T0343461

ADVANCES IN CELL AGING AND GERONTOLOGY
VOLUME 4

The Role of
DNA Damage and Repair
in Cell Aging

ADVANCES IN CELL AGING AND GERONTOLOGY
VOLUME 4

The Role of DNA Damage and Repair in Cell Aging

Volume Editors:

Barbara A. Gilchrest
Boston University, School of Medicine
Department of Dermatology
Boston, MA
USA

Vilhelm A. Bohr
Laboratory of Molecular Genetics
National Institute on Aging
NIH
Baltimore, MD
USA

2001

ELSEVIER

Amsterdam - London - New York - Oxford - Paris - Shannon - Tokyo

ELSEVIER SCIENCE B.V.
Sara Burgerhartstraat 25
P.O. Box 211, 1000 AE Amsterdam, The Netherlands

© 2001 Elsevier Science B.V. All rights reserved.

This work is protected under copyright by Elsevier Science, and the following terms and conditions apply to its use:

Photocopying
Single photocopies of single chapters may be made for personal use as allowed by national copyright laws. Permission of the Publisher and payment of a fee is required for all other photocopying, including multiple or systematic copying, copying for advertising or promotional purposes, resale, and all forms of document delivery. Special rates are available for educational institutions that wish to make photocopies for non-profit educational classroom use.

Permissions may be sought directly from Elsevier Science Global Rights Department, PO Box 800, Oxford OX5 1DX, UK; phone: (+44) 1865 843830, fax: (+44) 1865 853333, e-mail: permissions@elsevier.co.uk. You may also contact Global Rights directly through Elsevier's home page (http://www.elsevier.nl), by selecting 'Obtaining Permissions'.

In the USA, users may clear permissions and make payments through the Copyright Clearance Center, Inc., 222 Rosewood Drive, Danvers, MA 01923, USA; phone: (+1) (978) 7508400, fax: (+1) (978) 7504744, and in the UK through the Copyright Licensing Agency Rapid Clearance Service (CLARCS), 90 Tottenham Court Road, London W1P 0LP, UK; phone: (+44) 207 631 5555; fax: (+44) 207 631 5500. Other countries may have a local reprographic rights agency for payments.

Derivative Works
Tables of contents may be reproduced for internal circulation, but permission of Elsevier Science is required for external resale or distribution of such material. Permission of the Publisher is required for all other derivative works, including compilations and translations.

Electronic Storage or Usage
Permission of the Publisher is required to store or use electronically any material contained in this work, including any chapter or part of a chapter.

Except as outlined above, no part of this work may be reproduced, stored in a retrieval system or transmitted in any form or by any means, electronic, mechanical, photocopying, recording or otherwise, without prior written permission of the Publisher.
Address permissions requests to: Elsevier Global Rights Department, at the mail, fax and e-mail addresses noted above.

Notice
No responsibility is assumed by the Publisher for any injury and/or damage to persons or property as a matter of products liability, negligence or otherwise, or from any use or operation of any methods, products, instructions or ideas contained in the material herein. Because of rapid advances in the medical sciences, in particular, independent verification of diagnoses and drug dosages should be made.

First edition 2001

Library of Congress Cataloging in Publication Data
A catalog record from the Library of Congress has been applied for.

ISBN: 0-444-50494-X

⊚ The paper used in this publication meets the requirements of ANSI/NISO Z39.48-1992 (Permanence of Paper).
Printed and bound by CPI Antony Rowe, Eastbourne
Transferred to digital printing 2006

TABLE OF CONTENTS

v

INTRODUCTION

Nearly 40 years ago it was demonstrated that aging occurs at the level of individual cells. In the intervening decades, a complex interplay between intrinsic "programming" and exogenous "wear and tear" has become apparent, with genetically-determined cellular capacity to repair environmentally-induced DNA damage playing a central role in the rate of aging and its specific manifestations. In 12 chapters, "The Role of DNA Damage and Repair in Cell Aging" provides an intellectual framework for aging of mitotic and post-mitotic cells, describes a variety of model systems for further studies, and reviews current concepts of DNA damage responses and their relationship to the phenomenon of aging.

As part of a series entitled "Advances in Cell Aging and Gerontology," this volume also summarizes seminal recent discoveries such as the molecular basis for Werner syndrome (a mutant DNA helicase), the complementary roles of telomere shortening and telomerase activity in cell senescence versus immortalization, the role of apoptosis in the homeostasis of aging tissue, and the existence of an inducible SOS-like response in mammalian cells that minimizes DNA damage from repeatedly encountered injurious environmental agents. Insights into the relationship between cellular aging and age-associated diseases, particularly malignancies, are also provided in several chapters.

Whereas most of the early work on DNA repair pathways involved DNA lesions caused by environmental factors such as UV irradiation, there is now very active research on the DNA repair processes involved in the removal of oxidative DNA lesions that result, for example, from intracellular metabolism and presumably impact on "the rate of living." Many studies have shown that the frequency of oxidative DNA lesions increase with age in mammalian systems, and this could be due to increased formation and/or decreased repair of these lesions. Persistent lesions in the genome could explain many of the characteristics of the molecular phenotype of aging, such as the changes in transcription and the significantly increased genomic instability, issues touched upon throughout the volume.

The molecular biology of aging has surely come of age, lending its powerful techniques to the dissection of pathways responsible cellular growth, growth arrest, and differentiated function. This new knowledge confirms the long-presumed key role of DNA damage and repair in normal and pathologic aging processes. Building on this strong foundation, the field of cellular gerontology should make rapid progress in the new millennium.

Barbara A. Gilchrest
Vilhelm A. Bohr

© 2001 Elsevier Science B.V. All rights reserved.
The Role of DNA Damage and Repair in Cell Aging
B.A. Gilchrest and V.A. Bohr, volume editors.

AGING IN MITOTIC AND POST-MITOTIC CELLS

Judith Campisi[1] and *Huber R. Warner*[2]

[1]Lawrence Berkeley National Laboratory, Berkeley, CA;
[2] Biology of Aging Program, National Institute on Aging, Bethesda, MD

1. Introduction

Aging in most multicellular organisms entails distinctive changes in both extracellular and cellular components. Extracellular components include the soluble and insoluble molecules that contribute to tissue function and structure. Cells, of course, are the units that allow expression of the genome that defines the characteristics of the organism. Here we consider some of the fundamental mechanisms by which individual cells age, and speculate on the consequences of cellular aging for tissue function and, ultimately, the longevity of the organism. Our intention is not to ignore the importance of age-dependent changes in extracellular components. It is now well established that tissue function depends on a continuous, dynamic, and reciprocal interaction between the cells that comprise that tissue and the tissue microenvironment in which they reside; therefore it is important to understand aging of both cellular and extracellular components (Fig. 1).

Cells can be (very simply) classified into two fundamental types: mitotic and post-mitotic. Mitotic cells, or more accurately mitotically competent cells, are those that retain the capacity to proliferate (used here interchangeably with growth), independent of their differentiation (ability to perform specialized functions). Some mitotic cells may proliferate relatively often during the life span of an organism -- for example, basal epidermal keratinocytes. Others, such as hepatocytes and stromal fibroblasts, may proliferate rarely or only when the need for cell replacement arises. Regardless of their proliferative status at any given time, mitotic cells share with each other the capacity to undergo a complete cell cycle when appropriately stimulated. Mitotic cells also share the property of being susceptible to tumorigenic transformation. Post-mitotic cells originate from mitotic stem cells. During the process of differentiation, post-mitotic cells irreversibly lose their ability to proliferate. Despite their inability to initiate a cell cycle, post-mitotic cells may persist and function for long periods of time within the organism. Examples of post-mitotic cells include mature muscle and neuronal cells. Post-mitotic cells cannot be stimulated to proliferate by any physiological stimulus, nor by non-physiological stimuli such as carcinogens, irradiation or oncogene expression. In contrast to mitotic cells, post-mitotic cells never undergo tumorigenic transformation. Both post-mitotic and mitotic cells are critically dependent upon extracellular components for their viability and their phenotype (ability to express the genes that specify their specialized functions).

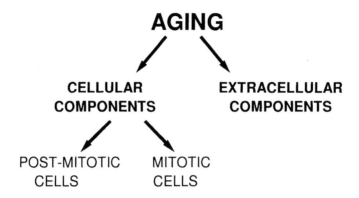

Figure 1. Aging components in higher eukaryotic organisms. A simplified view of the aging process considers that both cells and extracellular components undergo age-dependent changes. Cells may also be viewed as belonging to one of two classes: post-mitotic cells, which cannot divide, and mitotic cells, which are competent to divide when the need arises.

Because of their contrasting proliferative potentials, it is possible that mitotic and post-mitotic cells age by fundamentally different mechanisms. It is also possible that mitotic and post-mitotic cells differ in their sensitivity to stimuli that contribute to or accelerate aging, and/or to age-related changes in factors such as hormones, neighboring cells or the adjacent extracellular material (the tissue microenvironment). Here, we review some of the differences and similarities in the aging of mitotic and post-mitotic cells.

2. Impact of Aging on Cells

There are two broad classes of hypotheses that have been proposed to explain the initiating causes of aging. The first invokes extrinsic or intrinsic factors that damage intracellular (or extracellular) macromolecules. Damage may be repaired, but if repair is imperfect a cell may die or cease to function properly. If the damage and imperfect repair occur in DNA, and the cell is mitotic, a heritable mutation may result. The second category of hypotheses invokes changes in gene expression that are a programmed response or epigenetic in origin. Some age-related changes in gene expression may result from basic cellular processes, such as repeated cell division, whereas others may be a cellular response to extrinsic or intrinsic damage. Whatever the initiating cause, aging results in changes (increases or decreases) in cell proliferation, cell death and/or cell function. The response of any particular cell to damage or stress depends on its phenotype and genotype. Moreover, the response may differ depending on whether a cell is post-mitotic or mitotic.

Aging is characterized by a loss of post-mitotic cells -- for example, neurons in certain age-related pathologies, and muscle atrophy during "normal" aging. Cell loss in neurodegenerative disease is presumably due to apoptosis, although this remains somewhat controversial. Aging may also cause post-mitotic cells to malfunction. For example, neuronal synapses are commonly lost during aging, although neuronal cell bodies may persist. Results using cell cultures, as well as experimental animal models, suggest that agents, especially oxidants, that damage macromolecules frequently induce post-mitotic cells to die.

Because there is no way to replace lost or dysfunctional post-mitotic cells, cell death or malfunction can severely compromise tissues composed largely or entirely of such cells. Some tissues contain stem cells that can replace post-mitotic cells as they are lost due to damage, trauma or differentiation. In the skin, for example, the terminally differentiated keratinocytes that comprise the outer epidermal layers are continually shed, but they are replaced by stem cells in the basal epidermis. Thus, an age-related increase in the death or malfunction of terminally differentiated keratinocytes may not have a large impact on tissue function. In skeletal muscle, by contrast, stem cells (myoblasts) are also available to replenish damaged or wasted muscles, but the supply of myoblasts appears to be limited and this occurs only when the need arises. Nervous tissue may be particularly vulnerable to age-related cell loss or malfunction. Although there is increasing evidence for neuronal stem cells (neuroblasts) in adults, these cells do not appear to be very efficient at replenishing neuronal loss due to trauma or age.

Aging is also characterized by a loss of mitotic cells (for example, the increased acellularity of stromal tissue). Mitotic cells, like post-mitotic cells, can die in response to damaging agents. However, in contrast to the potentially detrimental effects of cell death in post-mitotic tissue, cell death may have, within limits, a positive effect on the health of mitotic tissues. Because mitotic cells can proliferate, dead cells can be readily replaced. Moreover, there is now substantial evidence that the ability of mitotic cells to die in response to damage constitutes a powerful tumor suppressive mechanism. Nonetheless, this and other tumor suppressive mechanisms eventually fail or are inadequate, because an additional characteristic of aging is an increase in hyperplasia and neoplasia.

3. Aging of Mitotic Cells

Replicative senescence - what is it?

Mitotic cells, with few exceptions, do not divide indefinitely (Hayflick, 1965; reviewed in Stanulis-Praeger, 1987; Cristofalo and Pignolo, 1993; Campisi et al., 1996). This trait has been termed the finite replicative life span of cells, and the process that limits cell division potential has been termed cellular or replicative senescence. Replicative senescence may be a fundamental primitive cellular trait because it has been documented in simple single celled organisms such as the budding yeast *Saccharomyces cerevisiae* (Jazwinski, 1993). It has been shown to occur in a large number of higher eukaryotic cell types from a variety of vertebrate and some invertebrate species. Here, we confine our discussion to higher eukaryotic cells, and principally to mammalian cells.

In higher eukaryotes, cellular senescence is thought to be a powerful tumor suppressive mechanism (Sager, 1991; Campisi, 1996, 1997a; Smith and Pereira-Smith, 1996; Wright and Shay, 1996). This notion stems from a large body of cellular, molecular and organismal biology. Among the most persuasive evidence for this idea is the finding that the p53 and pRB (retinoblastoma) tumor suppressor proteins, which are the most commonly lost functions in human cancers, are essential for establishing and maintaining the senescent state (Hara et al., 1991; Shay et al., 1991; Sakamoto et al., 1993; Dimri and Campisi, 1994; Hara et al., 1996a). In addition, there are mouse models in which inactivation of the p53 and p16 tumor suppressor proteins renders the cells refractory to senescence, and the organism is destined to die at an early age from cancer (Donehower et al., 1992; Harvey et al., 1993; Serrano et al., 1996). Cellular senescence has also been proposed to contribute to organismic aging (Campisi, 1996; Smith and Pereira-Smith, 1996; Campisi, 1997b). The evidence for this idea is less compelling, as discussed below, but support for it is slowly building.

The idea that the same process (cellular senescence) may have both beneficial (tumor suppression) and detrimental (aging) effects may, at first glance, seem inconsistent. However, the evolutionary idea of antagonistic pleiotropy predicts that some traits selected by evolution for their beneficial effects early in life may have *unselected* adverse effects late in life. Because the force of natural selection declines with age, late life adverse effects will not be selected against. Thus, cellular senescence may be an example of an antagonistically pleiotropic trait.

Replicative senescence results in an essentially irreversible block to cell proliferation. It is exceedingly stringent in human cells, but rather leaky in cells from many rodent species (Ponten, 1976; Sager, 1984; McCormick and Maher, 1988; Sager, 1991). Human cells rarely spontaneously escape and acquire an indefinite or immortal replicative life span. Rodent cells, by contrast, spontaneously immortalize with frequencies in the range of one cell in 10^5-10^7. The number of doublings at which mitotic cell populations senesce depends on the species and genotype of the donor. Thus, fetal human fibroblasts generally senesce after 50-80 doublings, whereas fetal mouse fibroblasts generally senesce after 10-15 doublings. Cells from patients with hereditary premature aging syndromes senesce much more rapidly than cells from age-matched wild-type controls. On the other hand, cells in which the p53 or pRb tumor suppressors have been inactivated by mutation or neutralized by viral or cellular oncogenes show an extended replicative life span relative to wild-type cells (reviewed in Stanulis-Praeger, 1987; Cristofalo and Pignolo, 1993; Campisi et al., 1996; Campisi, 1996; Smith and Pereira-Smith, 1996; Wright and Shay, 1996; Campisi, 1997b).

There are two types of cells that do not undergo replicative senescence. The first is the germ line, which must be immortal in order for higher eukaryotic species to persist. Second, a majority of tumor cells, whether experimentally induced or naturally occurring, do not undergo replicative senescence (reviewed in Sager, 1991). It is also possible that selected pluripotent stem cells in adult organisms do not undergo replicative senescence, but this has yet to be unambiguously demonstrated.

SENESCENT PHENOTYPE OF MITOTIC CELLS

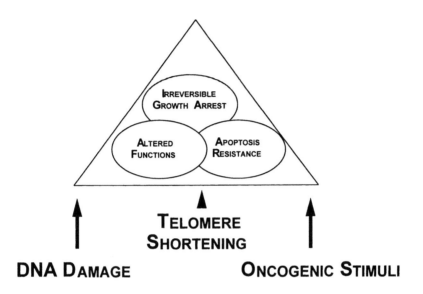

Figure 2. The senescent phenotype of mitotic cells. Mitotically competent cells irreversibly arrest growth, become resistant to apoptotic stimuli, and acquire altered differentiated functions when they adopt the senescent phenotype. The senescent phenotype can be induced by certain types of DNA damage, the acquisition of one of more critically short telomere, or certain types of potentially oncogenic stimuli.

The Senescent Phenotype

It is now clear that the irreversible block to cell proliferation that is the hallmark of replicative senescence is accompanied by two additional phenotypic changes. These additional phenotypic characteristics are resistance to apoptotic death and selected, sometimes striking, changes in differentiated functions (reviewed in Campisi et al., 1996) (Figure 2). Thus, upon completing a finite number of divisions, cells adopt an altered phenotype, which we refer to here as the senescent phenotype. As discussed below, it is also now clear that repeated cell division is not the only stimulus that can induce the senescent phenotype.

An irreversible growth arrest is a universal feature of the senescent phenotype — that is, all cells, regardless of cell type or species of origin, irreversibly lose the capacity to proliferate when they complete a finite number of cell divisions or express the senescent phenotype for other reasons. Thus, senescent cells resemble post-mitotic cells in that they are permanently incapable of cell division. The mechanisms responsible for the senescence-associated irreversible growth arrest are incompletely understood. Some of the events that undoubtedly are critical for the growth arrest of senescent cells are the overexpression of at least two inhibitors of cyclin-dependent protein kinases (p21 and p16) (Noda et al., 1994; Alcorta et al., 1996; Hara et al., 1996b), and the repression of at least three growth-regulatory transcription factors (c-fos, Id, and E2F) (Seshadri and Campisi, 1990; Hara et al., 1994; Dimri et al., 1994; Good et al., 1996). Resistance to apoptosis may also be a universal characteristic of senescent cells (Wang et al., 1994; Effros, 1996), but this phenomenon has been less widely studied. The mechanisms responsible for the relative unresponsiveness of senescent cells to apoptotic stimuli are poorly understood.

All senescent cells show functional changes. However, the nature of these changes are cell type-specific. For example, upon reaching replicative senescence, dermal fibroblasts switch from a matrix-producing to a matrix-degrading phenotype. At a molecular level, this switch entails a decline in the production of extracellular matrix proteins such as collagen and elastin, and a marked increase in the expression and activity of proteases which degrade matrix proteins (West et al., 1989; Millis et al., 1992; Wick et al., 1994; reviewed in Campisi et al., 1996). On the other hand, adrenal cortical epithelial cells produce an altered profile of steroid hormones upon replicative senescence (Hornsby et al., 1987). Thus, the senescent phenotype also entails changes in differentiated cell functions.

It is very likely that the growth arrest associated with senescence is critical for its function as a tumor suppressive mechanism. Thus, the growth arrest is very likely the feature of the senescent phenotype that was selected during evolution. The resistance to apoptosis and altered function of senescent cells, by contrast, may be unselected phenotypes, and thus responsible for the adverse effects of senescent cells.

What is the evidence that senescent cells have adverse effects that contribute to aging? At present, this idea is speculative, supported essentially by two lines of evidence: 1) the limited, albeit highly suggestive, evidence that senescent cells accumulate in aged tissues; and 2) the substantial body of evidence that, at least in culture, senescent cells secrete molecules that are potentially harmful to tissues. The idea that senescent cells exist *in vivo* and accumulate with age derives from several *ex vivo* studies of cells from young and old donors, which show substantial scatter in the data, and variable conclusions, depending on the data set (reviewed in Stanulis-Praeger, 1987; Campisi et al., 1996). More direct evidence derives from a limited number of *in situ* examinations of cells in young and aged tissue samples, showing the existence and age-dependent accumulation of cells expressing markers of the senescent phenotype (Dimri et al., 1995; Pendergrass et al., 1999). In culture, it is well-established that senescent cells secrete molecules that can have far-ranging and deleterious effects in tissues (reviewed in Campisi et al., 1996; Campisi, 1997a). These molecules include matrix-degrading enzymes, as discussed above, as well as pro-inflammatory cytokines, anti-angiogenic factors, and pro-thrombotic factors. Overexpression of any of these secreted molecules, much less several of them, would be expected to compromise the integrity and function of tissues. Because such molecules can act at a distance within tissues, only a few senescent cells may be needed for them to have adverse effects on tissue function and/or integrity. Moreover, because the molecules secreted by senescent cells also include growth factors, they may also contribute to the increase in hyperplasia and neoplasia that is the hallmark of aging in many mammalian species, including humans (Campisi, 1997a).

Causes of the senescent phenotype

Telomere Shortening. The driving force behind the decline in proliferative potential that occurs as a consequence of repeated cell division is telomere shortening (Harley et al., 1990; Bodnar et al., 1998). Telomeres are the ends of eukaryotic chromosomes, composed of several kilobases of a repetitive DNA sequence (TTAGGG in vertebrates) and specialized proteins. Telomeres are essential for chromosome stability. Chromosomes that lack a telomere are extremely susceptible to random fusion, breakage and recombination (reviewed in Blackburn, 1991). Owing to the biochemistry of DNA replication, each cell cycle leaves 50-200 bp of telomeric DNA unreplicated at the 3' ends of linear chromosomes (Levy et al., 1992). Thus, telomeres shorten progressively with each cell division (Harley et al., 1990). There is now very solid evidence that cells express the senescent phenotype when they acquire one or more telomere that reaches a critically short length (Bodnar et al., 1998). It is not known exactly how short a telomere must be to induce the senescent phenotype, but it is likely that senescence-inducing chromosomes retain substantial telomeric sequences. Thus, in human cells, where the length of the terminal (telomeric) restriction fragment (TRF) averages 15-20 kb, replicative senescence occurs when the TRF reaches an average of 5-7 kb (reviewed in Wright and Shay, 1996). Cells that do not undergo replicative senescence express the enzyme telomerase, a ribonucleoprotein complex that can add telomeric repeat sequences *de novo* to the 3' ends of chromosomes (reviewed in Blackburn, 1992; Lingner and Cech, 1998). Telomerase is expressed by germ cells and early embryonic cells, but is repressed during fetal development such that somatic cells generally do not express the enzyme. In the few somatic cells that do express telomerase, the enzyme is tightly regulated and serves to retard but not completely prevent telomere shortening. The exception to stringent control over telomerase expression and activity is, of course, tumor cells, the vast majority of which constitutively express the enzyme (reviewed in Holt et al., 1997).

It is not known how short telomeres signal normal cells to arrest growth with a senescent phenotype, although studies in yeast have provided several potential mechanisms. These include a DNA damage response induced by a short telomere, the release of transcription-modulating factors from short telomeres, and changes in heterochromatin induced by a short telomere (reviewed in Campisi, 1997b). These potential mechanisms are not mutually exclusive. Unfortunately, at present, there are very few data to suggest which, if any, of these mechanisms operate in mammalian cells.

Recent evidence suggests that the senescent phenotype is also induced by at least two stimuli in addition to telomere shortening. These stimuli are certain types of DNA damage, and inappropriate mitogenic or oncogenic signals (Figure 2).

DNA damage. Agents that produce oxidative lesions in DNA or cause DNA double-strand breaks induce normal human cells to irreversibly arrest growth with associated phenotypic changes that resemble those of replicatively senescent cells (DiLeonardo et al., 1994; Chen et al., 1998; Robles and Adami, 1998). Although many studies show that moderate levels of DNA damage often result in apoptosis, most of these studies utilized immortal cells, and frequently immortal rodent cells. Normal cells, particularly normal human cells, often do not undergo apoptosis in response to moderate DNA damage. Rather, they respond by adopting a senescent phenotype.

The ability of certain types of DNA damage to induce a senescent phenotype may explain the premature replicative senescence of cells from donors with the Werner syndrome (WS). WS is a hereditary, adult-onset premature aging syndrome of man (reviewed in Goto, 1997). Patients with WS are asymptomatic until puberty, and even then symptoms are mild. However, WS individuals in their 20's and 30's develop a cluster of age-related pathologies, including cancer, atherosclerosis, type II diabetes, cataracts, osteoporosis, loss and graying of hair, and skin atrophy. WS is not a exact phenocopy of normal aging, but rather is characterized by the development of a subset of age-related pathologies 20-30 years prematurely. The average life span of individuals with WS is approximately 45 years, with cardiovascular disease and cancer being the leading cause of death.

WS is caused by homozygous inactivation of the recently cloned *WRN* gene, which is located on human chromosome 8 (Yu et al., 1996). *WRN* encodes a large protein that has intrinsic DNA helicase and 3' to 5' exonuclease activities (Gray et al., 1997; Huang et al., 1998). Although the precise function of the WRN protein is not yet known, its biochemical activities, and its striking homology to the helicase domain of the *E. coli* RecQ gene, strongly suggest that it participates in one or more DNA repair pathway. WS cells, which lack WRN function, accumulate a variety of mutations, with an unusually high proportion of chromosomal deletions and translocations (Fukuchi et al., 1989). It is well-established that WS cells undergo replicative senescence after many fewer doublings than cells from age-matched controls (Martin et al., 1970; Salk et al., 1985). There is also suggestive evidence that WS cells senesce with telomere lengths that are demonstrably longer than those typical of senescent normal cells (7-9 kb, compared to 5-7 kb) (Schultz et al., 1996). Thus, accumulated DNA damage, rather than a critically short telomere, may cause the premature senescence of WS cells. The fact that loss of a single gene function (*WRN*) causes both the acceleration of aging phenotypes *in vivo* and accelerated senescence of cells in culture, support the idea that cellular senescence may contribute to aging, or at least to several age-related pathologies.

Inappropriate oncogenic or mitogenic stimuli. Very recent data suggest that normal cells may respond to potentially oncogenic stimuli by adopting a senescent phenotype. The first evidence for this idea derived from studies in which an activated (oncogenic) form of the *RAS* gene was introduced into normal human fibroblasts.

RAS is a well-characterized proto-oncogene that transduces signals from certain growth factor receptors. Upon binding ligand, RAS-responsive receptors stimulate the RAS protein to bind GTP. GTP-bound RAS then transmits the mitogenic signal by activating the mitogen-activate protein kinase (MAPK) pathway. The signal is terminated by

RAS GTPase activity. Mutations that convert RAS into an oncoprotein inactivate GTPase, but not GTP-binding activity, causing the protein to transmit a continuous mitogenic signal. Activated forms of RAS stimulate the growth of many rodent cells, and transform immortal cells into tumorigenic cells (reviewed in McCormick, 1989). It was therefore unexpected when normal human fibroblasts responded to introduction of an oncogenic *RAS* gene by arresting growth with characteristics of senescent cells (Serrano et al., 1997). Subsequent studies showed that activated forms of two downstream effectors of RAS activity, the RAF and MEK protein kinases, also elicited a senescence response in normal human fibroblasts (Lin et al., 1998; Zhu et al., 1998). By contrast, activated forms of RAS and MEK stimulated proliferation in immortal cells or cells in which the p53 tumor suppressor protein was inactive. Thus, inappropriate and potentially oncogenic mitogenic signals stimulated proliferation in cells with compromised p53 function, but induced a senescent phenotype in normal human cells.

The senescence response of normal human cells is not limited to hyperstimulation of growth factor signal transduction pathways. Our recent findings indicate that overexpression of the E2F1 transcription factor also induces a senescence response from normal human fibroblasts (Dimri et al., 2000). E2F1 is a multifunctional transcription factor that is negatively regulated by the retinoblastoma tumor suppressor protein, and is important for the transcription of many genes that are required for DNA synthesis (reviewed in Helin, 1997). The ability of overexpressed E2F1 to induce a senescence response also required uncompromised p53 function, and depended on its ability to stimulate transcription. Of the many genes that are E2F1 target genes, the p14/ARF tumor suppressor (Bates et al., 1998) is the most likely candidate for effecting the E2F1-induced senescence response. p14/ARF is highly expressed by senescent human fibroblasts, and expression of p14/ARF in presenescent cells to the level found in senescent cells induces a senescent phenotype.

Taken together, these recent findings suggest that normal mitotic cells respond to potentially oncogenic stimuli by adopting a senescent phenotype. Thus, the senescence response may be the first line of defense against tumorigenesis for many normal cells.

4. Aging of Post-Mitotic Cells

In mitotic cells, damage-responsive checkpoints exist to provide a means for deciding whether to repair and then proliferate, or to undergo apoptosis (Weinert, 1998). Known checkpoint proteins include p53 (Levine, 1997), ATM (Westphal et al., 1997), and possibly BRCA1 and 2 (Zhang et al., 1998). DNA damage stimulates phosphorylation of p53 by DNA-dependent protein kinase (Woo et al., 1998), implicating this enzyme in checkpoint function. In post-mitotic tissues proliferation is no longer possible, so if cell replacement from a stem cell is not an option, the possible outcomes are either complete repair, continued survival of a dysfunctional cell, or apoptosis (Figure 3). The final outcome may eventually be significant tissue pathology.

The major source of most cell damage is assumed to be oxidative stress due to mitochondrial production of reactive oxygen species (ROS) such as the superoxide anion

FIGURE 3

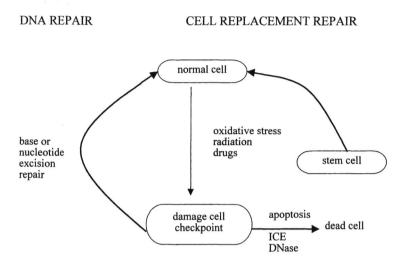

DNA REPAIR   CELL REPLACEMENT REPAIR

Figure 3. *Options for dealing with damaged post-mitotic cells. If the damage cannot be repaired, and cell replacement is not possible, the result may be a net loss in cell number as in the case of neurodegenerative diseases. Although this figure refers to DNA damage, protein damage could also lead to apoptosis as discussed in the text.*

(O_2^{-}), hydrogen peroxide (H_2O_2), and hydroxyl radical (HO·). The initial source of these ROS is the superoxide anion generated as a byproduct of oxidative phosphorylation. This superoxide can then be converted into hydrogen peroxide by the mitochondrial manganese-containing superoxide dismutase (SOD-2). The cytosolic form of this enzyme, SOD-1, is distinct from the mitochondrial form, and it contains copper and zinc rather than manganese. Because the mitochondria contain no catalase activity, the hydrogen peroxide formed by the activity of SOD-2 either leaks out of the mitochondria, or is converted to the very reactive hydroxyl radical by the Fenton reaction, which is catalysed by various forms of heavy metal ions such as ferrous ion (Fe^{++}). The hydrogen peroxide can also react directly with unsaturated lipids, particularly those in the mitochondrial membrane to produce a variety of reactive products.

The various forms of ROS can interact with, and thereby damage, nucleic acids, proteins, and lipids. Because DNA damage and repair is dealt with thoroughly in many of the other chapters in this volume, this review will focus on protein and lipid damage and their implications for age-related pathology. An obvious place to start is in the mitochondrion itself, because it is the source of most of the oxidative stress imposed on the cell.

Two mitochondrial proteins have been singled out for study because they appear to be readily oxidized. These two proteins are mitochondrial aconitase, an essential component of the citric acid cycle within the mitochondrion, and adenine nucleotide translocase, a mitochondrial membrane protein.

Oxidized aconitase protein can be detected in mitochondria through the carbonyl groups formed, and oxidized aconitase levels increase with age, and in response to increased oxidative stress (Yan et al., 1997). Because iron catalyses the Fenton reaction, thus generating hydroxyl radical from hydrogen peroxide, and because this enzyme contains an iron-sulfur cluster, it appears to be an obvious target for attack by ROS. They also showed that as the level of oxidation of this protein increases in house flies, the specific activity drops. Older flies are no longer able to fly and these flies contain more oxidized protein than older flies still able to fly, suggesting a possible causal association among aging, physical activity, oxidation of cis-aconitase, and citric acid cycle activity. Lowered citric acid cycle function would result in a lower rate of production of reduced nicotinamide adenine dinucleotide (NADH), and subsequently a lower rate of adenosine triphosphate (ATP) production by the electron transport system inside the mitochondria. Mitochondrial aconitase activity is also deficient in Friedreich's ataxia, which is characterized by accumulation of iron in mitochondria (Rötig et al., 1997), which would be expected to lead to increased protein oxidation in mitochondria through the Fenton reaction.

Once ATP is produced inside the mitochondria, it must be transported to the cytosol by adenine nucleotide translocase. This is accomplished by mitochondrial membrane adenine nucleotide translocase which exchanges mitochondrial ATP for cytosolic ADP. The oxidation of this protein also increases with age in houseflies, and the specific activity decreases with age (Yan and Sohal, 1998). The loss of flying ability is associated with increased oxidation of this protein, again suggesting an association among aging, diminished mitochondrial ATP production, oxidation of adenine nucleotide translocase, and reduced physical activity. The absence of adenine nucleotide translocase I activity has also been examined in mice in which the gene for this enzyme has been knocked out (Graham et al., 1997). Mice lacking this enzyme activity exhibit muscle mitochondrial myopathy, ragged red muscle fibers, cardiac hypertrophy, defects in respiration, metabolic acidosis, and premature death. Of particular interest is the recent observation that bcl-X_L inhibits apoptosis by facilitating ATP/ADP exchange (Heiden et al., 1999), suggesting the importance of maintaining an appropriate ATP/ADP for cell survival.

Another approach to understanding the possible role of mitochondrial oxidative stress in aging *in vivo* is to examine the impact of altering the antioxidant defense systems in mitochondria. This has been accomplished by knocking out the gene for the mitochondrial superoxide dismutase in mice (Li et al., 1995). The result is very early death with cardiomyopathy, usually by 10 days after birth, although the age of death is dependent on the mouse strain used (Huang et al., 1997). These authors also found that there is severe reduction in not only aconitase activity in heart, liver and brain, but also succinate dehydrogenase, another iron-containing protein. These results suggest that the mitochondria in these newborn mice are being exposed to severe oxidative stress due to absence of mitochondrial superoxide dismutase, resulting in cellular death and premature death of the mouse. Melov et al. (1998) found that a superoxide dismutase mimetic, manganese 5, 10, 15, 20-tetrakis (4-benzoic acid) porphyrin prolongs the life of these mice by 2 to 3-fold. However, these mice show signs of progressive neuropathology, presumably because the mimetic fails to cross the blood brain barrier.

A more modest level of stress was used by Williams et al. (1998) through use of mice heterozygous for mitochondrial superoxide dismutase. Such mice are viable, although it is not yet known whether life span is normal or slightly truncated. These mice have a 30% decrease in mitochondrial superoxide dismutase activity in liver, and show signs of elevated oxidative stress as evidenced by 50% reduction in glutathione levels in liver, which is accompanied by decreased levels of mitochondrial aconitase activity, decreased mitochondrial function, and increased oxidation of DNA and mitochondrial proteins, and increased rate of the mitochondrial permeability transition. This latter could lead to loss of cytochrome C, a stimulator of the apoptotic process, leading eventually to death of cells where this occurs (Green and Reed, 1998).

The results summarized briefly above suggest one route by which oxidative stress could lead to cell death through a pathway including oxidation of critical proteins, reduction in cellular ATP levels, and loss of membrane integrity. While internal oxidative stress occurring slowly, but continuously, in this manner may not have the catastrophic impact associated with myocardial infarction or stroke, the gradual loss of cells in postmitotic tissues with aging could conceivably ultimately have significant impact on tissue function. Gradual loss of neuronal cells through apoptosis in particular areas of the brain, is associated with a variety of age-related neurodegenerative diseases (Warner et al., 1997), and could be significant in loss of muscle mass during aging if cell replacement is insufficient to maintain normal tissue function in muscle.

The above model is clearly oversimplified and does not include a variety of other known events which could contribute in either a negative or positive way. For example, the oxidative interaction between hydrogen peroxide and unsaturated fatty acids leads to the production of dialdehydic intermediates such as 4-hydroxy-2-nonenal, which can interact with, and crosslink, proteins (Lucas and Szweda, 1998). Such crosslinked proteins may become not only dysfunctional, but also resistant to degradation. Also, membranes which contain significant levels of unsaturated fatty acids may lose their functional integrity through such oxidative damage and subsequent post-translational modification of the proteins.

A different, but related process is glycation of proteins which involves the non-enzymatic reaction between reducing sugars, such as glucose, and the amino groups of proteins (Monnier, 1990). Such complexes can undergo a variety of intramolecular rearrangements, producing intermediates which are also sensitive to oxidative modification (Elgawish et al., 1996). Evidence has been obtained linking the presence of glycation endproducts to Alzheimer disease pathology (Smith et al., 1994), and showing an age-related increase of these endproducts in human extracellular matrix (Sell and Monnier, 1989). A critical factor may be the ability of the cell to recognize and repair and/or remove such altered proteins, especially if such repair systems decline with age. Much of the data accumulated so far with regard to whether the activities of repair systems are maintained with increasing age are mixed, and incomplete in the sense of not covering a wide range of tissues. Dietary aminoguanidine may provide a promising intervention to reduce accumulation of glycation endpoints (Li et al., 1996).

Summary

In summary, the progressive loss of post-mitotic tissue during aging may be due to the death of damaged post-mitotic cells which cannot be replaced, or possibly their malfunction. The damage may derive from either intrinsic or extrinsic sources, and may consist of damage to proteins as well as DNA. The loss of neuronal cells is particularly problematic, since they are least readily replaced (if at all). Mitotic cells may also respond to intrinsic or extrinsic damage by cell death, but this generally is not problematic in mitotic tissues. In contrast, however, damage or repeated cell division can cause mitotic cells to express a senescent phenotype, which may compromise the function and integrity of the surrounding tissue. Finally, imperfect repair of DNA damage may cause the accumulation of potentially oncogenic mutations, which can cause mitotic, but not post-mitotic, cells to express a hyperplastic or neoplastic phenotype.

References

Alcorta, D.A., Xiong, Y., Phelps, D., Hannon, G., Beach, D., and Barrett, J.C. (1996). Involvement of the cyclin-dependent kinase inhibitor p16 (INK4a) in replicative senescence of normal human fibroblasts. Proc. Natl. Acad. Sci. USA 93, 13742-13747.

Bates, S., Phillips, A.C., Clark, P.A., Stott, F., Peters, G., Ludwig, R.L., and Voudsen, K.H. (1998). p14[ARF] links the tumor suppressors RB and p53. Nature 395, 124-125.

Blackburn, E.H. (1991). Structure and function of telomeres. Nature 350, 569-573.

Blackburn, E.H. (1992). Telomerases. Ann. Rev. Biochem. 61, 113-129.

Bodnar, A.G., Ouellette, M., Frolkis, M., Holt, S.E., Chiu, C.P., Morin, G.B., Harley, C.B., Shay, J.W., Lichtsteiner, S., and Wright, W.E. (1998). Extension of life span by introduction of telomerase into normal human cells. Science 279, 349-352.

Campisi, J. (1996). Replicative senescence: An old lives tale? Cell 84, 497-500.

Campisi, J, Dimri, G.P., Hara, E. (1996). In Handbook of the Biology of Aging, Schneider E, Rowe J (eds): "Control of replicative senescence." Academic Press: New York, pp 121-149.

Campisi, J. (1997a). Aging and cancer: The double-edged sword of replicative senescence. J. Am.Geriatrics Soc. 45, 1-6.

Campisi, J. (1997b). The biology of replicative senescence. Eur. J. Cancer 33, 703-709.

Chen, Q., Bartholomew, J.C., Campisi, J., Acosta, M., Reagen, J.D., and Ames, B.N. (1998). Molecular analysis of H_2O_2-induced senescent-like growth arrest in normal human fibroblasts: p53 and Rb control G(1) arrest but not cell replication. Biochem. J. 332, 43-50.

Cristofalo, V.J, Pignolo, R.J. (1993). Replicative senescence of human fibroblast-like cells in culture. Physiol. Rev. 73, 617-638.

DiLeonardo, A., Linke, S.P., Clarkin, K., and Wahl, G.M. (1994). DNA damage triggers a prolonged p53-dependent G1 arrest and long-term induction of Cip1 in normal human fibroblasts. Genes Dev. 8, 2540-2551.

Dimri, G.P., and Campisi, J. (1994). Molecular and cell biology of replicative senescence. Cold Spring Harbor Symposia on Quantitative Biology. Molec. Gen. Cancer 54, 67-73.

Dimri, G.P., Hara, E., Campisi, J. (1994). Regulation of two E2F related genes in presenescent and senescent human fibroblasts. J. Biol. Chem. 269, 16180-16186.

Dimri, G.P., Lee, X., Basile, G., Acosta, M., Scott, G., Roskelley, C., Medrano, E.E., Linskens, M., Rubelj, I., Pereira-Smith, O., Peacocke, M., and Campisi, J. (1995). A novel biomarker identifies senescent human cells in culture and aging skin in vivo. Proc. Natl. Acad. Sci. USA 92, 9363-9367.

Dimri, G.P., Acosta, M., Hahana, K. and Campisi, J. (2000). Regulations of a senescense checkpoint response by the E2F1 transcription factor and p14/ARF tumor suppressor. Mol. Cell. Biol. 20: 273-285.

Donehower, L.A., Harvey, M., Slagke, B.L., McArthur, M.J., Montgomery, C.A. Jr., Butel, J.S., and Bradley, A. (1992). Mice deficient for p53 are developmentally normal but susceptible to spontaneous tumors. Nature 356, 215-21.

Effros, R.B. (1996). Insights on immunological aging derived from the T lymphocyte cellular senescence model. Exp. Gerontol. 31, 21-27.

Elgawish, A., Glomb, M., Friedlander, M., and Monnier, V.M. (1996). Involvement of hydrogen peroxide in collagen cross-linking by high glucose in vitro and in vivo. J. Biol. Chem. 271, 12964-12971.

Fukuchi, K., Martin, G.M., and Monnat, R.J. 1989. Mutator phenotype of Werner syndrome is characterized by extensive deletions. Proc. Natl. Acad. Sci. USA 86, 5893-5897.

Good, L.F., Dimri, G.P., Campisi, J., and Chen, K.Y. (1996). Regulation of dihydrofolate reductase gene expression and E2F components in human diploid fibroblasts during growth and senescence. J. Cell Physiol. 168, 580-588.

Goto, M. (1997). Hierarchical deterioration of body systems in Werner's syndrome: Implications for normal ageing. Mech. Ageing and Dev. 98, 239-254.

Graham, B.H., Waymire, K.G., Cottrell, B., Trounce, I.A., MacGregor, G.R., and Wallace, D.C. (1997). A mouse model for mitochondrial myopathy resulting from a deficiency in the heart/muscle isoform of the adenine nucleotide translocator. Nature Genetics 16, 226-234.

Gray, M.D., Shen, J.C., Kamath-Loeb, A.S., Blank, A., Martin, G.M., Oshima, J., and Loeb, L.A. (1997). The Werner syndrome protein is a DNA helicase. Nature Genetics 17, 100-103.

Green, D.R., and Reed, J.C. (1998). Mitochondria and apoptosis. Science 281, 1309-1312.

Hara, E., Tsuri, H., Shinozaki, S., and Oda, K. (1991). Cooperative effect of antisense-Rb and antisense-p53 oligomers on the extension of lifespan in human diploid fibroblasts, TIG-1. Biochem. Biophys. Res. Comm. 179, 528-534.

Hara, E., Yamaguchi, T., Nojima, H., Ide, T., Campisi, J., Okayama, H., and Oda, K. (1994). Id related genes encoding helix loop helix proteins are required for G1 progression and are repressed in senescent human fibroblasts. J. Biol. Chem. 269, 2139-2145.

Hara, E., Uzman, J.A., Dimri, G.P., Nehlin, J.O., Testori, A., and Campisi, J. (1996a). The helix-loop-helix protein Id-1 and a retinoblastoma protein binding mutant of SV40 T antigen synergize to reactivate DNA synthesis in senescent human fibroblasts. Dev. Genetics 18, 161-172.

Hara, E., Smith, R., Parry, D., Tahara, H., and Peters, G. (1996b). Regulation of p16 (CdkN2) expression and its implications for cell immortalization and senescence. Molec. Cell Bio. 16, 859-867.

Harley, C.B., Futcher, A.B., and Greider, C.W. (1990). Telomeres shorten during ageing of human fibroblasts. Nature 345, 458-460.

Harvey, M., Sands, A.T., Weiss, R.S., Hegi, M.E., Wiseman, R.W., Pantazis, P., Giovanella, B.C., Tainsky, M.A., Bradley, A., and Donehower, L.A. (1993). In vitro growth characteristics of embryo fibroblasts isolated from p53-deficient mice. Oncogene 8, 2457-67.

Hayflick, L. (1965). The limited in vitro lifetime of human diploid cell strains. Exp. Cell Res. 37, 614-636.

Heiden, M.G., Chandel, N.S., Schumacker, P.T., and Thompson, C.B. (1999). Bcl-X_L prevents cell death following growth factor withdrawal by facilitating mitochondrial ATP/ADP exchange. Mol. Cell 3, 159-167.

Helin, K. (1997). Regulation of cell proliferation by E2F transcription factors. Curr. Opin. Genet and Dev. 8, 28-35.

Holt, S.E., Wright, W.E., and Shay, J.W. (1997). Multiple pathways for the regulation of telomerase activity. Eur. J. Cancer 33, 761-766.

Hornsby, P.J., Hancock, J.P., Vo, T.P., Nason, L.M., Ryan, R.F., and McAllister, J.M. (1987). Loss of expression of a differentiated function gene, steroid 17a-hydroxylase, as adrenocortical cells senesce in culture. Proc. Natl. Acad. Sci. USA 84, 1580-1584.

Huang, T.T., Yasunami, M., Carlson, E.J., Gillespie, A.M., Reaume, A.G., Hoffman, E.K., Chan, P.H., Scott, R.W., and Epstein, C.J. (1997). Superoxide-mediated cytoxicity in superoxide-deficient fetal fibroblasts. Arch. Biochem. Biophys. 15, 424-432.

Huang, S., Li, B., Gray, M.D., Oshima, J., Mian, S., and Campisi, J. (1998). The premature aging syndrome protein WRN is a 3' to 5' exonuclease. Nature Genetics 20, 114-116.

Jazwinski, S.M. (1993). The genetics of aging in the yeast Saccharomyces cerevisiae. Genetica 91, 35-51.

Levine, A.J. (1997). p53 the cellular gatekeeper for growth and division. Cell 88, 323-331.

Levy, M.Z., Allsopp, R.C., Futcher, A.B., Greider, C.W., and Harley, C.B. (1992). Telomere end-replication problem and cell aging. J. Mol. Biol. 225, 951-960.

Li, Y., Huang, T.T., Carlson, E.J., Melov, S., Ursell, P.C., Olson, J.L., Noble, L.J., Yoshimura, M.P., Berger, C., Chau, P.H., Wallace, D.C., and Epstein, C.J. (1995). Dilated cardiomyopathy and neonatal lethality in mutant mice lacking manganese superoxide dismutase. Nature Genetics 11, 226-381.

Li, Y.M., Steffes, M., Donnelly, T., Liu, C., Fuh, H., Basgen, J., Bucala, R., and Vlassara, H. (1996). Prevention of cardiovascular and renal pathology of aging by the advanced glycation inhibitor aminoguanidine. Proc. Natl. Acad. Sci. USA 93, 3902-3907.

Lin, A.W., Barradas, M., Stone, J.C., van Aelst, L., Serrano, M., and Lowe, S.W. (1998). Premature senescence involving p53 and p16 is activated in response to constitutive MEK/MAPK mitogenic signaling. Genes and Dev. 12, 3008-3019.

Lingner, J., and Cech, T.R. (1998). Telomerase and chromosome end maintenance. Curr. Opin. Genet. Dev. 8, 226-232.

Lucas, D.T., and Szweda, L.I. (1998). Cardiac reperfusion injury: Aging, lipid peroxidation, and mitochondrial dysfunction. Proc. Natl. Acad. Sci. USA 95, 510-514

Martin, G.M., Sprague, C.A., and Epstein, C.J. (1970). Replicative life span of cultivated human cells. Effect of donor's age, tissue and genotype. Lab. Invest. 23, 86-92.

McCormick, J.J., and Maher, V.M. (1988). Towards an understanding of the malignant transformation of diploid human fibroblasts. Mutation Res. 199, 273-291.

McCormick, F. (1989). ras GTPase activating protein: Signal transmitter and signal terminator. Cell 56, 5-8.

Melov, S., Schneider, J.A., Day, B.J., Hinerfield, D., Coskun, P., Mirra, S.S., Crapo, J.D., and Wallace, D.C. (1998). A novel neurological phenotype in mice lacking mitochondrial manganese superoxide dismutase. Nature Genetics 18, 159-163.

Millis, A.J.T., Hoyle, M., McCue, H.M., and Martini, H.. (1992). Differential expression of metalloproteinase and tissue inhibitor of metalloproteinase genes in diploid human fibroblasts. Exp. Cell Res. 201, 373–379.

Monnier, V.M. (1990). Nonenzymatic glycosylation, the Maillard reaction and the aging processes. J. Gerontol. 45, B105-111.

Noda, A., Ning, Y., Venable, S.F., Pereira-Smith, O.M., and Smith, J.R. (1994). Cloning of senescent cell derived inhibitors of DNA synthesis using an expression screen. Exp. Cell Res. 211, 90-98.

Pendergrass, W.R., Lane, M.A., Bodkin, N.L., Hansen, B.C., Ingram, D.K., Roth, G.S., Yi, L., and Wolf, N. (1999). Cellular proliferation potential during aging and caloric restriction in Rhesus monkeys (Macaca mulatta). J. Cellular Phys. 180, 123-130.

Ponten, J. (1976). The relationship between in vitro transformation and tumor formation in vivo. Biochem. Biophys. Acta. 458, 397-422.

Robles, S.J., and Adami, G.R. (1998). Agents that cause DNA double strand breaks lead to p16[INK4a] enrichment and the premature senescence of normal fibroblasts. Oncogene 16, 1113-1123.

Rötig, A., de Lonlay, P., Chretien, D., Foury, F., Koeing, M., Sidi, D., Munnich, A., and Rustin, P. (1997). Aconitase and mitochondrial iron-sulfur protein deficiency in Friedreich ataxia. Nature Genetics 17, 215-217.

Sager, R. (1984). Resistance of human cells to oncogenic transformation. Cancer Cells 2, 487-493.

Sager, R. (1991). Senescence as a mode of tumor suppression. Env. Health Persp. 93, 59-62.

Sakamoto, K., Howard, T., Ogryzko, V., Xu, N.Z., Corsico, C.C., Jones, D.H., and Howard, B. (1993). Relative mitogenic activities of wild-type and retinoblastoma binding defective SV40 T antigens in serum deprived and senescent human fibroblasts. Oncogene 8, 1887-1893.

Salk, D., Bryant, E., Hoehn, H., Johnston, P., and Martin, G.M. (1985). Growth characteristics of Werner syndrome cells in vitro. Adv. Exp. Biol. Med. 190, 305-311.

Schultz, C.P., Zakian, V.A,. Ogburn, C.E., McKay, J., Jarzebowicz, A.A., and Martin, G.M. (1996). Accelerated loss of telomeric repeats may not explain accelerated replicative decline of Werner syndrome cells. Human Genet. 97, 750-754.

Sell, D.R.. and Monnier, V.M. (1989). Structure elucidation of a senescence cross-link from human extracellular matrix. J. Biol. Chem. 264, 21597-21602.

Serrano, M., Lee, H., Chin, L., Cordon-Cardo, C., Beach, D., and DePinho, R.A. (1996). Role of the INK4A locus in tumor suppression and cell mortality. Cell 85, 27-37.

Serrano, M., Lin, A.W., McCurrach, M.E., Beach, D., and Lowe, S.W. (1997). Oncogenic ras provokes premature cell senescence associated with accumulation of p53 and p16^{INK4a}. Cell 88, 593-602.

Seshadri, T., and Campisi, J. (1990). c-fos repression and an altered genetic program in senescent human fibroblasts. Science 247, 205-209.

Shay, J.W., Pereira-Smith, O.M., Wright, W.E. (1991). A role for both Rb and p53 in the regulation of human cellular senescence. Exp. Cell Res. 196, 33-39.

Smith, M.A., Taneda, S., Richey, P.L., Miyata, S., Yan, S.-D., Stern, D., Sayre, L.M., Monnier, V.M., and Perry, G. (1994). Advanced Mallaird reaction end products are associated with Alzheimer disease pathology. Proc. Natl. Acad. Sci. USA 91, 5710-5714.

Smith, J.R., and Pereira-Smith, O.M. (1996). Replicative senescence: implications for in vivo aging and tumor suppression. Science 273, 63-67.

Stanulis-Praeger, B. (1987). Cellular senescence revisited: A review. Mech. Aging Dev. 38, 1-48.

Wang, E., Lee, M.J., and Pandey, S. (1994). Control of fibroblast senescence and activation of programmed cell death. J. Cell Biochem. 54, 432-439.

Warner, H.R., Hodes, R.J., and Pocinki, K. (1997). What does cell death have to do with aging? J. Am. Geriatrics Soc. 45, 1140-1146.

Weinert, T. (1998). DNA damage and checkpoint pathways: Molecular anatomy and interactions with repair. Cell 94, 555-558.

West, M.D., Pereira-Smith, O.M., and Smith, J.R. (1989). Replicative senescence of human skin fibroblasts correlates with a loss of regulation and overexpression of collagenase activity. Exp. Cell Res. 184,138–147.

Westphal, C.H., Rowan, S., Schmaltz, C., Elson, A., Fisher, D.E., and Leder, P. (1997). atm and p53 cooperate in apoptosis and suppression of tumorigenesis, but not in resistnce to acute radiation toxicity. Nature Genetics 16, 397-401.

Wick, M., Burger, C., Brusselbach, S., Lucibello, F.C., and Muller, R. (1994). A novel member of human tissue inhibitor of metalloproteinases (TIMP) gene family is regulated during G1 progression, mitogenic stimulation, differentiation and senescence. J. Cell Biochem. 269, 18953-18960.

Williams, M.D., Van Remmen, H., Conrad, C.C., Huang, T.T., Epstein, C.J., and Richardson, A. (1998). Increased oxidative damage is correlated to altered mitochondrial function in heterozygous manganese superoxide dismutase knockout mice. J. Biol. Chem. 273, 28510-28515.

Woo, R.A., McClure, K.G., Lees-Miller, S.P., Rancourt, D.E., and Lee, P.W.K. (1998). DNA-dependent protein kinase acts upstream of p53 in response to DNA damage. Nature 394, 700-704.

Wright, W.E., and Shay, J.W. (1996). Mechanism of escaping senescence in human diploid cells. In Modern Cell Biology Series-Cellular aging and cell death. Holbrook, N.J., Martin, G.R., and Lockshin, R.A. eds. Wiley and Sons, Inc., pp. 153-167.

Yan, L.-J., Levine, R.L., and Sohal, R.S. (1997). Oxidative damage during aging targets mitochondrial conitase. Proc. Natl. Acad. Sci. USA 94, 11168-11172.

Yan, L.-J., and Sohal, R.S. (1998). Mitochondrial adenine nucleotide translocase is modified oxidatively during aging. Proc. Natl. Acad. Sci. USA 95, 12896-12901.

Yu, C.E., Oshima, J., Fu, Y.H., Wijsman, E.M., Hisama, F., Alisch, R., Matthews, S., Nakura, J., Miki, T., Ouais, S., Martin, G.M., Mulligan, J., and Schellenberg, G.D. (1996). Positional cloning of the Werner's syndrome gene. Science 272, 258-262.

Zhang, H., Tombline, G., and Weber, B.L. (1998). BRCA1, BRCA2, and DNA damage response: Collision and Collusion? Cell 92, 433-436.

Zhu, J., Woods, D., McMahon, M., and Bishop, J.M. (1998). Senescence of human fibroblasts induced by oncogenic raf. Genes and Dev. 12, 2997-3007.

© 2001 Elsevier Science B.V. All rights reserved.
The Role of DNA Damage and Repair in Cell Aging
B.A. Gilchrest and V.A. Bohr, volume editors.

AGE-ASSOCIATED CHANGES IN DNA REPAIR AND MUTATION RATES

Lawrence Grossman[1], Shin-Ichi Moriwaki,[2] Satyajit Ray,[1] Robert E. Tarone,[3] Qingyi Wei,[4] Kenneth H. Kraemer[2]

[1]Department of Biochemistry, The Johns Hopkins University School of Hygiene and Public Health, Baltimore, MD; [2] Laboratory of Molecular Carcinogenesis, National Cancer Institute, NIH, Bethesda, MD; [3] Biostatistics Branch, National Cancer Institute. National Institutes of Health, Bethesda, MD; [4] Department of Epidemiology, The University of Texas M.D. Anderson Cancer Center, Houston, TX

1. Introduction

Human populations typically display a range of inherent sensitivities to radiation and chemical carcinogens. Given a common carcinogenic insult, some individuals develop associated neoplasms, some may develop only an associated pre-neoplastic lesion, while others remain clinically-free from all related effects of the exposure. Even within specific cancer populations, age of onset, extent and severity of neoplasia often vary between patients. Such variability in host response, in part, may be due to inherent differences between individuals to monitor and repair damaged sites induced in their genetic material by exogenous and endogenous genotoxic agents. In most cases mutations at specific loci appear to provide the necessary signal releasing these genes from a cryptic or quiescent state eventuating in cancer formation. The linkage between persistent DNA damage and oncogene activity suggests that such long-lived DNA damage is a reflection of the diminished involvement of DNA repair or surveillance activities in tumorigenesis.

A human model supporting such assumptions exists with the rare, cancer-prone inherited disorder (xeroderma pigmentosum) (XP) (Kraemer and Slor, 1985 and Cleaver and Kraemer 1980). XP patients experience a greater than 2,000-fold excess frequency of sunlight related skin cancers. Coupled to this marked susceptibility is the consistent laboratory finding that all cells tested from the XP patients are defective in repairing DNA damage induced by ultraviolet (UV) and other UV-mimetic agents. Since the UV component of solar radiation exists as the predominant environmental risk factor for skin cancer, a causal association between UV exposure, defective repair of UV-induced DNA photoproducts, and skin cancer is inferred. The link is further strengthened by clinical reports showing the occurrence of skin tumors is practically ameliorated in XP patients afforded early and lifelong protection from sunlight exposure (Pawsey *et al* 1979 and Lynch *et al* 1977).

Although the rarity of XP precludes any particular significance as a public health concern, the clear etiologic model of cancer susceptibility it provides may have relevance to the general population. This is made apparent through three basic observations. The first is that the defect in XP is not complete, but is expressed phenotypically as a range of diminished repair proficiencies estimated to encompass residual capabilities of <2% up to approximately 80% of "normal". The second is that besides UV damage, XP cells are equally defective in repairing genetic damage induced by a variety of "bulky"

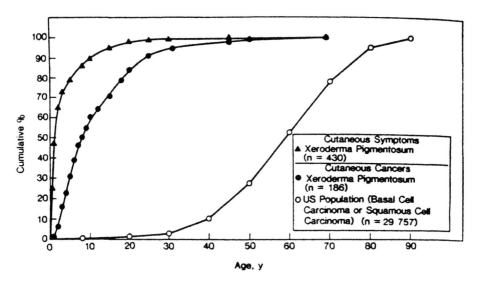

Figure 1. Age at onset of xeroderma pigmentosum symptoms. Age at onset of cutaneous symptoms (generally sun sensitivity or pigmentation) was reported for 430 patients. Age at first skin cancer was reported for 186 patients and is compared with age distribution for 29, 757 patients and is compared with basal cell carcinoma or squamous cell carcinoma in the US general population (Kraemer et al, 1987).

genotoxic drugs and chemical carcinogens. Lastly, like many other human phenotypic traits, individual variability in DNA repair capacity DRC has been demonstrated (Setlow, 1985).

Taken together, these observations show that DRC could exist as an etiologic correlate of cancer risk outside of XP. If XP is considered to represent the lower range of repair capabilities in humans, those individuals expressing a somewhat reduced repair response within the upper quadrant of the shoulder of a dose-response curve may be at increased risk for skin cancer or internal neoplasms, given an appropriate exposure. Because it can be anticipated that human populations may include individuals showing only marginal damage vulnerability, an assay was developed capable of detecting the level of DRC found in the heterozygotes of autosomal recessive DNA repair diseases with extreme precision and minimal intra-assay variation. This methodology has been substantiated with a pilot project (Athas *et al*, 1991) of 38 patients with BCC and 27 controls, which was further extended to a larger population study permitting stratification of data capable of determining the role of repair as a susceptibility factor in disease (Wei et al 1987)

If XP is considered to represent the lower range of repair capabilities in humans, those individuals expressing a somewhat reduced repair response may be at increased risk for skin cancer or internal neoplasms given an appropriate exposure. It can be anticipated that human populations may include individuals showing only marginal damage vulnerability, therefore, any DNA repair assay must be able to detect the level of DRC found in the heterozygotes of autosomal recessive DNA repair diseases with extreme precision and minimal intra-assay variation given that 5 percent of the population may be carrying *xp* genes (Friedberg 1985).

Further, the incidence of sunlight-induced skin cancer in the general population increases with increasing age, Figure 1, (Kraemer *et al* 1987). Sun exposure results in DNA damage in the form of photoproducts such as cyclobutane dimers, 6-4 pyrimidine-pyrimidone adducts and other lesions (Cleaver and Kraemer, 1989) that influence cellular functions including replication, transcription and DNA repair. Characteristic mutations in human skin cancers are closely associated with UV type photoproducts (Pawsey *et al*, 1979, Brash *et al* 1991). DNA repair and cell cycle regulation are major cellular systems that cope with UV-induced DNA damage Bredberg *et al* 1986) Aging in humans is associated with an accumulation of cellular mutations in vivo (Hartwell and Kastan 1994, Cole *et al* 1988, Tates *et al*, 1989 Cole *et al* 1992, Robinson *et al*, 1994). The mechanism by which these mutations accumulate is not understood. Several human diseases with DNA repair deficiencies are associated with accelerated aging (Scotto *et al* 1982, King *et al* 1994, and Bernstein and Bernstein, 1991). For example, patients with the rare inherited disease xeroderma pigmentosum have photodamaged skin as teenagers that is similar to farmers or sailors with many years of sun exposure. Xeroderma pigmentosum patients develop sun induced skin cancers about 50 years earlier than in the normal population - Figure 1 (Kraemer et al 1997). Xeroderma pigmentosum patients have defects in their DNA repair capacity (DRC) (Cleaver and Kraemer, 1995, Protic-Sabljic and Kraemer, 1985, Athas *et al*, 1991 Wei *et al* 1993). The life spans of organisms have been compared with the overall efficiencies of their DNA repair systems (Bernstein and Bernstein, 1991, Hart and Setlow, 1974). Results from such studies have provided no clear consensus in support of a progressive loss of DRC in connection with either aging or senescence in human populations (Hart and Setlow, 1974, Lambert *et al* 1979 and Nette *et al* 1984).

2. Previous Findings

In our previous study (Wei *et al*, 1993) it was indicated that fresh circulating lymphocytes from young people with skin cancers have significantly lower post-UV DRC than age-matched controls. The post-UV DRC was examined by measuring the expression level of a non-replicating UV-damaged recombinant DNA plasmid (pCMV*cat*) which codes for the reporter gene chloramphenicol acetyltransferase (*cat*, Kraemer *et al* 1994; Protic-Sabljic and Kraemer 1985). We observed that the post-UV DRC of fresh circulating T-lymphocytes in the control population declined at an estimated 0.6% per annum over four decades from 20 to 60 years of age (Wei et al 1993; Moriwaki, *et al* 1996).

Table 1   *Multiple linear regression modelling of DNA repair related to risk factor among basal cell carcinoma cases (BCC) and controls*

Parameter	Estimate[a]	T-value[b]	p-value[c]
Model 1 (control only, n=135); F-test for the model: F-value=6.36; R^2=0.088; p=0.002			
Intercept	11.725	10.23	0.000
Age in years	-0.071	-3.47	0.000
Sex	-0.409	-1.11	0.271
Model 2 (BCC case only, n=88); F-test for the model: F-value=2.17; R^2=0.049; p=0.120			
Intercept	10.445	6.96	0.000
Age in years	-0.047	-1.85	0.067
Sex	-0.547	-1.26	0.211
Model 3 (BCC case only, n=88); F-test for the model: F-value=3.51; R^2=0.111; p=0.019			
Intercept	11.374	7.77	0.000
Age in years	-0.134	-3.09	0.003
Sex	-0.360	0.84	0.404
Age of onset	0.879	2.44	0.017
Model 4 (All the subjects, n=223); F-test for the model: F-value=7.44; R^2=0.092; p=0.0001			
Intercept	11.569	12.81	0.000
Age in years	-0.067	-4.24	0.000
Sex	-0.423	-1.51	0.133
BCC FH	-0.645	-1.93	0.055

[a] Least square estimate of regression coefficient from multiple linear regression models.
[b] Student's t test for the null hypothesis that the estimate is equal to zero. |[c] Two-sided Student's t test.

Notes: All the models but model 2 are significantly different from a model with no risk factors according to the F-test. The variables included in these linear regression models are as follows: dependent variable is percent CAT activity at UV dose of 700 J/m^2 (continuous variable); independent variables: Age (at assay, in years); Age of onset (age at first BCC, in group): 1≤35, 2=35-44, 3=45-54, 4=55-60; Sex: 1=male, 2=female; BCC FH (family history): 1=without family history of BCC, 2=with BCC family history. Reprinted with permission from Wie et al, Proc Natl Acad Sci 1993; 90:1614-1618.

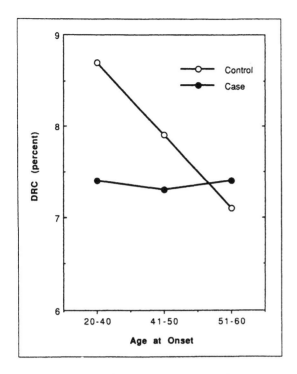

Figure 2. Relationship between age at first basal cell carcinoma and DNA repair capacity. The age related decline in DNA repair capacity among controls in comparison with that of age- matched cases is displayed. The linear regression modeling and statistical tests are presented in Table 1 (Wei et al 1993).

Multiple linear regression analysis shows that the DNA repair capacity of subjects declined with age over the 20 to 60 year span studied. Based on the estimates from the linear regression model for the 106 controls who had neither family history of skin cancer nor the presence of any premalignant skin lesions, the decline is 0.63 percent per year between 20-60 years of age (Table 1, model 1). This would amount to about a 25 percent decrease in cumulative DNA repair capacity over a 40 year period. This age-related decline occurred in both basal cell carcinoma cases (BCC) and controls, although it only reached a significant decline in the controls (Fig. 2, Table 1, models 1 and 2). These findings suggest that the basal cell carcinoma seen in the cases reflects the expected immortality associated with the transformed state.

In keeping with the XP paradigm, BCC cases with their first skin cancers at an early age, repair DNA photoproducts poorly when compared to controls or to those cases expressing BCC at a later age of onset (Fig. 2). This suggests that poor DNA repair is associated with the precocious skin aging as manifested by an early age of the first BCC. After adjusting for age at onset, the age-related decline in the repair of UV damage amongst the cases was at least as sharp as that of controls ($p=0.003$). When controlling for current age, the age at onset of BCC was positively correlated with DNA repair ($p = 0.017$), indicating the earlier age of onset, the lower the DNA repair (Table 1, model 3).

The group mean comparison data proposes that the mean DNA repair capacity of all BCC cases ($n=88$) is 5 percent lower than that of all control ($n=135$) and is of border-line significance ($p=0.103$). This difference is as high as 8 percent and statistically significant ($p=0.047$) when the controls with a familial history for the disease and with identifiable premalignant skin lesions are removed from comparison (Table 1, model 4). These findings suggest that heredity may also be associated with reduced DNA repair in the general population. After further adjustment for age and sex, prior family history of BCC is a statistically significant indicator of the individuals' DNA repair capacity regardless of whether they were BCC cases or control subjects (Table 1, model 4).

In the present study we sought to measure the ability of cells from individuals of different ages to repair UV-induced DNA damage and to establish a cellular model for age-related changes in the processing of damaged DNA. We examined post-UV DRC in undamaged primary skin fibroblast cultures obtained from donors from the first to the ninth decade of life. Post-UV cellular mutagenesis was determined by measuring the mutations induced by passage of a UV-damaged replicating recombinant DNA plasmid through undamaged human lymphoblastoid cells and assessing mutations in indicator bacteria (Moriwaki *et al* 1996). A donor age related decline in post-UV DNA repair capability and a corresponding increase in post-UV mutagenesis in these cultured cells was found.

3. Methods

A. *DNA Repair Capacity Skin fibroblasts.* The primary skin fibroblasts used in this study were obtained from 20 apparently normal donors of age 3 years to 96 years. All of the donors except 48BR, were male. Twelve of the cell lines were obtained from the subjects participating in the Baltimore Longitudinal Study of Aging conducted by the Intramural Research Program, National Institute on Aging at the Gerontology Research Center, Baltimore, USA and maintained by the Human Cell Repository, Camden, NJ. A uniform biopsy site (mesial aspect of the mid-upper left arm) and method for primary culture was used. Five cell lines were obtained from the National Institute of the General Medical Sciences Human Cell Repository, Camden, NJ, USA. Cell lines 1BR3, 48BR and 155BR were generously provided by Dr. C.F. Arlett of the MRC Cell Mutation Unit, University of Sussex, Brighton. All of the donors were Caucasian.

The fibroblasts were maintained in Dulbecco's modified Eagle's media (Gibco) supplemented with 10% fetal calf serum, non-essential amino acids and 100 units/ml of penicillin, 100 µg/ml streptomycin, and 0.25 µg/ml of fungazone. The passage numbers were determined by plating 10,000 cells per cm2 and then splitting at a density of 100,000 per cm^2 or confluence. This represents 2.5 doublings per plating. The primary skin fibroblasts employed in these studies were from similar passages and doubling times.

B. *Plasmid Assay.* Briefly, the plasmid DNA, pCMV*cat* harboring the*cat* gene was irradiated *in vitro* with 350 and 700 Jm^{-2} of 254 nm UV before transfection. Transfection with plasmid DNA was performed by a modification of the DEAE-dextran procedure of Athas *et al* (17). For the DRC assay 2×10^5 cells were inoculated into 60 mm tissue culture dishes at the same time as the plating efficiency of the flasks was determined. After 24 hrs, the cells were washed twice with 5 ml of TBS plus 0.1% dextrose Buffer. Two hundred and fifty microliter aliquots of a mixture containing 2 μg/ml DNA and 200 μg/ml DEAE-dextran in TBS plus 0.1% dextrose buffer were added to each plate and cells were transfected for an hour in a 37°C incubator. The cells were then washed with 2 ml culture medium and fresh media was added. After 40 hrs of incubation the standard assay for gene expression of CAT activity by measurement of [^{3}H]diacetylchloramphenicol, the product of the CAT reaction, in a Beckman Scintillation counter (LS 3801).

The DRC was calculated based on scintillation counts as the percentage of the residual CAT gene expression after repair of damaged DNA compared with undamaged DNA which is considered as 100%. In all cases, the undamaged gene generated ~30,000 to 200,000 dpm which reflects between 40 and 60 percent of the substrate converted to [^{3}H]-diacetylchloramphenicol. For each assay dose curves ranging from 0, 350 and 700 Jm^{-2} were performed in triplicate. The repair capacity at 700 Jm^{-2} was taken as the repair capacity for each subject since the dose curves are linear in this range of dosimetry (Athas *et al* 1991; Wei *et al* 1992). The variability of the assay (intra-individual variability), determined by using SV40 transformed control cell lines, was less than 0.05 of the mean.

C. *Mutagenesis Lymphoblastoid cell lines.* Fifteen normal control Epstein Barr virus transformed lymphoblastoid cell lines from donors ranging in age from 4 to 98 years were obtained from the Human Cell Repository, Camden NJ and one control line was a generous gift from Dr Alan Bale (Yale University, New Haven, CT). Cells were cultured in RPMI1640 (Gibco, BRL, Bethesda, MD) medium supplemented with 15% fetal bovine serum (S&S Media, Rockville, MD) and 2 mM L-glutamine (Gibco, BRL) at 37°C in a 5% CO_2 atmosphere as described (Moriwaki *et al* 1994).

The shuttle vector plasmid, pSP189, (Parris and Seidman 1992) was UV treated, transfected, recovered and assayed as previously described (Moriwaki *et al* 1994). Briefly, the 5.5-kilobase pair plasmid contains SV40 origin of replication, enhancer, and large-T antigen sequences that permit replication in human cells and the pBR327 origin of replication that leading to growth in bacteria. An approximately 150-base pair bacterial *sup* F suppressor tRNA gene serves as a marker for mutations and a gene for ampicillin resistance which permits selection in bacteria. pSP189 was treated with 1000 Jm^{-2} UV from a germicidal lamp and transfected into the human cells by use of DEAE dextran. Our previous study (Moriwaki *et al* (1994) showed a linear relationship between UV dose to the plasmid and the plasmid mutation frequency in lymphoblastoid cells from normal donors. Untreated plasmid pZ189K (with the gene for kanamycin resistance) was co-transfected as an internal standard. After 2 days the replicated plasmids were harvested from the cells and used to transform indicator bacteria. The bacteria were plated on agar containing ampicillin or kanamycin and colorless dye (IPTG plus X-

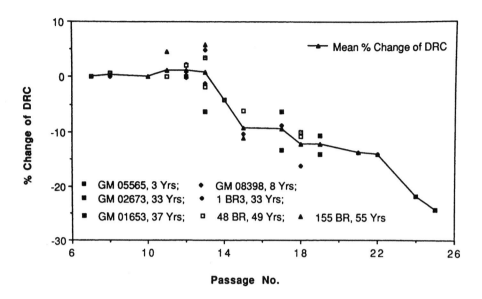

Figure 3. Relationship between fibroblast passage number and UV DNA repair capacity. Each point represents the % difference between the mean of the triplicate determinations of post-UV pCMVcat activity at the indicated passage number and the mean at the lowest passage level tested for each primary fibroblast cell line. GM05565, 3 yr (▣); GM08398, 8 yr (♦); GM02673, 33 yr (▪); 1BR3, 33 yr (◇); GM01653, 37 yr (▪); 48BR, 49 yr (□); 155BR, 55 yr (▲). The line indicates the mean of all values at a given passage level. The SEM represents + 0.02 thus the error bars are smaller than the individual points.

gal). Bacteria containing a plasmid with wild type *sup* F gene yield blue colonies while those containing plasmids with inactivating mutations in the *sup* F gene yield white or light blue colonies. The total number of colonies reflect plasmid survival while the proportion of white or light blue colonies is a measure of the plasmid mutation frequency. Plasmid mutation frequency was calculated as described (Moriwaki *et al* 1994) and the mean and standard error of the mean were calculated from three or more independent transfections with each donor line.

D. *Statistical Analysis*. The data was fitted to least-squares linear regression curves and the 95% prediction intervals and the mean of triplicate determinations from a single individual have a 95% chance of falling within the 95% prediction interval and was plotted using the SlideWrite Plus computer program. Two sided **p** values are reported.

4. Results

The effect of passage number of the primary fibroblasts on DRC was examined in order to determine the level of cellular senescence of the cells in culture. The post-UV DRC of

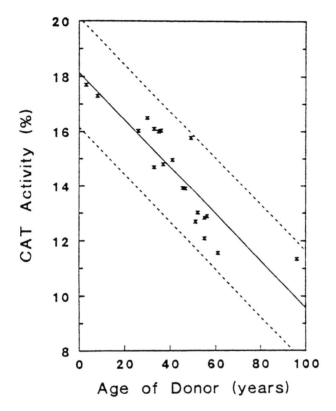

Figure 4. *Relationship between donor age and the post-UV DNA repair capacity. Post-UV plasmid pCMVcat activity was determined in fibroblast cell lines from 20 normal donors. The mean +/- SEM of triplicate determinations is shown for each donor. The solid line is the least squares linear regression line and the dashed lines indicate the 95% prediction interval. Correlation coefficient r =0. 88.*

7 primary fibroblast lines from normal donors of age 3 years to 55 years was determined in a range of cell passages from passage 7 to passage 25 (Figure 3). The results showed that the post-UV DRC of these populations did not change substantially prior to passage number 13. However, an approximately 10% decline in the post-UV DRC was observed after passage number 12–13 and an additional 10% decline between passages 15 and 25. Thus, cellular senescence in culture was associated with decline in post-UV DRC. Consequently, the post-UV DRC was measured only with fibroblasts of passage numbers less than 14.

The post-UV DNA repair capacity was determined for primary fibroblast cultures from 20 normal subjects of age 3 to 96 years (Figure 4). Post-UV DRC was found to decrease with donor age. Linear regression analyses of the DRC data indicated that approximately 80% of the variability in post-UV DRC can be explained by age. Linear regression analyses of the estimated post-UV DRC at birth was 18.1% and that at age 100 years was 9.6%. This corresponds to a rate of decrease of the post-UV DRC of

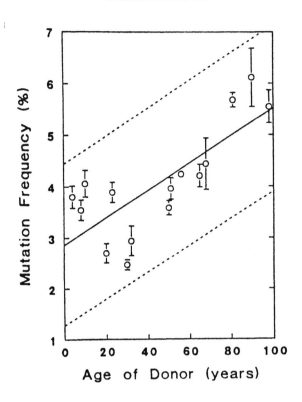

Figure 5. *Relationship between donor age and post-UV mutation frequency. Plasmid pSP189 was treated with UV and transfected into lymphoblastoid cell lines from 15 normal donors. Replicated plasmids were assayed for mutation frequency by transformation of indicator bacteria. The mean +/- SEM of triplicate determinations is shown for each donor. The solid line is the least squares linear regression line and the dashed lines indicate the 95% preselection interval. Correlation coefficient r=. 77*

0.6% per year (**p**=0.0001). The linearity of the age-related decline in post-UV DRC indicates a steady decline in the inherent DNA repair capability with age. Figure 4 also shows the 95% prediction interval. All 20 normal donors tested had values within this 95% prediction interval.

The relationship of donor age with cellular growth rate (doubling time) was determined for 5 of the skin fibroblast cell lines: cell line-AG09605, donor age-30 yr, doubling time 26 hr; 1BR3, 33 yr, 33 hr; AG07724, 46 yr, 20 hr; AG07136, 51 yr, 34 hr; 155BR, 55 yr, 32 hr. Correlation analyses were performed and the Pearson correlation coefficient was found to be 0.18 (**p**=0.77, not significant). For these 5 cell lines the Pearson correlation coefficient between donor age and post-UV DRC was -0.95 (**p**=0.012). Thus, as was the case with fresh human lymphocytes (Wei et al 1993), the decrease in post-UV DRC was not linked to changes in cellular growth rate but was related to donor age.

A. *Post-UV Plasmid Mutation Frequency*. UV-treated plasmid was passed through lymphoblastoid cell lines from 15 normal donors ranging in age from 4 years to 98 years and assessed for mutation frequency (Figure 5). The post-UV plasmid mutation frequency was found to increase with donor age. Linear regression analysis indicated that approximately 60% of the variability in post-UV mutation frequency can be explained by age. Using linear regression, the estimated post-UV mutation frequency at birth was 2.9×10^{-2} and that at age 100 years was 5.5×10^{-2}. This corresponds to a rate of increase of the post-UV plasmid mutation frequency of 0.6% per year ($p=0.001$). Figure 5 also shows the 95% prediction interval. All 15 normal donors tested had values within this 95% prediction interval.

B. *Post-UV Effects at the Chromosome Level* (Van der Riet et al 1994). Unlike squamous cell carcinoma of the skin, BCC is generally indolent, non-invasive and rarely metastatic. To study the involvement of tumor suppressor genes in these neoplasms, we analyzed 36 BCCs for p 53 mutations and a subset of these tumors for loss of chromosome 17p and 9q. Sixty-nine percent of sporadic BCCs had lost a 9q allele with the common area of loss surrounding the putative gene for nevoid BCC or Gorlin's syndrome. Forty-four % (16 of 36) of BCCs had a mutated p 53 allele, usually opposite pyrimidine tracts, which is consistent with UV-induced mutations. Surprisingly, only one tumor had lost a 17p allele, and all the BCCs only one p 53 allele was inactivated. This is in direct contrast to other epithelial tumors, which usually progress by the inactivation of both p 53 alleles.

5. Discussion

A. *Shuttle vectors and DNA damage processing*. Shuttle vectors have been used to measure the ability of cells to repair damaged DNA. This host cell reactivation assay is dependent on cellular systems to repair the damage in plasmid DNA. Plasmids have been developed to measure DNA repair, UV hypersensitivity and UV mutagenesis in human cells (Protic-Sabljic and Kraemer, 1991, Athas *et al*, 1991, and Wei *et al*, 1993). Use of plasmids damaged *in vitro* ensures that the DNA damage is the same in all cells, avoids radiation exposure to the cells, and permits a rapid assessment of post-UV plasmid survival and mutagenesis. Abnormalities have been reported in cells from patients with xeroderma pigmentosum Bredberg *et al* 1986, Protic-Sabljic and Kraemer, 1985, Athas *et al*, 1991, Seetharam *et al*, 1987, Yagi *et al* (1993 and Waters *et al*, 1993), Cockayne syndrome (Barrett *et al*, 1991 and Parris and Kraemer 1993), familial melanoma (Moriwaki *et al* 1994) and in apparently normal individuals with non-melanoma skin cancer (Wei *et al* 1993).

B. *DNA Repair, mutagenesis and aging*. Data exist demonstrating accumulation of mutations with increasing age. The studies of Cole *et al* (1988), Tates *et al* (1989), Cole *et al* (1992) and King *et al* (1994) revealed an increased frequency of spontaneous hypoxanthinephopsphoribosyl transferase (HPRT) mutations with donor age in fresh peripheral blood T-lymphocytes from normal individuals. Both groups reported a rate of increase of 1.3% per year in the frequency of accumulated mutations in the HPRT gene in adults. Such an increase can result from an accumulation of DNA damage over time, a decrease in the ability to process new DNA damage with increasing age, or a combination of both. However, studies of *in vivo* HPRT mutations cannot provide information as to the effect of age on cellular processing of newly damaged DNA.

There have been very few studies that measured age-related changes in the ability of human cells to process new DNA damage. Nette *et al* 1984 and Lambert *et al* 1979 reported a decrease in post-UV unscheduled DNA synthesis equivalent to about 1.3% per year from age 17 to 77 years in dissociated human epidermal cells. Previously we observed that the post-UV DRC of fresh circulating T-lymphocytes in a control population declined at an estimated rate of 0.6% per year over four decades from 20 to 60 years of age (Wei *et al* 1993). The present study confirms this finding in a different cell type: we found a decrease in post-UV DNA repair capacity of 0.6% per year in cultured skin fibroblasts (Figures 2). This study also found a "mirror image" (perfect inverse correlation) increase of 0.6% per year in the post-UV plasmid mutation frequency in cultured lymphoblastoid cells (Figure 5). Examination of DRC and mutation data (Figures 4 and 5) suggest that the slopes of both curves may change with age, with both curves possibly becoming steeper at older ages. However, multiple piece-wise linear regression analyses (Neter and Wasserman, 1974) indicate that the changes in slope with age are not significant for either DRC or mutation data. The changes in DRC and mutation frequency with age suggest that an important consequence of the age-related decline in DRC is the cumulative increase in persistent DNA damage that escaped DNA repair. The physiological consequence of enduring such DNA damage is its influence on those mutational events required for tumor progression (Peto *et al* 1975, and Magee 1978) as well as on the rate of spontaneous mutations of the HPRT locus associated with the progression of the aging process (Cole *et al* 1988, Tates *et al*, 1989 and Cole *et al* 1992) and King *et al* (1994). Importantly, there is a strong inverse correlation between the loss of DNA repair capacity (Figure 4) and the increase in mutations in UV treated pSP189 (Figure 5). These findings are consistent with hypotheses invoking the persistence of DNA damage arising from reduced DNA repair (Bernstein and Bernstein 1991) or alterations in cell cycle regulation (Hartwell and Kastan, 1994) and the resulting impaired ability of cells to deal with damaged DNA as people age. Despite our lack of understanding of the detailed mechanism of these age-associated cellular alterations, they may play a role in the greatly increased rate of sunlight related skin cancer occurrence with age in the normal population (Scotto *et al* 1982, 1994).

C. *Molecular model for aging.* Earlier studies utilizing fresh peripheral blood lymphocytes showed decreased DNA repair of damaged DNA (Athas *et al* 1991 and Wei et al, 1993) and increased accumulation of spontaneous mutations (Cole *et al* 1988, Tates *et al*, 1989 and Cole *et al* 1992) and King *et al* (1994) with age. Cellular transformation of fibroblasts or lymphocytes has been shown to produce similar post-UV plasmid DNA repair capacity (Protic-Sabljic and Kraemer 1985 and Athas *et al* 1991) or post-UV plasmid mutation frequency (Moriwaki *et al* 1996) as primary fibroblasts or lymphocytes. In the present study we used cultured primary skin fibroblasts and EB virus transformed lymphoblastoid cell lines to show a decreased ability to process new UV-induced DNA damage with increasing age. While the mechanism underlying these abnormalities is not known in detail, the finding of similar abnormalities in several types of cultured cells should provide a model for laboratory studies of the effect of age on DNA damage processing in humans.

Acknowledgements

Grant support from NIH Merit award GM-22846, P30 ES-02819 and CA-62924

References

Athas W. F, Hedayati MA, Matanoski et al. Development and field-test validation of an assay for DNA repair in circulating lymphocytes. Cancer Res 1991; 51:5786-579.

Bernstein, C., and H. Bernstein (1991) Aging, sex, and DNA repair, Academic Press, San Diego.

Brash, D. E., J. A. Rudolph, J. A. Simon, A. Lin, G. J. McKenna, H. P. Baden, A. J. Halperin and . Pontn (1991) A role for sunlight in skin cancer: UV-induced p53 mutations in squamous cell carcinoma, Proc. Natl. Acad. Sci. U. S. A. 88, 10124-10128.

Bredberg, A., K. H. Kraemer and M. M. Seidman (1986) Restricted ultraviolet mutational spectrum in a shuttle vector propagated in xeroderma pigmentosum cells, Proc. Natl. Acad. Sci. 83, 273-8277.

Cleaver, J. E. and Kraemer, K. H. Xeroderma pigmentosum. In Scrivner, C. R., Beuber, A. L., Sly. W. S. et al ed: The metabolic basis of inherited diseases II, New York, NY: McGraw Hill Inc. 1989 2949-2971

Cleaver, J. E. and K. H. Kraemer (1995) Xeroderma pigmentosum and Cockayne syndrome, in: C. R. Scriver, A. L. Beaudet, W. S. Sly and D. Valle (Eds.), The Metabolic and Molecular Bases of Inherited Disease, 7th Edn., McGraw-Hill, Inc. New York, pp. 4393-4419.

Cole, J., M. H. Green, S. E. James, L. Henderson and H. Cole (1988) A further assessment of factors influencing measurements of 6-thioguanine-resistant mutant frequency in circulating T-lymphocytes, Mutat. Res. 204, 493-507

Cole, J., C. F. Arlett, P. G. Norris, G. Stephens, A. P. Wah, D. M. Beare and M. H. Green (1992) Elevated hprt mutant frequency in circulating T-lymphocytes of xeroderma pigmentosum patients, Mutat. Res. 273, 171-178.

Friedberg, E. C. (1985) DNA Repair, W. H. Freeman, New York.

Hall, J., D. R. English, M. Artuso, B. K. Armstrong and M. Winter (1994) DNA repair capacity as a risk factor for non-melanocytic skin cancer—A molecular epidemiological study, Int. J. Cancer, 58, 179-184.

Hart, R. W. and R. B. Setlow (1974) Correlation between deoxyribonucleic acid excision-repair and life-span in a number of mammalian species, Proc. Natl. Acad. Sci. U. S. A. 71, 2169-2173.

Hartwell, L. H. and M. B. Kastan (1994) Cell cycle control and cancer, Science, 266, 1821-1828.

King, C. M., E. S. Gillespie, P. G. McKenna and Y. A. Barnett (1994) An investigation of mutation as a function of age in humans, Mutat. Res. 316, 79-90.

Kraemer, K. H., M. -M. Lee, A. D. Andrews and W. C. Lambert (1994) The role of sunlight and DNA repair in melanoma and non melanoma skin cancer: The xeroderma pigmentosum paradigm, Arch. Dermatol. 130, 1018-1021

Kraemer, K. H., Lee, M-M and Scotto, J. (1987) Xeroderma pigmentosum, cutaneous, ocular and neurologic abnormalities in 830 published cases, Arch. Dermatol 123:241-250

Lambert, B., U. Ringborg and L. Skoog (1979) Age-related decrease of ultraviolet light-induced DNA repair synthesis in human peripheral leukocytes, Cancer Res. 39, 2792-2795.

Lehmann AR. Ageing, DNA repair of radiation damage and carcinogenesis: Fact and fiction. In: Likhachev A et al. ed. Age-Related Factors in Carcinogenesis. WHO, IARC, Lyon, 1985:203-209.

Magee, P. N. (1978) Carcinogenesis and aging, Adv. Exp. Med. Biol. 97, 133-146.

Moriwaki, S. -I., R. E. Tarone and K. H. Kraemer (1994) A potential laboratory test for dysplastic nevus syndrome: Ultraviolet hypermutability of a shuttle vector plasmid, J. Invest. Dermatol. 103, 7-12.

Moriwaki, S-I., Ray, S., Tarone, R. E., Kraemer, K. H., and Grossman, L. (1996) The effect of donor age on the processing of UV-damaged DNA by cultured human cells: Reduced DNA repair capacity and increased DNA mutability Mutation Res. 364: 117-123

Nette, E. G., Y. P. Xi, Y. K. Sun, A. D. Andrews and D. W. King (1984) A correlation between aging and DNA repair in human epidermal cells, Mech. Ageing Dev. 24, 283-292.

Neter, J. and W. Wasserman (1974) Applied linear statistical models, Irwin Co. Homewood, Il

Parris, C. N. and K. H. Kraemer (1992) Ultraviolet mutagenesis in human lymphocytes: the effect of cellular transformation, Exp. Cell Res. 201, 462-469.

Parris, C. N. and M. M. Seidman (1992) A signature element distinguishes sibling and independent mutations in a shuttle vector plasmid, Gene, 117, 1-5.

Parris, C. N. and K. H. Kraemer (1993) Ultraviolet-induced mutations in Cockayne syndrome cells are primarily caused by cyclobutane dimer photoproducts while repair of other photoproducts is normal, Proc. Natl. Acad. Sci. U. S. A. 7260-7264.

Pawsey, S. A. Magnus, L. A., Ramsey et al. (1979) Clinical, Genetic and DNA repair studies on a consecutive series of patients with xeroderma pigmentosum. Quart. J. Med 48: 179-210

Protic -Sabljic, M. and K. H. Kraemer (1985) One pyrimidine dimer inactivates expression of a transfected gene in xeroderma pigmentosum cells, Proc. Natl. Acad. Sci. U. S. A. 82, 6622-6626.

Robinson, D. R., K. Goodall, R. J. Albertini, J. P. O'Neill, B. Finette, M. Sala-Trepat, E. Moustacchi, A. D. Tates, D. M. Beare, M. H. L. Green and J. Cole (1994) An analysis of in vivo hprt mutant frequency in circulating T-lymphocytes in the normal human population: A comparison of four data sets, Mutat. Res. Environ. Mutagen. Rel. Subj. 313, 227-247.

Peto, R., F. J. Roe, P. N. Lee, L. Levy and J. Clack (1975) Cancer and ageing in mice and men, Br. J. Cancer, 32, 411-426.

Seetharam, S., M. Protic -Sabljic, M. M. Seidman and K. H. Kraemer (1987) Abnormal ultraviolet mutagenic spectrum in plasmid DNA replicated in cultured fibroblasts from a patient with the skin cancer-prone disease, xeroderma pigmentosum, J. Clin. Invest. 80, 1613-1617.

Scotto, J., J. F. Fraumeni, Jr., T. R. Fears and K. H. Kraemer (1994) Skin cancer (other than melanoma), in: D. Schottenfeld and J. F. Fraumeni, Jr. (Eds.), Cancer Epidemiology, W. B. Saunders Co. Philadelphia, pp. (In Press)

Scotto, J., T. R. Fears and J. F. Fraumeni (1982) Incidence of non-melanoma skin cancer in the United States, U. S. Department of Health and Human Services, Bethesda, MD.

Tates, A. D., L. F. Bernini, A. T. Natarajan, J. S. Ploem, N. P. Berwoerd, J. Cole, M. H. Green, C. F. Arlett and P. N. Norris (1989) Detection of somatic mutants in man: HPRT mutations in lymphocytes and hemoglobin mutations in erythrocytes, Mutat. Res. 213, 73-82.

van der Riet, Karp, D., Farmer, E. V., Wei, Q., Grossman, L., Tokino, K. Ruppert, J. M. and Sidransky, D (1994) Progression of Basal Cell Carcinoma through loss of Chromosome 9q and Inactivation of a single Allele. Cancer Res. 54:25-27

Waters, H. L., S. Seetharam, M. M. Seidman and K. H. Kraemer (1993) Ultraviolet hypermutability of a shuttle vector Barrett, S. F., J. H. Robbins, R. E. Tarone and K. H. Kraemer (1991) Evidence for defective repair of cyclobutane pyrimidine dimers with normal repair of other DNA photoproducts in a transcriptionally active gene transfected into Cockayne syndrome cells, Mutat. Res. 255, 281-291.

Wei, Q., G. M. Matanoski, E. R. Farmer, M. A. Hedayati and L. Grossman (1993) DNA repair and aging in basal cell carcinoma: a molecular epidemiology study, Proc. Natl. Acad. Sci. U. S. A. 90, 1614-1618.

Yagi, T., J. Tatsumi-Miyajima, M. Sato, K. H. Kraemer and H. Takebe (1991) Analysis of point mutations in an ultraviolet-irradiated shuttle vector plasmid propagated in cells from Japanese xeroderma pigmentosum patients in complementation groups A and F, Cancer Res. 51, 3177-3182.

Ziegler, A., D. J. Leffell, S. Kunala, H. W. Sharma, M. Gailani, J. A. Simon, A. J. Halperin, H. P. Baden, P. E. Shapiro, A. E. Bale and D. E. Brash (1993) Mutation hotspots due to sunlight in the p53 gene of non melanoma skin cancers, Proc. Natl. Acad. Sci. U. S. A. 90, 4216-4220

© 2001 Elsevier Science B.V. All rights reserved.
The Role of DNA Damage and Repair in Cell Aging
B.A. Gilchrest and V.A. Bohr, volume editors.

RELATIONSHIP OF DNA REPAIR CAPACITY TO SPECIES LIFESPAN

R. W. Hart, A. Turturro, and S.Y. Li

National Center for Toxicological Research, 3900 NCTR Road, Jeferson, AR

1. Introduction

Historically science either attempts to discern the impact of an event, substance or condition on an organism or how different organisms, sexes, stains or species respond to such challenge. These two approaches are also reflexive of how we approach gerontological research. The former approach, which can be called the ontogenetic approach, seeks to account for the progressive age related changes that occur in an organism as a function of time. The latter approach can be termed the evolutionary-comparative approach and looks for differences in the genetically determined constitutive characteristics of individuals, sexes, strains and species that may be responsible for differences in response of the organism to challenge or, in gerontological research, the rate of aging. Until recently the ontogenetic approach has dominated the contemporary research scene. However, with the advent of gene sequencing this may be about to change.

The ontogenic approach while it produces important kinds of information, unfortunately, is limited in its ability to resolve issues of cause and effect. No matter how close to the fundamental biosynthetic processes an observed age change may be it is still an effect, for which there must be an antecedent cause. It is difficult, within the framework of the ontogenetic paradigm, to determine the causes of aging, i.e. those controlling parameters that determine whether or not a given molecular event will occur, and the rate at which it will occur.

The evolutionary-comparative paradigm is based on two postulates: First, a difference in the aging rates of two basically similar organisms is due to a difference in the physicochemical environments of the essential macromolecules that constitute the two organisms; and second, the parameters of the molecular environment within an organism are an expression of its genome. More specifically, the hypothesis is that the eventual fate of the organism, in terms of its rate of aging, disease susceptibilities, and life span, is primarily determined by specific genetically determined constitutive properties, both molecular and organizational, which can be measured in its youth, prior to the onset of aging and predict the impact of time on the organism, i.e. serve as biomarkers of aging (Sacher and Hart, 1978).

2. Comparative Approach

The evolutionary aspect of the evolutionary-comparative paradigm examines longevity in relation to the other aspects of population biology in order to determine the selective factors that contributed to the evolution of longevity. This approach is based on the belief that knowledge of these selective forces will assist in the discovery of the genetic mechanisms that effectuate evolutionary increase in lifespan, and will thereby contribute to an understanding of the underlying mechanisms that control the rate of aging. This is not to suggest that these mechanisms per se are immune to age change; indeed there is reason to assume they are not. Rather, it is to suggest that the nature and rate of the aging process for each species is essentially determined by its phenotypic constitution and can be observed in early life. It is also important that such genes have not to be either numerous nor specific to aging; rather it may be more important that they are fundamental to the initiation of numerous events in response of the organisms to its environment and critical to basic evolutionary principles such as survival of the species.

Thus, when attention is focused away from aging to survival till the age of reproduction, a shift in philosophical approach is achieved away from genes that cause death toward identification of positive, genically controlled mechanisms that ensure survival and longevity. The question of how and why aging evolved becomes meaningless and what becomes important are the nature of those longevity-assurance systems that are responsible for length of life and how do these differ between individuals, sexes, strains and species. This approach to aging places the field of gerontology on a strong foundation of investigating those systems by which organisms are able to maintain homeostasis in the face of a less than friendly environment. Two questions arise: How do these differences relate to the overall ability of an organism to survive? How and for what reasons have they changed so rapidly within the placental mammals, as represented by the rapid increase in primate and even human longevity over a relatively short period of time, with the latter rate approaching 1,000 milli-Darwins (Sacher, 1975)?

Over the course of the last century the comparative-evolutionary approach to the study of longevity has advanced from the study of organism and organ systems to identification of the differences between genetically defined populations at the biochemical and physiological levels. With the advent of gene sequencing, the probable next step will be both the identification of differences in specific gene sequences and organizational differences at the molecular level leading to differences in gene expression and functionality. Each of these stages in the comparative-evolutionary approach has or is lending credibility to the concept that a limited number of homeostatic systems serve as the rate limiting factors for longevity and that manipulation of these system alters both average and maximum longevity within individuals and species.

3. Model Development

Stage one. In gerontology the comparative-evolutionary approach began with Rubner (1908) who noticed an inverse relationship between mammalian life span and metabolic rate, but noted humans fell outside this relationship. In 1910 (Friedenthal, 1910) noted that the discrepancies in Rubner's calculation could be reduced by introducing an additional factor for brain size, with larger brained animals living longer than smaller brained ones. Little changed until 1959 when Sacher (Sacher, 1959) carried out a multiple regression analysis showing that the three variable (life span, metabolic rate and brain size) should be logarithmically transformed, as would normally be found for allometric relations between dimensions. He showed that the multiple regression found by least squares was:

Log L = 0.64 log E – 0.23 log S + 1.035

Where life span is denoted by L, brain weight by E and body weight by S. Since the squared multiple correlation of the above is 0.84 this means that 84% of the life span variance among the 63 species of mammals studied by Sacher (Sacher, 1959) could be accounted for by the allometric relation to brain weight and body weight. This relationship has been shown to hold both within as well as across orders of mammals as well as reptiles and birds (Mailouk, 1975). The above equation was further refined by taking temperature differences (due to differences in body mass) into account leading to the following:

Log L = 0.62 log E – 0.41 log S – 0.52 log M + 0.026T_b + 0.894

where T_b is deep body temperature in degrees Celsius. In other words it would appear that at least within mammals and maybe within other classes of tetrapods over 80% of all the variance in maximum achievable life span of the species can be accounted for by regression on two constitutional variables, brain weight and body weight and on two metabolic variables, specific metabolic rate and body temperature. These relationships, taken together with information on the relation of these constitutional variables to mammalian growth rate and reproduction (Sacher and Staffeldt, 1974) support the hypothesis that longevity is a component of fitness and ability to maintain homeostasis. It is especially interesting that brain, which is the principle organ responsible for neuroendocrine response and coordination of reproductive function and growth, is so closely tied to longevity and the capacity of the organism to respond to stress.

Stage Two. The second phase of study using the comparative-evolutionary approach started with the discovery of Hart and Setlow (1974) who first demonstrated a direct relationship between species maximum achievable life span and the ability of cells from that species to perform DNA excision repair. The importance of this and similar initial studies were not fully appreciated in the evolutionary-comparative paradigm. It had independently been speculated that the capacity for coordinated and sometimes rapid selective modification of the survival characteristics of a population implies that either longevity is governed by a relatively small number of genetically controlled longevity-assurance mechanisms, or that the overall longevity-assurance system, however complex, is controlled by a comparatively small number of genes potentially located very close to each other (Sacher, 1975). If the latter is correct then it is also a reasonable speculation that longevity may be under the same regulatory gene control or regulation as those systems that govern an organism size and/or homeostatic capability as a whole. This is especially the case since increased body size would (all other factors being equal) lend to an increased risk of death as result of clonal diseases such as cancer. This does not occur, indeed with an increase in body size there is an increase in the species longevity. For example across the placental mammals there is a fifty (50) fold increase in the maximum achievable life span and a ten thousand (10,000) – fold increase in the number of cells per organism with larger species generally living longer than smaller ones. This is an increase of approximately five hundred thousand (500,000) – difference in risk per susceptible cell per unit time for spontaneous transformation. Larger organisms are generally more competitive. Regardless of class, within the vertebrates there is always a drift as a function of time toward larger species, which are more dominant within various ecological niches. In order to achieve this condition successfully, selection may have allowed only those organisms capable of enhanced cellular homeostasis to achieve increased body size. Longevity, therefore, may be a by-product of the genetic stability and enhanced information processing systems responsible for preventing the onset of clonal diseases such as cancer prior to reproduction.

If indeed longevity and cancer/mutagenesis are related as a result of natural selection for more competitive (and in many cases larger) species via a requirement to maintain cellular homeostasis until the time of reproduction, then it is reasonable to assume that: 1) a direct measurement of changes in the target macromolecule (DNA) as a function of species life span should be observable; 2) that conditions that slow the rate of accumulation of genetic damage should be enhanced; 3) that conditions that uniformly alter the maximum achievable life span such as caloric restriction should alter functions protective of integrity, such as DNA repair (Turturro and Hart, 1991); 4) that strains, species and individuals defective in repair of genetic damage should exhibit shorter life span and 5) that conditions that improve homeostasis should lengthen life span. Sufficient data now exist to say that genomic integrity fulfills each and everyone of the conditions cited above and thus failure to maintain genomic integrity may be fundamental as a causative factor in longevity (Turturro, et al., 1999).

4. Interspecies Relationship Between DNA Repair and Longevity

DNA, RNA, protein, and lipid membranes are each a possible target for molecular senescence. Each is a major component of the cell, and each shows alterations with age. However, DNA not only appears as a unique copy within a cell, but is also larger than any other macromolecule and the template for all other cellular components. In addition DNA is also the primary target of evolution change, thus alteration in this molecule are of particular significance. The only direct proof that DNA damage is causative in aging and that DNA repair reversed this effect came from studies in *Paramecium* where Smith-Sonneborn (1979) showed that the life span of *Paramecium* could be shortened with UV- (254 nm) light and restored by photoreactivation (exposure to white light). Since photoreactivation is specific for the monomerization of UV (254 nm) light induced cyclobutane-type pyrimidine dimers, this was the first study to demonstrated that a specific form of DNA damage was causal in aging. While a number of correlation and direct genetic studies covered in this volume strongly suggests that the same may be true in higher organisms, there is no direct evidence that genetic damage is the cause of aging and that DNA repair prevents this event.

DNA can be damaged by physical, chemical or biological agents generated internally or externally. These agents can cause DNA breaks (breakage of one or both strands), distortions (kinking or opening up of the intertwined strands), adducts (addition of extra molecules to the DNA) and loss of bases such as a lose of a purine or pyrimidine.

Example of internal factors producing DNA damage are: 1) normal cellular metabolism, which generates free radicals and reactive metabolites capable of cross-linking DNA to proteins, RNA, and itself; 2) metabolism of certain externally originating molecules, producing excited molecules which form DNA adducts; 3) heat from body temperature, which causes a loss of bases and subsequently single-strand breaks and single-stranded regions in cellular DNA; and 4) the action of enzymes that can degrade DNA (Hart and Modak, 1980).

Examples of external factors producing DNA damage include: 1) ultraviolet light that produces, dependent upon wavelength, a number of forms of damage including the joining of adjacent bases to form bulky products that distort the DNA helix; 2) gamma rays and x-rays, that generate free radicals and thereby alter or remove bases; 3) chemical mutagens and carcinogens, both man made and naturally occuring toxins, which either bind to DNA and form adducts, produce DNA strand breaks, or slip between bases to form nonbound intercalations; 4) viral DNA that can insert into host DNA, thereby altering its information content; and 5) interaction between these factors such as the interaction of ultraviolet light with a number of chemical agents causing chemically mediated DNA-DNA, DNA-RNA and DNA-protein cross-links (Hart and Modak, 1980).

In non-replicating tissue, many of these types of DNA damage, have been shown to accumulate with age: DNA-protein cross links, single-strand breaks, single-stranded regions, endonuclease-sensitive sites and chromosomal aberrations among others. In dividing or mitotically active cells, minor DNA damage or small viral incorporations often appear to have little or no effect on DNA replication or cell division and may be passed on to one or both of the daughter cells. Other forms of damage in mitotically active cells, such as adducts, dimers, cross-links etc can either permanently block DNA replication or lead to mutation.

Of the various forms of DNA repair the most studied using the evolutionary-comparative paradigm is DNA excision repair. It has been evaluated using UV-induced unscheduled DNA synthesis or the endonuclease-sensitive site assay. In the broadest sense DNA excision repair involves the enzymatic recognition of a site, its removal, insertion of new bases and their reannealing to the old strand. There are several separate endonucleases, each classified based upon its ability to recognize various forms of DNA damage. A given cell type may be more proficient in one of these types of repair than in others, and several disease syndromes, including those associated with premature aging, may have deficiency in one repair system while being normal for other forms of DNA repair. Hart and Setlow (1974) examined UV-induced (254 nm) DNA repair in fibroblasts, taken from animals of different species that had completed approximately the same percentage of their maximum achievable lifespan. They maintained the primary cell cultures under identical conditions

Table 1. Relative DNA repair to rat (*Rattus norvegicus*) across five independent studies: Hart and Setlow (1) Treton and Courtois (2), Francis et al (3), Kato et al (4), Hall et al (5). Adapted from Cortopassi and Wang (1996). Only data included were those collected in fibroblasts of terrestrial mammals with an independent measurement in at least two studies

Species	1	2	3	4	5	Mean	MLS
Shrew	0.7	-	0.8	1.1		0.9	1.5
Mouse	0.8	-	0.5	1.3		0.9	2
Rat	1.0	1.0	1.0	1.0		-	3.3
Hamster	2.1	-	-	2.6		2.4	3.5
Rabbit	-	2.0	-	3.8		2.9	7.0
Dog	-	2.7	-	2.3		2.5	20
Cat	-	3.3	-	3.1		3.2	28
Cow	3.9	3.5	-	-		3.7	32
Horse	-	5.5	-	2.6		4.0	44
Elephant	4.2	-	2.1	-		3.2	50
Gorilla	-	-	2.9	-	6.0	4.5	51
Human	4.5	-	3.2	5.5	8.0	5.3	120

and showed a log-linear correlation of unscheduled DNA synthesis with species maximum achievable life span (MAL) across a number of placental mammals. Shortly thereafter a relationship was also formed in two rodent species of differing MAL chosen for their similarity in most other physiological variables except MAL (Sacher and Hart, 1978). Unstimulated lymphocytes and fibroblasts derived from primates have also been shown to have a log-linear correlation with MAL (Hall et al., 1984). Thus, even among a relatively closely related group of placental mammals such as the primates, across a twenty-fold difference in MLS there appears to be a close correlation between DNA excision repair and MAL. Francis et al. (1981), using a different set of species and a different technique, also showed a correlation between DNA repair capacity and MAL, although they contended that a better linear fit is achieved when total incorporated thymidine analogue is compared to species MAL.

The extent of agreement in the measurements from different laboratories is exceptional, especially when one normalizes repair to a single species. For examples, when Cortopassi and Wang (1996) analyzed all data for which there existed at least two measurements of DNA repair in terrestrial mammals, and expressed these values relative to DNA repair activity of an internal standard, the common laboratory rat, they found substantial agreement among studies and generally good agreement with the rank order of repair activity (Table 1). They found that for rat, shrew and mouse, across all studies, DNA excision repair was lower than for the longer-lived hamster and rabbit.

Following the Hart and Setlow (1974) paper, a number of studies confirmed their initial observation using a variety of cell types, animal species, and repair assays (Sacher and Hart, 1978; Hall et al, 1984; Walford 1979; Hall et al., 1984; Francis et al., 1981; Treton and Courtois, 1982; Paffenholz, 1978; Yagi, 1982; Walford and Bergman, 1979; Pegg et al, 1982). Nevertheless, despite the apparent correlation achieved in a number of laboratories between mean life span and DNA repair, these studies must be viewed with caution in that generally the range of life spans in an interspecies comparison is small, and the possibility that a strain has inherited some genetic defect causing it to be susceptible to some disease not related to DNA repair is great. Indeed, a number of strains used in the above studies were inbred with the express purpose of developing a model susceptible to a particular type of disease. Finally, few data exist relative to the relationship between DNA excision repair and MAL in cold-blooded vertebrates (Woodhead et al., 1980, and Regan et al., 1982). Interpretation of these data is compromised by the small number of species tested (a turtle and four species of fish), the different tissue types used, and the fact that only a single individual from each species was studied.

References

Altman, P.L. and Dittmer, D.S. (1962). Growth. Washington, D.C., Fed. Am . Soc. Exp. Biol., pp 445-449.

Altman, P.L. and Dittmer, D.S. (1972). Biology Data Book. Bethesda, MD: Fed. Am. Soc. Exp. Biol., pp 229-230.

Busbee, D., Miller, S., Schroeder, M., Srivastava, V., Guntupalli, G., Merriam, E., Holt, S., Wilson, V., and Hart, R. (1995) DNA polymerase α function and fidelity: dietary restriction as it affects age-related enzyme changes. In: Dietary Restriction: Implications for the Design and Interpretation of Toxicity and Carcinogenicity Studies. (Hart, R., Neuman, D., Robertson, R. Eds.) pp 245-270, ILSI Press, Washington, D.C.

Cleaver, J.E., Thomas, G.H., Trosko, J.E., and Lett, J.T. (1972). Excision repair (dimer excision, strand breakage and repair replication) in primary cultures of eukaryotic (bovine) cells. Exp. Cell Res. 74, 67-80.

Cleaver, J.E. and Trosko, J.E. (1970). Absence of excision of ultraviolet-induced cyclobutane dimers in xeroderma pigmentosum. Photochem. Photobiol. 11, 547-550.

Cortopassi, G.A. and Wang, E. (1996). There is substantial agreement among interspecies estimates of DNA repair activity. Mech. Age. Devel. 91, 211-218.

Cutler. R.G. (1975). Evolution of human longevity and the genetic complexity governing aging rate. Proc. Natl. Acad. Sci. USA. 72, 4664-4668.

Francis, A.A., Lee. W.H. and Regan, J.D. (1981). The relationship of DNA excision repair of UV-induced lesions to the maximum life span of mammals. Mech. Age. Devel. 16, 181-189.

Friedenthal, H. (1910). Ueber die Giltigkeit des Mussenwirkung fur den Energieumsatz der lebendigen Substanz. Zbl Physiol. 24, 321-317.

Hall, K., Hart, R., Benirschke, A., and Walford, R. (1984) Correlation between ultraviolet- induced DNA repair in primate lymphocytes and fibroblasts and species maximum achievable lifespan. Mech. Age. Devel. 24, 163-173.

Hart, R.W. and Setlow, R.B. (1974). Correlation between deoxyribonucleic acid excision-repair and lifespan in a number of mammalian species. Proc. Natl Acad. Sci. U.S.A. 71, 2169-2173.

Hart, R.W., and Modak, A. (1980). Aging and changes in genetic information. Adv. Exp. Med. Biol. 129, 123-137.

Hart, R.W. and Turturro, A. (1981). Evolution and longevity-assurance processes. Naturwissenschaften 68, 552-557.

Jones, M.L. (1982). Longevity of captive mammals. Zool. Gart. 52, 113-129:

Kato, H. Harado, M., Tsuchiya, K., and Morawaki, K. (1980). Absence of correlation between DNA repair in UV-irradiated mammalian cells and life span of the donor species. J. Genet. 55, 99-108.

Lipman, J., Turturro, A. and Hart, R. (1989) The influence of dietary restriction on DNA repair in rodents: A preliminary study. Mech. Age. Devel. 48, 135-143.

Mailouk, R.S. (1975). Longevity in vertebrates is proportional to relative brain weight. Fed. Proc. 34: 2101-2103.

Paffenholz, H. (1978). Correlation between DNA repair of embryonic fibroblasts and different life span of three inbred mouse strains. Mech. Age. Devel. 7, 131-150.

Pegg, A.E., Roberfroid, M., von Bahr, C., Foote, R.S., Mitra, S., Bresil, H., Likhochev, A., and Montesano, R. (1982). Removal of 0^6-methylguanine from DNA by human liver fractions. Proc. Natl Acad. Sci. U.S.A. 79, 5162-5166.

Rubner, M. (1908). Das Problem der Lebensdauer und seine Beziehungen zum Wachstum und Ernährung Oldenbourg, Munich.

Regan, J.D., Carrier, W. L., Samet, C., and Olla, B. L. (1982). Photoreactivation in two closely related marine fishes having different longevities. Mech. Age. Devel. 18, 59-65.

Sacher, G.A. (1959). Relation of lifespan to brain weight and body weight in mammals. In:.CIBA Foundation Colloquia on Aging. The Lifespan of Animals. (Wolstenholme, GEW, O'Connor, M. Eds) pp 115-133, Churchill, London.

Sacher, G. A., and Staffeldt, E. F. (1974). Relation of gestation time to brain weight for placental mammals. Implications for the theory of vertebrate growth. Am. Naturalist. 108, 593-615.

Sacher, G.A. (1975). Maturation and longevity in relation to cranial capacity in hominid evolution. In: Antecedents of Man and After. I. Primates: Functional Morphology and Evolution. (Tuttle, R. Ed.), pp. 417-441. Mouton Publishers, The Hague.

Sacher, G.A. (1976). Life table modification and life prolongation. In: Handbook of Aging, (Finch, C.E., Hayflick Eds.), pp. 582-587. Von Nostrand, New York.

Sacher, G. A. and Hart, R.W. (1978). Longevity, aging and comparative cellular and molecular biology of the house mouse, *Mus musculus*, and the white-footed mouse, *Peromyscus leucopus*. In: Birth Defects: Original Articles Series. The National Birth Defects Foundation, New York. pp. 71-96.

Setlow, R. B., Regan, J.D., and Currier, W .L. (1972). Different levels of excision repair in mammalian cell lines. Biophys. Soc. 12, 19a (Abstract)

Smith-Sonneborn, J. (1979). DNA repair and longevity assurance in *Paramecium tetraurelia*. Science 203, 1115-1117.

Srivastava, V.K., Miller, S., Schroeder, M.D., Hart, R., and Busbee, D. (1993) Age-related changes in expression and activity of DNA polymerase: Some effects of dietary restriction. Mutat. Res. 295, 265 -280.

Treton, J.A. and Courtois, Y. (1982). Correlation between DNA excision repair and mammalian lifespan in lens epithelial cells. Cell Biol. Int. Rep. 6, 153-158.

Turturro, A. and Hart, R.W. (1984) DNA repair mechanisms in aging. In: Comparative Pathobiology of Major Age-related Diseases: Current Status and Research Frontiers. (Scarpelli,,D., Migaki, G., Eds.). pp. 19-45, A.R. Liss, New York.

Turturro, A. and Hart, R.W. (1991) Caloric restriction and its effects on molecular parameters, especially DNA repair. In: Biological Effects of Dietary Restriction, (Fishbein, L., Ed.), pp. 185-192, ILSI Press, Wash., D.C..

Turturro, A., Duffy, P.H., and Hart, R.W. (1999). Antioxidation and evolution: dietary restriction and alterations in molecular processes. In: Antioxidants in Human Health and Disease. (Bosu, T., Temple, N., Gary, N., Eds.), CAB International, Oxford, In press.

Walford, R. L. (1979). Multigene families, histocompatibilities, transformation, meiosis, stem cells and DNA repair. Mech. Age. Devel. 9, 9–26.

Walford, R. L., and Bergman, K. (1979). Influence of genes associated with the main histocompatibilities complex on deoxyribonucleic acid excision repair capacity and bleomycin sensitivity in mouse lymphocytes. Tissue Antigens 14, 336-343.

Weraarchakul, N., Strong, R., Wood, W.G., and Richardson, A. (1989) The effect of aging and dietary restriction on DNA repair. Exp. Cell Res. 181, 197-204.

Woodhead, A.D., Setlow, R.B., and Grist, E. (1980). DNA repair and longevity in three species of cold-blooded vertebrates. Exp. Gerontol. 15, 301-307.

Yagi, T. (1982). DNA repair ability of cultured cells derived from mouse embryos in comparison with human cells. Mutat. Res. 96, 89-97.

© 2001 Elsevier Science B.V. All rights reserved.
The Role of DNA Damage and Repair in Cell Aging
B.A. Gilchrest and V.A. Bohr, volume editors.

DNA BASE MODIFICATIONS IN AGING

James Warren Gaubatz

Department of Biochemistry and Molecular Biology, University of South Alabama,
College of Medicine, Mobile, AL

1. Introduction

The arrangement of bases along the length of a DNA strand determines the genetic information for all cells. If one defines DNA damage as a change in the structure of DNA, then there are a multitude of possible forms of damage. There can be alterations of the bases, sugars, and phosphodiester linkages. Modifications can come in the form of covalent bond breakage and rearrangement or through the addition of groups to components of the repeating units of the polydeoxynucleotide. These covalent additions may range from small aliphatic groups and large bulky aromatic compounds to macromolecules. Amongst all the modifications to all the components of DNA, the bases appear to be exceptionally interesting because of their diversity, abundance and possible functional consequences. The primary focus of this chapter is to summarize and discuss data about the effects of aging on covalent modifications to the bases in DNA.

The DNA bases are chemically reactive and therefore susceptible to covalent modification. Although related in structure, each DNA base has unique properties, such as relative chemical reactivity. For example, purines are generally more reactive than pyrimidines (Singer and Kusmierek, 1982). Base modifications includes deaminations (Karran and Lindahl, 1980), oxidative adducts and ring openings (Dizdaroglu, 1991), alkylation adducts (Gaubatz, 1986), reactions with reducing sugars (glycation) (Bacala et al., 1984), bond formation with aldehydes (Beckman and Ames, 1998), and reactions with proteins or other macromolecules (Cutler, 1976). In addition to adducts formed through chemical reactions, vertebrate chromosomes are modified by DNA cytosine-5-transmethylases on C5 of cytosine positioned in CpG dinucleotides (Bird, 1992).

Base modifications result in several possible biological outcomes, depending on the base, on the modification, on where the damage is located and on how a cell responds to the damage. Various base modifications can block transcription and replication (Holmes et al., 1992). Other DNA base adducts miscode during replication and if not repaired properly will cause mutations (Wang et al., 1998). Some DNA lesions are very efficiently repaired; while other forms of DNA alterations can persist for extended intervals (Bohr and Anson, 1995). Two obvious biological end points resulting from base damage are cell death and cell mutations. A third potential outcome is an incompetent, dysfunctional cell. For instance, non-replicating, fixed postmitotic cells might exhibit altered metabolism and growth characteristics due to unrepaired (possibly unrepairable) DNA lesions that have set up residency in the genome (Bernstein and Gensler, 1993).

Modified bases can be classified as either benign, destabilizing, miscoding or lethal. Different methyl-DNA adducts will be used to illustrate each category. A benign lesion

41

is one with no apparent effect on replication/repair, gene expression, and genomic stability. Methylphosphotriesters likely represent benign modifications to DNA. In some cases, bond breakage is coupled to and follows bond addition. For instance, covalent modification of the nitrogens in purines destabilizes the base-sugar bond, causing increased frequencies of abasic sites which are promutagenic lesions (Loeb, 1985). N7-methyldeoxyguanosine is an example of such a destabilizing adduct. In contrast, O^6methylguanine is an alkylation adduct that can mispair with thymine during replication and thereby induce mutations, predominately G→A transitions (Pegg, 1983). Lastly, 3-methyladenine is considered a lethal lesion based on its ability to strongly block DNA and RNA polymerases (Xiao et al., 1996).

DNA base modifications and other changes to DNA have been proposed to be involved in the decline of cellular function with chronological age (Alexander, 1967; Curtis, 1971; Harmon, 1981; Gensler and Bernstein, 1981). Indeed, many attempts have been made to measure DNA alterations as an animal or population of animals grows old. There is now solid experimental evidence that some base modifications, such as methyl-DNA adducts (Tan et al., 1990) and oxidative base adducts (Park et al., 1989), can be measured at uninduced, background levels in somatic cells. As discussed in the following sections, the indigenous levels of some bulky base modifications have been reported to change with age (Randerath, et al. 1986). There is evidence that 5-methylcytosine decreases in DNA with aging (Cooney, 1993) and circumstantial evidence that some other types of lesions, such as advanced glycosylation end-products (AGEs) of DNA, occur in vivo and may accumulate during senescence (Vassara, et al., 1994). Lastly, there are base modifications that might potentially fluctuate as a function of aging but have not been measured or were measured by methods that were not sensitive enough to detect low levels of genomic damage. To fit base modification into a larger picture of genomic instability of postmitotic mammalian cells, the reader is referred to a more extensive review (Gaubatz, 1995).

2. Sources of DNA Base Adducts

Although DNA is an extraordinary molecule because of its very long length and very narrow width, DNA is made up of rather ordinary components arranged in a highly repetitious manner. The molecule is not endowed with any great physical resistance to modification. Indeed, many sorts of structural modifications to DNA have been described. Many forms of DNA damage arise spontaneously as a consequence of the intrinsic chemical instability of DNA or through reactions with normal cellular constituents (Singer and Kusmierek, 1982; Ames and Gold, 1991). Other DNA alterations occur from environmental and dietary sources (Ames et al., 1995; Gold et al., 1995). Cellular temperature and hydrogen ion concentration, nucleases, DNA methylases, reactive metabolites such as Maillard products of sugars, endogenous oxidants and alkylating agents are examples of inherent sources of genetic instability, whereas dietary mutagens, inhalation mutagens, UV irradiation, ionizing radiation, and viruses are ex-

Table 1. Sources of DNA base modification

Spontaneous chemical changes (excluding base loss)
> Deamination

Reactions with other molecules

Endogenous agents	*Examples*
Reducing sugars	Glucose
Alkylating agents	S-adenosylmethionine
Reactive oxygen species	Hydroxyl radical
Aldehydes	Formaldehyde
DNA methylases	5-Methylcytosine

Exogenous agents	*Examples*
Dietary mutagens/promutagens	Heterocyclic amines
Inhaled pollution/cigarette smoke	Benzo(a)pyrene
N-nitroso compounds	N-nitrosamines
UV-irradiation	Sunlight
Ionizing radiation	Diagnostic X-rays
Medicines/food supplements	Cis-platinum

Sources of DNA base modifications were adapted from Singer and Kusmierek (1982).

ogenous examples of DNA damaging agents. Some of these sources of DNA damage, along with specific examples, are shown in Table 1.

Amino groups on the bases of DNA may be spontaneously lost by hydrolytic deamination (Lindahl and Nyberg, 1974; Karran and Lindahl, 1980). Alternatively, deamination may proceed from cellular oxidants. Another category of DNA base modification is alkyl adduction. Alkylation damage to DNA can arise through interactions with N-nitroso compounds that are potential mutagens and carcinogens (Pegg, 1983). Individuals may be exposed to N-nitrosoamines through the diet (Aucher, 1982), through inhalation (Hecht and Hoffman, 1989) and through endogenous formation (Mirvish, 1973; Tannenbaum, 1987). In addition to these exposures, endogenous sources of DNA methylating compounds, such as S-adenosylmethionine, can modify the bases in DNA nonenzymatically (Barrows and Magee, 1982; Ryberg and Lindahl, 1982).

Randerath and associates (Li and Randerath, 1989; Randerath et al., 1989a,1989b; Gupta et al., 1990) have argued that the background indigenous damage (I-compounds) detected by ^{32}P-postlabeling are unlikely to originate from exogenous environmental sources but rather appear to be formed from indigenous electrophilic DNA-reactive compounds. What these endogenous bulky, non-polar groups might be has not been resolved. On the other hand, the metabolic processing of heterocyclic amines, poly-cyclic aromatic hydrocarbons, and other bulky, non-polar compounds that are common in the environment, are known to generate DNA-reactive intermediates and thus to induce DNA base modifications (Doll and Peto, 1981; Miller and Miller, 1981).

Glycation is the addition of sugar groups to macromolecules. The nonenzymatic re-action of proteins with reducing sugars, such as glucose, is termed the Maillard reaction. The Maillard reaction begins with the formation of a Schiff base between the aldehyde of a reducing sugar and amino group of a protein or a nucleic acid base (Reynolds, 1965). The product of this reaction is an aldosylamine that undergoes an Amadori rear-rangement (Vlassara et al., 1994). Amadori products are transformed through a series of rearrangements and dehydrations to form yellow-brown fluorescent compounds (advan-ced glycosylation endproducts or AGEs) that can crosslink proteins, and potentially crosslink DNA (Vlassara et al., 1994).

Oxygen radicals are forms of oxygen containing an unpaired electron and can arise from the single electron reduction of oxygen during cellular respiration. The Fenton reaction is the reduction of H_2O_2 by reduced transition metals (e.g., Fe^{+2}) to produce the exceptionally reactive hydroxyl radical and related oxidants (Ames, 1989; Weindruch et al., 1993) The normal background of hydroxyl radicals can be increased by exogenous factors, such as ionizing radiation. Oxidative damage to DNA results from the interplay of reactive oxygen species with the components of DNA, mostly the nitrogenous bases. Reactive oxygen species have been implicated in the creation of single- and double-strand breaks, ring-opened sugars, abasic sites, protein-DNA crosslinks, and dozens of oxidative base adducts (Henner et al., 1982; Hutchinson, 1985; Dizdaroglu, 1991). Oxi-dants generated by the mitochondria are believed to be the predominant source of reac-tive oxygen species leading to oxidative lesions that accumulate in the genome with aging (Bernstien and Gensler, 1993). Indeed, hydrogen peroxide and superoxide radical are oxidants continuously produced during aerobic metabolism. Under normal circum-stances, the cellular production of reactive oxygen species is balanced in the cell by antioxidant protective systems, such as superoxide dismutase, catalase, and 9-tocopherol (Halliwell, 1996). Age-dependent increases in mitochondria deficiencies may be contributing to greater oxidative stress and thus to cellular aging of the animal (Sohal and Dubey, 1994).

3. DNA Base Modifications

A. *Adduction of Bulky Non-polar Groups*

The ^{32}P-postlabeling assay for detecting DNA damage involves the enzymatic transfer of radioactive phosphorous from ATP to hydrolysis products of DNA that can retro-spectively determine in vivo DNA damage, such as animal exposure to experimental carcinogens (Reddy et al., 1984). Randerath et al. (1986) first described DNA modifica-tions in untreated rats using ^{32}P-postlabeling assays. Tissue-specific patterns of ^{32}P-labeled spots were observed following autoradiography of thin-layer chromatograms.

These spots were first termed I-spots, then later renamed as I-compounds based on the assumption that they are non-polar, covalent DNA modifications that arise indigenously in cells (Randerath et al., 1986; Randerath et al., 1989). It was therefore postulated that I-compounds were background genomic damage derived from endogenous DNA-reactive compounds generated during normal metabolism. To detect background levels of DNA damage, the original postlabeling procedure has been modified several times to enhance the sensitivity of the assay (Reddy and Randerath, 1986). Detection limits are routinely claimed in the range of one adduct per 10^8 to 10^{10} nucleotides for various enhancement procedures (Randerath and Randerath, 1994). In addition to non-polar types of I-compounds, polar types of I-compounds have also been described (Randerath et al., 1993b).

I-compounds appear as a number of spots with differing intensities and locations on autoradiograms of polyethyleneimine cellulose thin layers following multidimensional chromatography. Randerath and associates have shown that the types and quantities of I-compounds vary with tissue, sex, age, species, (Randerath et al., 1986, 1989, 1990, 1993c) hormones, diet, individual nutrients and minerals (Li and Randerath, 1990a, 1990b, 1992; Li et al., 1990a, 1990b, 1992), physiological and pathological processes, (Randerath et al., 1988, 1990) and time-of-day (circadian rhythm) (Nath et al., 1992). In the context of aging, I-compounds associated with DNA isolated from tissues of rats appeared to increase early in the life span then level off after maturity (Randerath et al., 1986, 1989). Other postlabeling analyses examined a possible accumulation of indigenous bulky non-polar DNA adducts in the genome of aging mouse heart (Gaubatz, 1989). The results showed that several adducts were more abundant in ^{32}P-maps of senescent heart DNA. One adduct in particular showed a striking 10-fold higher content in 39-month-old mice, compared to 17-month-old mice. The relationship between these heart DNA adducts and other I-compounds is not straight forward because different enhancement versions of the postlabeling procedure were used, e.g., phase transfer to 1-butanol enhancement versus nuclease P1 enhancement (Gaubatz, 1989). Age-dependent increases of I-compounds have been reported for human brain (Randerath et al., 1993a). Two autoradiogram spots out of a pattern of three abundant spots increased in human brain DNA after the age of 60 years. Considerable inter-individual variation was observed in human brain I-compound levels that grew greater after the age of 60 years. Rat brain DNA, moreover, exhibited two similar spots that also increased from 1-month-old to 10-months-old (Randerath et al., 1993a). A subsequent study by Cai et al. (1996) determined the amounts of I-compounds in brains of male Fischer rats at ages 1, 6, 12, 18, and 24 months. The results indicated that I-compounds increased with aging from 6 to 24 months. However, the highest level of adducts was observed for 1-month-old rats. The reasons for different adduct levels in young rat brain between these two studies is not known.

Many studies have established that dietary restriction of total calories increases mammalian longevity. It was therefore surprising that dietary restriction actually increased I-compounds in the tissues of male rats (Randerath et al., 1993b). Yet other evidence indicated that treatment of rats with various carcinogens and toxins depressed the age-related increases of I-compounds in liver tissues (Randerath et al., 1988, 1990).

These results suggested a protective function for I-compounds. It would therefore follow that senescent animals have more protection since they exhibited greater levels of I-compounds. However, it is difficult to see how base or nucleotide modifications, other than the normal methylation of cytosine, can have a beneficial effect on the genome. It is an axiom that covalent modification of DNA induces genomic and chromosomal instability which in turn decreases organismal survival. Nevertheless, Randerath et al. (1990, 1993c; Randerath and Randerath, 1994) proposed that I-compounds may have an anti-aging role in cells. In studying biomarkers of mammalian aging, Randerath et al. (1993a, 1993b) correlated I-compound abundance with life spans of different rat strains maintained on different diets; therein, adducts that increased with caloric restriction and were elevated in longer-lived rodent strains were designated type I, or reverse, I-compounds with the notion that they exemplify "healthy" biomarkers of aging versus type II I-compounds that represent endogenous DNA lesions, whose levels would adversely affect cell function. However, there is no direct evidence that I-compounds have either beneficial or detrimental roles in cellular physiology.

In consideration of I-compounds, it is important to critically examine the methodology. There are a number of features of I-compounds and the methods used to measure them that are troublesome. To detect background levels of damage, at the highest levels of detection with the ^{32}P-postlabeling assay, just about every component of the assay, and materials that come into contact with reactions, may contribute to background spots on autoradiograms. These contributors include (γ-^{32}P-ATP, several nucleases, T4 polynucleotide kinase, buffer components, plastic pipettes and microcentrifuge tubes (Adams et al., 1994; Moller and Hofer, 1997). Many times extra spots on autoradiograms are simply incomplete digestion products, e.g., dinucleotides, and recently, one safrole-induced DNA-adduct was shown to alter T4 polynucleotide kinase specificity (Randerath et al., 1993d). Careful analyses have uncovered differences in efficiency of labeling, in adduct stability and in adduct recovery between different sorts of base modification (Gupta and Earley, 1988; Segerback and Vodicka, 1993). Numerous studies have now demonstrated potentially very large differences in comparing the quantity of a single adduct as measured by two or more variations of postlabeling assays (Gallagher, et al., 1989; Whong et al., 1992). It is also worrisome that natural dietary ingredients, such as oats and alfalfa meal, induced I-compounds in rat liver and kidney (Li and Randerath, 1992; Li et al., 1992). Indeed, the degree of concern is heightened by the finding that a single adduct from the same tissue, species/strain, environment, etc. can vary by 8- or 9-fold during different times of the day (Nath et al., 1992). These latter two observations alone seem to argue that it will always be very difficult if not impossible to compare background DNA I-compound data between laboratories.

Of course, an ultrasensitive method should reveal the minutia of DNA damage metabolism, and metabolism varies with time-of-day, nutritional status, physiologic stress, and age to name only a few factors. The original notion that I-compounds represent covalent modifications to DNA, and are therefore DNA-adducts, was based on the following: (1) polynucleotide kinase catalyzed the transfer of (γ-^{32}P-phosphate to the compounds; (2) the association of the labeled phosphate to its receptor molecule was DNA-dependent, and; (3) the chromatographic properties of I-compounds resembled nucleotides containing bulky, hydrophobic carcinogen adducts (Randerath et al., 1986, 1990; Randerath and Randerath, 1994). In an effort to characterize the chemical nature of I-compounds, Cai et al. (1996) incubated different deoxynucleosides with the lipid per-oxidation product malondialdehyde (MDA). The results demonstrated only dGMP-MDA adducts overlapped chromatographically with I-compounds. A total of five dGMP-MDA adducts were linked to I-compounds in rat brain. Cai et al. (1996) con-cluded that I-compounds may be a biomarker of indirect oxidative damage to DNA. Sequence-specific, bulky, non-polar DNA adducts have also been produced by incuba-tion of DNA in the presence of oxygen radical generating systems (Randerath et al., 1996). However, the chemical structure of a single I-compound has yet to be deter-mined, and I-compounds remain molecules of largely unknown structures and func-tions. This issue will need to be resolved before any type of definitive statement can be made on whether mammalian genomes accumulate bulky DNA adducts during aging and what effects such lesions might impart on senescent cells.

B. *Alkylation*

Some studies have focused on alkylating damage to DNA during aging because alkylat-ing agents are common in the environment (Mirvish, 1975; Aucher, 1982; Tannenbaum, 1987), and alkylation is one of the more frequent types of covalent modifications of DNA (Ryberg and Lindahl, 1982; Saul and Ames, 1986; Ames, 1989). Alkylating agents produce a spectrum of lesions at a dozen different DNA sites; eleven of which are base modifications (Margison and O'Connor, 1979). Various alkylated bases in DNA have the potential to be cytotoxic, mutagenic, and carcinogenic (Pegg, 1983). These agents modify DNA leading to promutagenic lesions such as O^6-methyl-2'-deoxyguanosine (O^6methyl-dG), but the low background levels of this lesion challenge existing methods of detection. Alternatively, 7-methyl-2'-deoxyguanosine (7-methyl-dG) is a much more abundant modified base (normally more than 70% of the total methyl adducts produced in DNA) and thus is more easily detected. Although not a miscoding lesion, 7-methyl-dG is frequently used as a surrogate for overall genome alkylation (Tan et al., 1990; Blomeke et al., 1996). 7-Methyl-dG is often considered an innocuous adduct but depurinates more rapidly than unmodified dG to create cytotoxic and mutagenic abasic lesions (Loeb, 1985; Loeb and Preston, 1986). The released base is 7-methylguanine. 7-Methylguanine is often assayed as a biomarker for total alkyla-tion damage to the genome and therefore is used as an indicator of exposure to exoge-nous and endogenous methylating agents. The work of Gaubatz (1985) and Tan et al. (1990) demonstrated that the steady-state levels of 7-methylguanine increased in nu-clear DNA of postmitotic rodent cells during aging. The first study examined alkylated nucleotides in mouse heart tissue with aging (Gaubatz, 1985). ^{32}P-postlabeling methods were used to separate and quantify adducts. One nucleotide adduct increased in an age-dependent manner. The chromatographic properties of the modified nucleotide were essentially the same as those of authentic 7-methyl-dG and a methylated nucleotide induced by treatment of DNAs, polymers and oligos containing dG (but not those lack-ing dG) with the alkylating agent N-methyl-N-nitrosourea. It was therefore assumed that the modified nucleotide detected by postlabeling represented indigenous steady-state levels of 7-methyl-dG. In young mice 2- to 10-months-old, there were approximately 10 adducts per 10^6 nucleotides. This level increased 2.5-fold in 15-to 17-month-old mice, and grew to a 4-fold increase by 29-months-of-age (Gaubatz, 1985). In the terminal portion of their life span between 29 months and 39 months of age, there was another doubling in 7-methyl-dG. Therefore, there was 9-times more modified base in senescent mouse cardiac muscle than in young heart (Gaubatz, 1985). These results suggested an exponential rise in this DNA damage past maturity.

Park and Ames (1988a) reported an analysis by HPLC with electrochemical detection (HPLC-EC) that suggested 7-methylguanine adducts are normally present at high levels in nuclear DNA and increased in rat liver on aging. The level of 7-methylguanine increase about 2- to 2.5-fold in old rats compared to young rats. This same study indicated that indigenous amounts of 7-methylguanine were 3 times higher in mt-DNA than those levels in nuclear DNA (Park and Ames, 1988a). These workers subsequently discovered that under their solvent elution conditions of 10% methanol, the HPLC peak previously quantified for 7-methylguanine consisted of 3 components that could be resolved by 2% methanol elution into 2 minor peaks that coelute with adenine and 7-methylguanine, and one major peak that was an unidentified contaminant (Park and Ames, 1988b). Therefore, uninduced 7-methylguanine was detected in rat liver DNA, but accurate estimates on adduct levels were obscured.

Using 2 independent HPLC systems and 2 different methods of detection, Tan et al. (1990) observed that low levels of 7-methylguanine are present in nuclear DNA of normal brain, liver, and kidney mouse tissues. To determine whether this adduct increased with aging, 7-methyl-dG was quantified by HPLC separation coupled with electrochemical detection. The results showed that steady-state levels of 7-methylguanine increased about 2-fold between 11 months and 28 months-of-age (Tan et al., 1990). The chemical identity of the age-associated adduct was confirmed by fast atom bombardment mass spectrometry. Therefore, 7-methylguanine has been connected with aging processes on the basis of higher adduct levels being present in tissues from old rodents, such that 7-methylguanine appears to accumulate as a function of age in the nuclear DNA of mice.

The loss of 7-methylguanine from DNA occurs by both spontaneous release and enzymatic release (Pegg, 1983). The enzymatic pathway involves DNA alkylglycosylases, such as 3-methyladenine-DNA-glycosylase, that are part of base excision repair (Mitra and Kaina, 1993). The steady-state levels of 7-methylguanine might increase during aging of postmitotic tissues if base excision repair declined with age. To test for the possibility that DNA repair of 7-methylguanine might decrease in senescent animals, methyl adducts were induced in young and old mice with the direct alkylating agent MNU, the persistence of 7-methylguanine was measured at different times over 8 days (Gaubatz and Tan, 1993; Gaubatz and Tan, 1994). The time course of adduct removal from DNA was biphasic. There was a rapid phase during the first 24 h following MNU administration and a slow phase that appeared to be a first-order decay process with half-lives consistent with the spontaneous hydrolysis of 7-methylguanine from DNA. This biphasic loss of 7-methylguanine from DNA was evident in both young and old brain, kidney and liver tissues. The pattern of damage induction and repair was tissue-specific and varied by more than 2-fold between tissues of mice of the same age. However, comparing repair between the same tissues, the results indicated that old tissues removed 7-methylguanine as rapidly as young tissues (Gaubatz and Tan, 1994). To obtain more information about the active phase of 7-methylguanine removal from DNA in both young and old cells, 3-methyladenine-DNA glycosylase mRNA abundance was measured in the relevant tissues at different ages. The data indicated that 3-methyladenine-DNA glycosylase mRNA levels did not change during aging of mouse brain, heart, kidney, and liver tissues (Gaubatz and Tan, 1994). The results from these studies suggested that the repair enzymes responsible for excising 7-methylguanine from DNA are not compromised in senescent tissues (Gaubatz and Tan, 1993; Gaubatz and Tan, 1994).

To determine whether there were chromosome sectors that changed with age to make 7-methylguanine repair more resistant, the distribution and elimination of 7-methylguanine from different fractions of mouse liver chromatin was studied after treating young and old mice with 25 mg/kg and 50 mg/kg doses of MNU (Gaubatz and Tan, 1997). Chromatin was separated into Micrococcal nuclease-soluble, low-salt, high-salt and nuclear matrix fractions. All four fractions of young liver chromatin was methylated to the same extent. In contrast, there were differences in 7-methylguanine levels between different fractions of old mouse liver chromatin. DNA in the nuclease-soluble fraction was most heavily modified, whereas nuclear matrix nucleotides were alkylated the least. Nevertheless, removal of 7-methylguanine occurred at fairly uniform rates in all four chromatin fractions regardless of age (Gaubatz and Tan, 1997). In agreement with data from earlier studies, these results suggested that glycosylases accountable for eliminating 7-methylguanine from DNA were not deficient in senescent postmitotic mouse tissues (Gaubatz and Tan, 1993; Gaubatz and Tan, 1994), but there may be age-related changes in chromatin composition and structure that render some genomic sequences more accessible to methylating agents in liver tissue of older animals (Gaubatz and Tan, 1997). Additionally, it was suggested that those methyl-DNA adducts that were not rapidly removed represented a fraction of damage in which the sequence context of the lesion greatly reduced or even eliminated DNA alkyl-glycosylase activity.

From the foregoing discussion of methyl-DNA adduct levels in rodent tissues, it can be inferred that at least the alkyl-glycosylase step of base excision repair is not compromised with aging, and 7-methylguanine was removed from different chromatin fractions equally well in young and old tissues. Furthermore, it has been reported that endogenous pools of SAM decrease in mouse tissues during aging (Eloranta, 1977; Stramentinoli et al., 1977). So what determines the higher steady-state levels of methyl adducts in old rodent organs? It has been proposed that old mice will have higher levels of DNA damage because their cells are exposed to active methylating species longer than cells from young mice (Gaubatz and Tan, 1997). This hypothesis is based on studies that showed initial molar ratios of induced 7-methylguanine/nucleotide for weight-normalized, MNU-treated mice were higher in old tissues compared to young liver, brain, and heart tissues (Gaubatz and Tan, 1994). Kidney however was the exception to this observation; wherein methylated guanine was higher in young kidney than in old kidney (Gaubatz and Tan, 1993). This age-related difference in tissue specific damage and repair profiles is consistent with younger animals having a faster rate of mutagen clearance compared to older mice. Hence, senescent cells may have a longer, higher exposure to DNA alkylating agents if they are not removed efficiently in old animals. This idea has not been tested directly however.

7-Methylguanine levels have been measured in DNA isolated from human white blood cells using the ^{32}P-postlabeling approach (Mustonen and Hemminke, 1992). The range of methyl adduct levels observed for this replicating cell was 600 to 4,200 residues per diploid genome. Recently, Blomeke et al. (1996) showed that background levels of 7-alkyl-dG could be detected in human lung, but for a group of 10 individuals, steady-state adduct levels did not correlate with age, gender, race, tobacco exposure, or a variety of medical conditions. However, the large inter-individual variation in 7-alkyl-dG levels (10-fold differences for same sex and race and similar age) may preclude finding an age-associated trend unless a large population of individuals is analyzed. There were other complicating factors such as the distribution of adduct within the lung tissue and the heterogeneity of cell types in a tissue which might change with age or disease. In support of the notion that induction and distribution can vary significantly, the intra-individual variation among different lobes of lung for some individuals was high (Blomeke et al. 1996).

In summary, 7-methyl-dG is a major adduct formed by alkylating agents in vivo and in vitro. Presumably, the methylation observed in background DNA damage studies arose through the action of endogenous methylation reactions for which there is precedent (Barrows and Shank, 1981). Since 7-methyl-dG is not a totally stable adduct having a chemical half-life in vivo of roughly 4 or 5 days (Gaubatz and Tan, 1994), the steady-state levels suggest that this lesion is continually being introduced into mammalian nuclear genomes. It is interesting to note that 7-methylguanine levels in human cells measured by either ^{32}P-postlabeling (Mustonen and Hemminke, 1992) or by HPLC/^{32}P-postlabeling assays (Blomeke et al., 1996) are lower than the adduct levels determined for mice. Mouse adducts calculated as 7-methyl-dG per dG (Tan et al., 1990) are approximately an order of magnitude higher than observed for human samples, 10^{-6} versus 10^{-7}, respectively. DNA methylation of this sort is thought to arise in untreated genomes as a consequence of endogenous processes, perhaps as a consequence of nonenzymatic reactions involving S-adenosylmethionine as a methyl donor (Ryberg and Lindahl, 1982; Barrows and Magee, 1982). However the adducts originate, 7-methylguanine has mutagenic potential (Loeb, 1985) and the content of adduct is generally a good indicator of overall alkylation damage to DNA. Indeed, in vivo studies suggest a role for endogenous DNA alkylation damage, including 7-methylguanine, as a source of spontaneous mutations in eukaryotic cells (Xiao and Sampson, 1993). Therefore, an age-related increase in the amounts of 7-methylguanine in cellular genomes may be pertinent to basic mechanisms of aging.

C. Deamination

Mutations in DNA may contribute to the aging process and to the age-related incidence of cancer (Ames and Gold, 1991; Schmutte and Jones, 1998). One ubiquitous source of promutagenic damage is DNA base deamination, and it has been estimated that base deamination is a frequent event in the genome of a mammalian cell (Lindahl and Nyberg, 1974; Karran and Lindahl, 1980). Uracil is the deamination product of cytosine. Deoxyuridine may form two hydrogen bonds and base pair with adenine, thereby inducing transitional mutations. Deaminated adenine is the base inosine or the nucleoside hypoxanthine which can base pair with cytosine, thus inducing transitional mutations (Singer and Kusmierek, 1982; Sidorkina et al., 1997). Deamination of guanine results in xanthine which pairs with cytosine - the appropriate base. Thymine has no primary amino group, but hydrolytic deamination of 5-methylcytosine leads to a G/T mismatch and if not repaired, to a C→T transition mutation (Schmutte and Jones, 1998).

What are the physiological steady-state levels of deamination base modifications, and do these modifications accumulate with age? The answers to these questions are not known. The few investigation that have been reported have failed to detect deaminated bases in DNA. For example, Kirsh et al. (1986) failed to detect deoxyuridine (or 5-hydroxymethyldeoxyuridine, an oxidation product of thymine, see below) in DNAs from brain, liver, and small intestine tissues of mice, ranging in age from 7 months to 31 months. This investigation employed reversed-phase HPLC separation of deoxynucleosides, coupled with UV-detection/quantification. The limits of detection corresponded to approximately 10 pmol of modified deoxynucleosides/(mol of normal deoxynucleosides or 1 modification per 10^5 nucleosides. To compare this method with those that have been used to measure indigenous methyl adducts and oxidative damage (see below), one can estimate that UV-detection is about 10- to 20-fold less sensitive than electrochemical detection, depending on the electrochemical activity of the individual base or deoxynucleoside. Therefore if background, steady-state levels of deoxyuridine and 5-hydroxymethyldeoxyuridine in DNA are similar to those for 7-methylguanine or 8-hydroxydeoxyguanosine, minimum detection limits of 1×10^{-6} will be required to measure them.

Whether hydrolytic or oxidative in action, there seems to be little doubt that spontaneous deamination of the DNA bases occurs in vivo. In fact, data support spontaneous hydrolytic deamination of C and A in the heavy strand during replication as a critical determinant in the origin of asymmetric, biased base compositions of mammalian mitochondrial genomes (Reyes et al., 1998). The scarcity of deaminated deoxynucleosides in somatic cell DNA implies efficient repair of these lesions, and levels might be kept exceedingly low by a system of dedicated base excision enzymes, such as uracil DNA-glycosylase (Pogribny et al., 1997). In contrast to direct measurements made for deamination products in DNA, recombinant methods have surveyed genomic sequences for age-related changes in the status of 5-methylcytosine. In fact, several hypotheses propose a role for methylation/demethylation in the loss of differentiated function associated with biological aging (Holliday, 1984, 1987). These results are discussed in a subsequent section of this chapter.

D. *Glycation and Macromolecular Adduction*

The work by Cerami and coworkers (Bucala et al., 1984,1985; Lee and Cerami, 1987a, 1987b, 1990, 1991) suggests that reducing sugars contribute to steady-state levels of DNA damage by modifying nucleotide bases. Reducing sugars such as glucose and glucose-6-phosphate have been shown to nonenzymatically react with the amino groups of proteins. These covalently bound sugars are transformed through a series of bond rearrangements and dehydrations to form yellow-brown fluorescent products (Reynolds, 1965). This class of irreversibly-formed, stable complexes has been collectively referred to as advanced glycosylation endproducts or AGEs (Vlassara et al., 1994). For long-lived proteins, AGE-associated protein crosslinks accumulate with time and therefore accumulate with age (Sell et al., 1996). Importantly, AGEs have become a primary consideration in explaining some of the biological effects associated with diabetes, Alzheimer's disease, and aging (Vlassara et al., 1994). Much like the reactions observed with proteins, the amino groups of DNA bases are able to react chemically with reducing sugars. The modification of DNA by reducing sugars can alter both the physical characteristics and biological functions of DNA (Bucala et al., 1984, 1985; Lee and Cerami, 1990, 1991). The levels of modifications to DNA via reducing sugars are dependent upon sugar concentration and exposure time (Lee and Cerami, 1987b, 1991).

Glucose-derived addition products to DNA cause strand breakage and induce mutation in bacterial cells (Bucala et al., 1984, 1985). Furthermore, AGEs have been shown to induce DNA rearrangements and transpositions in both bacterial and mouse lymphoid cells (Lee and Cerami, 1991; Bucala et al., 1993). The mechanisms by which advanced glycosylation leads to transpositions is not known, but the ability of endogenous reducing sugars to react with DNA and induce large rearrangements suggests that this pathway may be a major source of genomic instability. It is also thought that since DNA is a long-lived molecule and AGE products are generally quite stable, various glycation compounds might accumulate in mammalian chromosomes with aging (Bucala et al., 1984). Indeed, the total tissue content of AGEs correlates strongly with aging (Beckman and Ames, 1998). On the other hand, it is not known if DNA modification by reducing sugars changes during aging, or if DNA-related AGE molecules accumulate as a function of age. The measurement of individual AGE-modified DNA components in vivo has remained elusive. One approach to the study of AGE accumulation in cells has been to use immunochemical detection (Mikita et al., 1992). However, what fraction of AGE molecules recognized by polyclonal antibodies from sugar-modified DNA has not been determined. Recently, several nucleotide-derived AGEs have been isolated from in vitro reactions and structurally characterized (Papoulis et al, 1995). It seems that with the chemical identity of DNA-linked AGEs in hand, experiments can now proceed to quantify these adducts using chemical/physical and immunochemical methods.

In other model systems, the chemical reactions between sugars, lysine and DNA suggest ways by which amino acids or proteins can become adducted to DNA, possibly leading to intra-strand and inter-strand crosslinks (Lee and Cerami, 1987a). Over the years there has been circumstantial evidence related to the possible accumulation of cross-linkages in DNA of non-dividing mammalian cells. Some indirect signs that crosslinks increase in somatic cell DNA with time involve age-related changes in the melting temperature or other physical-chemical properties of DNA and chromatin, the degree of difficulty in removing proteins bound to DNA, and in nuclear template activity for exogenous RNA polymerases. The results from various investigations have been reviewed by Tice and Setlow (1985) and by Cutler (1976). A different approach has been to purify DNA from a tissue, hydrolyze DNA in 6 N HCl to release the bases, then separate the individual bases from larger products using molecular exclusion chromatography (Sharma and Yamamoto, 1980). It was shown by this approach that age-related modifications of DNA in the liver of mice and rats exhibit an unusually high fluorescence indicative of DNA-DNA or DNA-protein crosslinks in which covalent bonds between ring systems produce larger fluorescent effects. Similar compounds are formed when solutions of DNA are irradiated with (γ-photons (Mandal and Yamamoto, 1986). Yamamoto et al. (1988) examined fluorescent modifications to DNA in liver tissue of the New Zealand while rabbit. Chromatographically separated modified bases, which were obtained in relatively large yields from old animals, were highly fluorescent, and mass spectrometry indicated the compounds were cross-linked base products by the secondary ion mass spectrometry method. The yield of highly fluorescent products from rabbit liver DNA increased with age (after 6-months-of-age) and in fact, could be fitted with an exponential function. Livers from 10-year-old rabbits had 4 times the amount of modified bases found for young rabbit livers (Yamamoto et al., 1988). Thus, these data provide some direct support for an age-dependent accumulation of DNA cross-linkages in mammalian liver tissue.

E. *Oxidation*

Oxidative DNA damage has been implicated in aging as well as in mutagenesis and carcinogenesis (Ames, 1989; Ames and Gold, 1991). Reactive oxygen species, including free radicals, are formed in normal cells by oxidative metabolism. Oxidative damage generates a plethora of DNA alterations (see Scources of DNA Adducts). Among oxidative damages, DNA base modifications constitute an important class of lesions with mutagenic potential in vivo. The spectrum of oxidative damage to DNA involves modifications to all four bases and includes more than 20 different products (Dizdaroglu, 1991; Henle et al., 1996). Among base modifications studied as a function of age, oxidative damage to DNA is the most extensively documented, and probably the most abundant, class of lesions continually introduced in the genome. There are several well established correlations between DNA repair, oxidant levels, and longevity. First, there is a direct correlation between the ability to repair DNA damage and a species life span (Hart and Stelow, 1974). Second, there is an inverse correlation between levels of endogenous oxidant production and life span, in that mitochondrial respiration rates correlate negatively with maximum species longevity (Ku et al.,1993). Third, there is a direct correlation between antioxidant defense systems and life span (Cutler, 1985a; Sohal et al., 1994a). Further support for the free radical theory of aging comes from Orr and Sohal (1994) who showed a gene dosage-dependent extension of maximum life span in transgenic Drosophila containing additional germ line copies of antioxidant catalase and superoxide dismutase genes.

Oxidative base damage may constitute either a mutagenic lesion or a lethal lesion (Wang et al., 1998). A lethal lesion usually represents a barrier to DNA polymerases, thereby leaving a gap in the replicated strand, whereas a mutagenic lesion has altered base-pairing properties that miscode for a new nucleotide. The oxidation of guanine to 8-oxoguanine (8-oxoG) and of thymine to 5,6-dihydroxy-5,6-dihydrothymine (thymine glycol) are two abundant oxidative adducts (Dizdaroglu, 1991). In fact aside from abasic sites, 8-oxo-2'-deoxyguanosine (8-oxo-dG) is the most frequent lesion found after DNA exposure to Fenton chemistry (Henle et al., 1996). It has been proposed that reactions are initiated by hydroxyl radical addition to C8 of deoxyguanosine (dG), yielding 8-hydroxy-2'-deoxyguanosine (8-OHdG). However, the preferred tautomeric form for this adduct is 8-oxo-dG. In this chapter, the two forms will be used inter-changeably. Chemical modification of these bases in DNA causes a change in their nucleotide hydrogen bonding properties. For example, 8-OHdG and thymine glycol have been shown to miscode, leading to G→T and T→C nucleotide substitutions, re-spectively (Shitbutani et al., 1991; Cheng et al., 1992; Basu et al., 1989). Furthermore, thymine glycol acts as a blocking lesion in primer extension experiments using purified DNA polymerases, suggesting thymine glycol may also be a lethal lesion if not re-paired. Adelman et al. (1988) have measured thymine glycol and thymidine glycol ad-ducts in urine samples of four different mammalian species. Urinary output of oxidative products of DNA are presumed to be modified bases or nucleosides released form cellu-lar DNA by excision repair and are thought to be a general indicator of overall oxidative damage to DNA. The results demonstrated that this biomarker of oxidative DNA dam-age correlated highly with the specific metabolic rates of mice, rats, monkeys and hu-mans. Accordingly, mice excreted 18 times more glycol adducts than did humans, and monkeys excreted 4 times more glycol adducts than humans on a normalized-weight basis (Adelman, et al., 1988). The rate of oxidative DNA damage excretion was there-fore inversely related to life span.

Does oxidative DNA damage accumulate or otherwise change with age? An age-related accumulation of oxidative damage in macromolecules such as DNA would be consistent with the free radical theory of aging (Harman, 1992). To detect and quantify DNA damage arising from endogenous oxidation, methylation, and deamination, Ames and coworkers (Cundy et al., 1988; Park et al., 1989) developed a high-performance liquid chromatography system with electrochemical detection (HPLC-EC). The high sensitivity, simplicity and rapidity of HPLC-EC have made this the method of choice for analyzing oxidative adducts of DNA (Beckman and Ames, 1997). Furthermore, since 8-OHdG is a major product of oxidative reactions with DNA and is electrochemi-cally active (Park et al., 1989), this adduct has been frequently used as a biomarker of oxygen-derived DNA damage. Richter et al. (1988) reported that background oxidative damage to mitochondrial and nuclear DNA of somatic cells is extensive. Rat liver 8-OHdG residues were 16 times higher in mitochondrial DNA (mt-DNA) than in nuclear DNA; a result in keeping with the former molecule's closer proximity to the sites of oxidative metabolism. These damage levels consisted of 1.25 8-OHdG molecules per

10^4 nucleosides in rat liver mt-DNA and 7.6 adducts/10^6 bases in rat nuclear DNA. Recently, Yakes and Van Houten (1997) showed that damage to mt-DNA following oxidative stress is more extensive and persists longer that nuclear DNA damage in human cells. This study used a quantitative polymerase chain reaction (PCR) to monitor formation and repair of H_2O_2-induced DNA damage.

Using HPLC-EC, Fraga et al. (1990) studied oxidative DNA damage in rat tissues as a function of age. The amounts of oxidized nucleoside 8-OHdG increased with age in Fischer 344 rat liver, kidney and intestine. On the other hand, the levels of 8-OHdG did not change with age in brain tissue and testes. In this study, damage levels were twice as high in kidney DNA as compared to liver, brain and testes (Fraga et al., 1990). This same investigation measured urinary excretion of 8-OHdG from 2-month and 24-month-old rats which declined 65% during aging. Fraga et al. (1990) suggested that DNA repair enzymes were removing less oxidative adducts from DNA thereby leading to an accumulation of genomic damage in senescent cells. However, it is not clear why the levels of oxidized bases increased with aging in liver, kidney and intestine tissues but did not increase in brain and testes. Sai et al. (1992) measured 8-OHdG levels in Fischer 344 rat brain, kidney, liver, lung, and spleen from 6-months to 30-months-of-age. In agreement with the earlier experiments, the results showed a significant age-related increase in liver and kidney tissues but not in the others. The age-related increase in adduct levels was observed for both male and female rats. Oxidative damage to mt-DNA and nuclear DNA in human brain showed marked age-dependent increases, and mt-DNA seems to be preferentially damaged (Mecocci et al., 1993). Therefore, these results indicate that changes in steady-state levels of oxidative DNA damage are tissue-dependent and may increase upon aging.

Calorie restriction (or dietary restriction) is known to retard the aging processes and delay the onset of age-related diseases (Yu, 1990). Although the mechanisms altering life span are not clear, restriction of food consumption may reduce the production of reactive oxygen species that are a byproduct of food metabolism (Weindruch, et al., 1993). To test the effects of dietary restriction on endogenous oxidative DNA damage, Chung et al. (1992) measured 8-OHdG concentrations in dietary-restricted and ad libitum-fed Fischer 344 rat liver at 3 months- and 24-months-of-age. Both nuclear and mt-DNAs were analyzed, and 8-OHdG was detected and quantified by HPLC-EC. It was observed that mt-DNA had about 15-fold more 8-OHdG than nuclear DNA, consistent with the findings of Richter et al.(1988) and Mecocci et al. (1993). Dietary restriction reduced oxidative DNA damage in both nuclear and mitochondrial compartments. The results indicated that 8-OHdG levels were lower in old, calorie-restricted animals than in their ad libitum-fed counterparts (or the controls). However, there was a greater effect in young 3-month-old restricted rat (36% reduction) than with calorie restriction in 24-month-old liver tissue — a 29% decrease of 8-OHdG. Surprisingly, there was no significant increase in 8-OHdG levels between the young and old rat livers which was in variance with earlier investigations (Fraga et a., 1988; Sai et al., 1992).

Sohal et al. (1994b) also looked at effects of age and calorie restriction on oxidative damage in nuclear DNAs of different tissues from C57BL/6 mice. The concentration of 8-OHdG was compared in skeletal muscle, brain, liver, heart, and kidney tissues from

young and old mice, fed either ad libitum or food restricted to 60% of ad libitum calories. The results showed that 8-OHdG content was greater in old mice for all five tissues. However, the levels of oxidative damage were much more pronounced in the fixed
post-mitotic muscle and brain tissues than in the slowly dividing liver and kidney tissues (Sohal et al., 1994b). A similar analysis was performed on 3-month and 27-month-
old Sprague-Dawley rats. In old rat liver, 8-OHdG increased approximately 45% in total
DNA and about 70% in mitochondrial DNA. Tellingly, there was approximately 7-fold
more oxidative damage in mitochondrial DNA than in total DNA at either age. Dietary
restriction lowered damage levels in all tissues and with aging but had the greatest effect in postmitotic tissues. Dietary restriction lowered 8-OHdG values 20% to 35%
(Sohal et al., 1994b). The effect of dietary restriction on 8-OHdG levels was not significant when the same tissues of dietary restriction and ad libitum-fed mice were compared
at the same age until 23-months-of-age when the effect became quite pronounced.

The combined studies of Fraga et al. (1990) and Sai et al. (1992) suggested that 8-
OHdG increased linearly with age in rat liver and kidney nuclear DNAs. In contrast,
Kaneko et al. (1996) presented evidence for a non-linear accumulation of 8-OHdG in
organs of rat during aging. The results of Kaneko et al. (1997) also indicated that the
accumulation of 8-OHdG in nuclear DNA of various rat tissues begins after 24 months-
of-age but that 8-OHdG varied little before senescence. In addition, old rats accumulated 8-OHdG in a tissue-specific manner. There were significant increases in 8-OHdG
in kidney at 24-months-of-age, in heart and liver at 27-months-of-age, and in brain at
30-months-of-age. Interestingly, dietary restriction did not abolish the increase in 8-
OHdG during aging of these tissues but simply delayed the onset of increased damage
by several months. No difference was observed between ad libitum fed and dietary
restricted rats in the 8-OHdG levels of any organ until 24-months of age. Compared to
2-month, or 24-month-old rats, 30 month-old rats exhibited a 2-fold increase in 8-
OHdG in all organs tested (liver, heart, brain, and kidney). Adduct levels reported in
these studies ranged from 0.75 to 2.2 adducts/10^5 dG, corresponding well with the
steady-state levels of 8-OHdG determined by Fraga et al. (1990) that ranged from 0.8 to
7.3 adducts/10^5 dGs. Hirano et al. (1996) determined 8-OHdG levels in Sprague-
Dawley rat liver, kidney, spleen, small intestine, and brain organs at 3-weeks, 5 months,
and 30-months. HPLC-EC results showed no significant differences between 5-and 30-
month levels in these tissues. Alternatively, there was a significant difference in spleen
between 3-week-old and 30-month-old and in kidney and brain between 3-week and 5-
month-old rats. Adduct levels in this study ranged from 0.7 to 4.9 adducts per 10^5 dGs
(Hirano et al., 1996). No likely interpretation was offered to explain why these results
were in disagreement with other studies (Fraga et al., 1990; Sai et al., 1992; Kaneko et
al., 1996 and 1997).

Instead of HPLC-EC, Wang et al. (1995) used gas chromatography coupled with
mass spectrometry (GC-MS) for identification and quantification of oxidative adducts.
When endogenous oxidative damage to DNA during aging was measured, thymine
glycol, 8-hydroxy-2'-deoxyadenosine (8-OHdA), and 8-OHdG increased roughly two-
fold in rat liver tissue between 1-month and 12-months-of-age. In lung tissue, the
amount of thymine glycol in 12-month-old Wistar rats was ten times greater than in

young rats; 3-times for 8-OHdA and 2-times for 8-OHdG. The results showed a clear age-dependent increase, but adduct levels seemed exceptionally high compared to those obtained with HPLC-EC. Estimates of DNA oxidation damage via HPLC-EC were about one-tenth of those attained by GC-MS. Presumably this difference in oxidation damage is due to oxidation during sample preparation for GC-MS (Finnegan et al., 1996). The GC-MS method is now known to entail artifactual oxidation, such that 8-oxoG and 8-oxo-dG levels were artificially and massively elevated by ex vivo oxidation during sample preparation and analysis (Halliwell and Dizdaroglu, 1992; Ravanat et al., 1995; and see below). Using technology that eliminates or lowers ex vivo artifacts, levels were an order to two orders of magnitude less than with the GC-MS methods.

Chen et al. (1995) obtained data that support the hypothesis that oxidative DNA damage contributes to replicative senescence of normal human fibroblasts. Senescent IMR-90 human fibroblasts removed 4 times more 8-oxoG per day than did early-passage fibroblasts, perhaps signifying greater oxidative stress in senescent fibroblasts. Nonetheless, the steady-state level of 8-oxoG was significantly higher (about 35%) in DNA of senescent cells than in young cells. Furthermore, IMR-90 cells grown at 3% oxygen achieved twice as many population doublings compared to cells grown in air with approximately 20% oxygen. Based on an analysis of the levels of 8-OHdG in the genome of human fibroblasts after irradiation with visible light in the presence of riboflavin, Homma et al. (1994) presented evidence that repair of oxidative DNA damage was more efficient in cells at lower population doubling (25 PDL) than in cells at higher population doublings (55 PDL) and concluded that repair activity was lost with increased cellular passage number. As a function of population doublings, Hirano et al. (1995) analyzed the amount of endonuclease (nicking) activity using total cellular lysate activity from human diploid fibroblasts on a substrate consisting of a double-stranded, 22-mer oligonucleotide with a single central 8-OHdG residue. Results showed nicking activity increased from 20 to 30 PDL, then declined about 60% over the span of 40 to 60 PLD. Therefore, circumstantial evidence indicates that repair of base modifications may decline during cellular aging.

In addition to specific modified bases, less defined oxidative damage to DNA has been surveyed. For example, Randerath et al. (1991) incubated DNA under Fenton reaction conditions that generated additional I-compounds in postlabeling maps. The authors claimed that the oxygen-radical induced I-compounds were bulky, non-polar adducts and not small polar hydroxylated nucleotides. Rat liver mt-DNA was shown by Gupta et al. (1990) to exhibit I-compounds in ^{32}P-postlabeling assays. Two of these non-polar I-spots were mitochondria-specific and increased about 8-fold from 1-month to 9-months-of-age. Gupta et al. (1990) speculated that mitochondria I-compounds were some kind of poorly defined of oxidative DNA damage. In an effort to characterize the chemical nature of I-compounds, Cai et al. (1996) incubated deoxynucleosides with the lipid peroxidation product malondialdehyde (MDA) and analyzed the reactions via postlabeling techniques. Based on this analysis, the authors concluded that a total of five I-compounds have been identified as dGMP-MDA adducts in rat brain and suggested that these or similar I-compounds might act as a biomarker of indirect oxidative damage to DNA.

Over the past several years, a cloud of confusion has arisen regarding the measurements of oxidative base adducts in aging studies because of difficulties inherent in the analysis of oxidative adducts of DNA bases (Beckman and Ames, 1996). DNA can become oxidatively damaged during isolation, sample preparation or adduct detection. For example, phenol is a pro-oxidant used frequently in DNA extractions that can contribute to the level of 8-oxo-dG when compared to non-phenol methods (Claycamp, 1992). Cadet et al. (1997) have summarized studies that demonstrate oxidative adducts were caused by various treatments in preparing samples for GC-MS and for ^{32}P-postlabeling. These workers also observed a 5-fold reduction in levels of 8-OHdG measured by HPLC-EC by incorporating more appropriate DNA extraction techniques. Using a modified, non-phenol, chaotropic NaI, DNA-extraction method, Ames and coworkers (Helbock, et al., 1998) have reduced their estimated levels of steady-state oxidative adducts, such as 8-oxo-dG, by at least an order of magnitude; new 8-oxo-dG estimates were 0.04 to 0.11 adducts/10^5 dGs. Therefore, preceding reports that suggested oxidative adducts occurred at a frequency that is 10-fold or more higher than non-oxidative adducts are now open to question (Beckman and Ames, 1996, 1997). These types of analytical problems become even more daunting when measuring oxidative damage to mt-DNA because of the relatively small amounts of mt-DNA per cell and thus, greater difficulty in purification (Cadet et al., 1997).

Because of artifacts associated with the measurement of oxidized DNA bases, results from previous studies are now suspect and questions of differences in amounts of oxidative base adducts between tissues, ages, organisms, and between nuclear DNA and mitochondrial DNA are open to debate. So how valid are conclusions from previous studies determining oxidative nucleoside or base adducts? One could make the argument that for a single study where are the methods of procedure are the same and possibly carried out simultaneously, then the major variable showing significant differences from "background" would be the experimental variable of chronological age; this however remains to be determined. To summarize, there is much correlative evidence that oxidative base modifications increase with age. Whether age-related increases are linear or non-linear, whether the differences are small or large, whether the steady-state levels are high or low, depend upon the source of the DNA (experimental animal, tissue, diet, etc.), upon the experimental methods, and upon the specific base modification under investigation.

4. DNA Cytosine-5-Methyltransferases

In contrast to DNA damaging agents that alkylate DNA bases inappropriately, there is a directed methylation of the base cytosine in mammalian genomes. Methylation involves the enzymatic conversion of cytosine in DNA to 5-methylcytosine (5-methyl-C) (Bird, 1992). This modification is carried out by DNA 5-cytosine-methyltransferases (MTase), using S-adenosylmethionine as a methyl donor (Zimmerman et al., 1997). The preferred substrate for DNA methylation in mammals is the dinucleotide CpG (Antequera and Bird, 1993). Indeed, about 60% of the cytosine residues within CpG dinucleotides are non-randomly methylated in higher cells. DNA methylation provides one mechanism for stably altering the local structure of a gene. The methyl group that is added projects into the major groove of DNA, and this strategic positioning can affect DNA protein interactions (Tate and Bird, 1993). In fact, DNA methylation can stably repress transcription, perhaps by interfering with DNA binding proteins. Methylation at the 5' end of genes in regions rich in CpG clusters (CpG islands) has been correlated with inhibition of gene expression (Cedar, 1988; Bird, 1992; Tate and Bird, 1993). Although there are exceptions to this general rule, an important determinant of repression is the density of methyl-CpGs near a promoter sequence (Antequera and Bird, 1993). Therefore, methyl modification of mammalian DNA is one avenue whereby mammalian cells may regulate gene expression.

In general, the inheritance of methylation patterns is stable and reproducibly transferred from generation to generation at each cell division. This may be achieved by a maintenance methylase which conserves the methylation pattern after replication following the action of a de novo methylase that initiates a new pattern of methylation (Razin and Cedar, 1993). There appears, however, to be only one MTase gene in the mammalian genome which when mutated abolished both de novo and maintenance DNA methylation (Zimmermann et al., 1997), suggesting additional factors are involved in gene-specific modification. Demethylation is also a prominent feature of cell- and tissue-specific development. Although many of the details of the demethylation reaction are not clear, it seems that RNA molecules are involved in removing methylated nucleotides from DNA (Weiss et al., 1996). Therefore, normal mammalian cells have a steady-state balance between sequence-specific DNA methylation and demethylation. An imbalance in genomic methylation can contribute to altered cellular growth properties and advance tumor progression (Baylin et al., 1998), and alterations in DNA methylation patterns have long been postulated to play a role in fundamental aging processes (Holliday, 1987; Cooney, 1993).

Specific methyl-acceptor sites have been found to be modulated during differentiation of the organism in a programmed manner such that transcriptionally active genes are undermethlyated in tissues in which they are expressed (Cedar, 1988; Tate and Bird, 1993). These observations are consistent with a model in which discrete changes in methylation patterns dictate the characteristics of differentiated cells. It has been suggested that if DNA methylation exemplifies a developmental program, then demethylation during senescence represents a loss of that program, and as such is a dedifferentiation or dysdifferentiation process (Mays-Hoopes, 1989). One of the signal ideas of dysdifferentiation is that genetic regulation will breakdown in old cells, and as a indicator of deregulation, cells start to express proteins and other markers that are not tissue-specific. Evidence for the dysdifferentiation theory of aging has been elaborated by Cutler (1985b). In the context of DNA damage and repair, it can be argued that inappropriate gene methylation or demethylation constitutes genomic damage. For senescent cells, it is not known if such damage is repaired.

Studies on mammalian DNA methylation have revealed that sequence-specific methylation patterns and total genomic methylcytosine levels change with age. Most organisms exhibit a global loss of 5-methyldeoxycytidine (5-methyl-dC) as a function of age in vivo and in vitro. Human fibroblasts in culture showed a progressive loss of 5-methyl-dC as they approach senescence (Holliday, 1986; Wilson and Jones, 1983). When normal human diploid fibroblasts from mice, hamsters, and humans were grown in culture, the 5-methyl-dC content of their genomes decreased significantly. The greatest rate of loss was observed with mouse cells which survived the least number of divisions. In contrast, immortal mouse cell lines had more stable levels of DNA methylation (Wilson and Jones, 1983). Senescent human lymphocytes and T cells show reduced levels of DNA methylation upon aging (Drinkwaater et al., 1989; Golbus et al., 1990).

Wilson et al. (1987) have presented data that showed significant age-dependent losses of 5-methyl-dC residues from DNAs of C57BL/6J mouse brain, liver, and small intestine mucosa tissues and from DNAs of Peromyscus leucopus (white-footed deer mouse) liver and small intestine tissues. Methylation levels in these experiments were quantified by ^{32}P-postlabeling techniques. Similar losses in genomic 5-methyl-deoxycytidine were also shown to correlate with donor age of cultured normal human bronchial epithelial cells. An important finding to emerge from these studies is that the rate of 5-methyl-dC loss from nuclear DNA appeared to be inversely proportional to maximum life span for 3 different mammalian species (Wilson et al., 1987). The results suggested that the age-related reductions in 5-methyl-dC were not connected to mitotic activity in vivo since 3 different tissues, ranging from non-dividing to actively dividing tissues, exhibited similar rates of loss. Analytical HPLC methods used by Singhal et al. (1987) also noted a gradual decline in mouse liver DNA 5-methyl-dC content up to 24-months-of-age. After 24 months the data indicated an increase in genomic 5-methyl-dC in the oldest animals. This latter result is puzzling because C57BL/6J male mice from the same colony were used in both Wilson et al. (1987) and Singhal et al. (1987) studies which suggests that these discordant results might stem from different experimental methods.

Negative data showing no change with aging have also been reported. Excluding studies before 1980 when less sensitive methods of analysis were used to investigate DNA methylation, several groups have found either no change or rather insignificant differences in DNA methylation as a function of age. No significant differences in total 5-methyl-dC levels were detected during aging of human liver (Tawa et al., 1992), rat liver (Zhavoronkova and Vanyushin, 1987), or mouse brain (Tawa et al., 1990) tissues. Besides using different methods of procedure, there are other reasons why various studies do not reach the same conclusions. To point out just one complicating factor, 5-methyl-dC levels can be modulated by diet or compounds that affect methylation metabolism (Cooney, 1993). Even considering the negative results, there is still a consensus in favor of an age-related genomic decrease in 5-methyl-dC content.

Sequence-specific methylation sites have been examined by comparing DNA fragment sizes generated by isoschizomer restriction enzymes. The most commonly used pair of restriction enzymes have been HpaII (methylation-sensitive) and MspI (methylation-resistant) which both recognize the sequence — CCGG. MspI cleaves whether or not CCGG is methylated at the internal C, whereas HpaII cleaves only unmethylated CCGG. Following enzyme digestion and Southern hybridization to the probe of interest, methylation sites can be deduced by comparing band patterns between the 2 DNA digests. Studies have shown decreased DNA methylation of L1Md and IAP repetitive sequence families with aging in one or more mouse organs (Mays-Hoopes et al., 1983, 1986). Endogenous murine retroviral genomes also become hypomethylated during aging (Ono et al., 1989). Satellite sequences that are normally highly methylated in postmitotic tissues become undermethylated in senescent organisms (Howlett et al., 1989). Furthermore, DNA methylation levels associated with various repetitive sequence families seem to correlate well with total genomic levels and decline with age at the same rate as the rest of the genome. In contrast to demethylation of repetitive sequence families during aging, studies of individual genes have indicated no general pattern of methylation changes with age (Slagboom and Vijg, 1989; Slagboom et al., 1990). For the most part, methylation patterns are generally well maintained in stably differentiated cell in vivo. In vitro methylation patterns on single-copy-sequences of normal human diploid fibroblasts, however, do show variability between and within subclones of such fibroblasts as a function of population doublings (Shmookler Reis and Goldstein, 1982; Shmookler Reis et al, 1990). More recent results using genomic sequencing to elucidate methylation sites, demonstrated the importance of establishing the sequence-specific changes in methylation patterns in relation to aging (Halle et al., 1995). For example, overall methylation of the c-myc gene was stable during cellular aging of human fibroblasts in vitro, but after many PDLs, a single cytidine became demethylated at the end of the cells' proliferative life span.

In addition to transcriptional regulation, other roles have been proposed for DNA methylation. DNA methylation has been implicated in genomic imprinting (Peterson and Sapienza, 1993), chromosome inactivation (Pagani et al., 1990), chromosome condensation during mitosis (Cooney and Bradury, 1990), and chromosomal stability (Engler et al, 1993). One of the best characterized aspects of DNA methylation is in mutagenesis. More than one-third of all point mutations that are associated with human genetic diseases are transitions of CpG to TpG, despite the relative rarity of CpG dinucleotides (Cooper and Krawczak, 1989; Rideout et al., 1990). Mechanistically, spontaneous deamination of 5-methylcytosine in DNA will create deoxythymidine which will base pair with deoxyadenosine during DNA synthesis to yield the transition mutation described above. Therefore, there is good reason to think that CpG sequences will be hot-spots for mutagenesis, and the frequency of events, i.e., deaminations, is time-dependent and will increase with age. There are considerable data that support this notion (Baylin et al, 1998).

Demethylation can be due to both active and passive systems, such as demethylases and spontaneous glycolytic release of 5-methyl-dC, respectively. 5-Methyl-dC is in fact more unstable than the normal nucleoside. If sequence-specific remethylation of these lesions is incomplete or missing, then genomic loss of methylated residues will ensue. As a component of the aging process, random loss of methyl groups in DNA may lead to aberrant epigenetic changes in gene expression and perhaps affect chromosomal stability (Holliday, 1984, 1987; Antequera and Bird, 1993). For genes that do exhibit age-related changes in methylation status, there have been no consistent correlations with changes in gene expression (Slagboom and Vijg, 1989; Cooney, 1993). One caveat to correlations between site-specific methylation and gene expression is the uncertainty of whether some sites exert a stronger influence on transcriptional regulation than others. Therefore until good correlations are in hand, the argument that transcriptional control is diminished with aging due to aberrant DNA methylation is incomplete.

5. Mitochondrial DNA Base Modification

Fleming et al. (1982) proposed that damage accumulation in mitochondrial DNA (mt-DNA) might be a mechanism for aging of postmitotic cells. Oxidative damage to mt-DNA in particular might lead to the physiological decline associated with aging due to accumulated mutations in the mitochondrial genome (Miquel and Fleming, 1984; Miguel, 1991). Two points to be promoted in this section are: (1) all of the base modifications described previously for nuclear DNA can reasonably be expected to exist at various steady-state levels in mt-DNA, and; (2) whenever mt-DNA base modifications have been quantified as a function of age, base alterations increased during aging, and mt-DNA adduct levels were higher than those in the nuclear DNA. More recent work has focused on mitochondrial mutations of the deletion type, and mammalian mt-DNA damage and repair are covered in detail by V. Bohr in this volume.

We are not aware on any studies that have directly measured deaminated bases or glycosylated DNA (AGEs) in mitochondria. Two mitochondria-specific I-compounds increased in rat tissues during maturation from 1-month to 12 months, but old animals were not examined; therefore, it is not known whether these bulky, non-polar compounds increase during senescence (Gupta et al., 1990). On the other hand, there are reports that alkyl adducts and oxidative adducts accumulate in old tissues. One study indicated that the alkyl adduct 7-methylguanine appeared to accumulate in rat liver mt-DNA with aging, but subsequent work indicated imprecise quantification due to technical difficulties in HPLC separation of bases (Park and Ames, 1988a, 1988b). Richter et al. (1988) reported that oxidative damage to mt-DNA is more extensive than to nuclear DNA. In fact, mature rat liver mt-DNA contained 16 times more 8-OHdG than nuclear DNA. Yates and Van Houten (1997) confirmed that oxidative damage of mt-DNA is more extensive than nuclear DNA and additionally observed that damage induced by oxidative stress persists longer in mt-DNA than in nuclear DNA. Adduct levels increased with aging in both nuclear and mitochondrial DNAs, in both human tissues (Hayakawa et al., 1991, 1992, 1993) and cultured cells (Homma et al., 1994). Oxidative damage to mt-DNA showed a marked age-dependent increase in human brain (Mecocci et al., 1993); 8-OHdG increased with aging in both nuclear DNA and mt-DNA. However, the magnitude of steady-state 8-OHdG (and the rate of increase with age) was much greater for mt-DNA.

It has been proposed that DNA damage plays a role in generating mt-DNA deletions, perhaps by inducing alterations in DNA structure. The age-correlates between oxidative adducts and deletions in mt-DNA are consistent with this idea. An age-associated exponential accumulation of 8-OHdG in mt-DNA of human diaphragm and heart tissues has been observed to correlate with multiple mt-DNA deletions in these muscles during aging (Hayakawa et al., 1991, 1992, 1993). In addition, chronic cardiac ischemia has been associated with an immense increase in a very common 4,977 base pair deletion (Corral-Debrinski et al., 1991).

To summarize, although recent results suggest that mt-DNA damage and mt-DNA mutations accumulate over the life span of an organism (Bernstein and Gensler, 1993; Weindruch et al., 1993), oxidative DNA adducts are the only documented base modification damage to increase in aging mammalian mitochondria, as described by several different laboratories. Even these data are somewhat shaky because of possible artifactual introduction of oxidative adducts during DNA sample preparation and analysis, as discussed in the Oxidation section. Nevertheless, the data to date suggest that mitochondrial genomic instability is a major factor in aging, and this instability may cause functional deficits in non-renewing tissues, e.g., brain and heart, that have high energy demands and oxygen consumption.

6. Conclusions

This review has focused on DNA base modifications during aging. In the beginning, various sources for genetic alterations were briefly discussed. The sources of genetic damage have been reviewed many times (Cutler, 1976; Ames and Gold, 1991; Cooney, 1993; Gaubatz, 1995; Beckman and Ames, 1998) and have changed little over the years because there seems to be general agreement on what sorts of things effect DNA stability; although there is not necessarily agreement on the relative importance of individual sources. The types of base modifications that can be found in the genome have been fairly well documented (Brash and Hart, 1978; Singer and Kusmierek, 1982; Mullaart et al., 1990; Holmes et al., 1992; Bernstein and Gensler, 1993; Gaubatz, 1995), but again, the relative significance of different types of lesions is controversial.

The results discussed in this review lead to the conclusion that several altered bases and nucleotide adducts can be detected in somatic cell genomes, and there is suggestive evidence that the levels of several forms of indigenous damage, increase in cells during senescence. Less clear perhaps is what is the true genomic content of modified bases. What is not clear at all is why DNA damage levels might vary with aging. Furthermore, we do not know if many of the possible base modifications change with aging. For example, deaminated bases, such as xanthine, have not been examined during mammalian aging. For those base modifications that do change with age, we do not know the distribution of lesions within the genome sequences or within and between individual chromosomes. We do not know the impact of aging of endogenous rates of DNA base damage formation, nor do we know if aging affects the totality of DNA repair enzymes that remove altered bases.

When sensitive methods have been used, studies have demonstrated that some modified bases and DNA adducts increase during aging of non-dividing tissues in vivo. The differences are most apparent for alkyl adducts, oxidative adducts, and I-compounds increasing with aging (Gaubatz, 1986; Randerath et al., 1986; Fraga et al., 1990; Tan et al., 1990). Base cross-links have also been reported for aging rodent liver tissue (Yamamoto et al., 1988). Whereas liver, muscle, and kidney tissues exhibited higher levels of modified bases with aging, the results have been less obvious with brain where it seems that oxidative adducts are unchanged until quite late in the life span when there appears to be a rapid accumulation of damages (Taneko et al., 1996, 1997). This trend suggests that other related lesions will also be found to accumulate in the genome of these tissues with age.

The consensus of the evidence points to an age-related decline in total genomic 5-methyl-dC in postmitotic mammalian cells. Whether or not a positive correlation exists between decreased methylation and aberrant gene regulation has not been established. Likewise, the effects of aging of the methylation status of individual genes has not supported any unified mechanism of aging. Nevertheless, the stability of DNA methylation as a function of aging is an area of molecular investigation that deserves closer scrutiny.

Although some lesions and types of insults are more common than others, it is difficult to predict the biological outcomes of many of the damages. Some changes to DNA appear to be more mechanistically fundamental in relation to aging. Many studies conclude that there is a correlation between oxidative DNA damage and aging (Sohal and Dubey, 1994; Sohal et al., 1994; Beckman and Ames, 1998), and the participation of free radicals in aging and in chronic degenerative diseases continues to gain strength (Beckman and Ames, 1998). Oxidative damage to cellular DNA, both nuclear and mitochondrial, is a ubiquitous insult that happens under normal physiological conditions and may be enhanced by environmental agents and aging (Ku et al., 1993). Perhaps, the relative stability of mt-DNA has exceptional potential to determine aging rates for cells, for individuals, and thus for different species. Base damages, point mutations, and various deletions of mt-DNA have been reported to increase in senescent tissues (Hayakawa et al., 1991, 1992, 1993; Munscher at al., 1993). The aging changes in mt-DNA appear to be more conspicuous in fixed postmitotic cells, such as cardiac and neural tissues. The levels of measured single lesions might appear to be insufficient to account for major functional decrements, given the genomic and cellular redundancies of mammalian cells, but the totality of all lesions combined might be necessary and sufficient to cause cellular senescence. It seems reasonable to assume that if 8-OHdG increases as a function of age then other forms of oxidative base modification will also increase during aging. Concerning informational gaps, we clearly need to know more about mt-DNA repair enzymes, and the distribution of mt-DNA deletions among different cell types in aging tissues is of importance. However, future studies should take lessons learned from the current fund of knowledge. One lesson involves the potential introduction of oxidative artifacts in isolation and analysis of oxidative adducts of DNA (Cadet et al., 1996; Helbock et al., 1998).

Finally, the experimental results do not show that DNA base damage causes senescence of somatic cells. What we have are numerous descriptions and correlations. Examples of base adduct levels in the nuclear and the mitochondrial genomes correlate with life span and more generally with decreasing functions of aging, non-dividing cells. These genome-based changes might be the causes of various physiological decrements characteristic of aging mammals, but there is no proof of this at the present time. Molecular gerontology needs more than correlative data. Thus, the available information does not allow firm conclusions to be drawn about the contribution of base alterations to the aging process. However, there is reason to be optimistic that more mechanistic experiments will be forthcoming soon. Aging research will benefit from the continued development of recombinant DNA technology, biomedical reagents, more powerful detectors, and transgenic mice. Improved methods of analysis including sample isolation and adduct detection would help move this field of investigation forward. With the use of transgenic animals and techniques to inactivate individual genes in the germ cells, the role of various components of the aging process (or anti-aging mechanisms) can be dissected. Development of interventions that either enhance or diminish either antioxidants or oxidants and development of transgenic rodents with design-specific repair systems will also be important.

Acknowledgements

I thank the American Heart Association, Southeast Affiliate, Inc. (9810196) and the National Cancer Institute (CA78510) for finacial support and Jeffrey Dubuisson for reading the manuscript.

References

Adams, S.P., Laws, G.M., Selden, J.R., & Nichols, W.W. (1994) Phosphatase activity in commercial spleen exonuclease decreases the recovery of benzo[a]pyrene and N-hydroxy-2-naphthylamine DNA adducts by ^{32}P-postlableing. Anal. Biochem. 219, 121-130.

Adelman, R., Saul, R. L., & Ames, B. N. (1988) Oxidative damage to DNA: relation to species metabolic rate and life span. Proc. Natl. Acad. Sci. USA 85, 2706-2708.

Alexander, P. (1967) The role of DNA lesions in processes leading to aging in mice. Symp. Soc. Exp. Biol. 21, 29-50.

Ames, B.N. (1989) Endogenous oxidative damage, aging, and cancer. Free Rad. Res. Commun. 7, 121-128.

Ames, B.N., & Gold, L.S. (1991) Endogenous mutagens and the causes of aging and cancer. Mutat. Res. 250, 3-16.

Ames, B.N., Gold, L.S., & Willet, W.C. (1995) The causes and prevention of cancer. Proc. Natl. Acad. Sci. USA 92, 5258-5265.

Antequera, F., & Bird, A. (1992) CpG islands. In: DNA Methylation: Mmolecular Biology and Biological Significance (Jost, J P. & Saluz, H. P., Eds.) Birkhauser Verlag, Basel, pp. 169-185.

Aucher, M. C. (1982) Hazards of nitrate, nitrite, and N-nitroso compounds in human nutrition, in Nutritional Toxicology. Hathcock, J. N. (Ed.) Academic Press, New York, pp. 328-367.

Barrows L.R., & Shank, R.C. (1981) Aberrant methylation of liver DNA in rats during hepatotoxicity. Toxicol. Appl. Pharmacol. 60, 334-345.

Barrows, L.R., & Magee, P. N. (1982) Nonenzymatic methylation of DNA by S-adenosyl methionine in vitro. Carcinogenesis 3, 349-351.

Basu, A.K., Loechler, L.E., Leadon, S.A. & Essigmann, J.M. (1989) Genetic effects of thymine glycol: site-specific mutagenesis and molecular modeling studies. Proc. Natl. Acad. Sci. USA 86, 7677-7681.

Baylin, S.B., Herman, J.G., Graff, J.R., Vertino, P.M., & Issa, J-P. (1998) Alterations in DNA methylation: a fundamental aspect of neoplasia. Adv. Cancer Res. 72, 141-196.

Beckman, K.B., & Ames, B.N. (1997) Oxidative decay of DNA. J. Biol. Chem. 272, 19633-19636.

Bechman, K.B., & Ames, B.N. (1998) The free radical theory of aging matures. Physiol. Rev. 78, 547-581.

Bernstien, H. & Gensler, H. (1993) DNA Damage and Aging. In: Free Radicals in Aging (Yu, B.P., Ed.) CRC Press, Boca Raton, FL, pp. 89-122.

Bird, A.P. (1986) CpG-rich islands and the function of DNA methylation. Nature 321, 209-211.

Bird, A. (1992) The essentials of DNA methylation. Cell 70, 5-8.

Blomeke, B., Greenblatt, M.J., Doan, V.D., Bowman, E.D., Murphy, S.E., Chen, C.C., Kato, S. & Shields, P.G. (1996) Distribution of 7-aklyl-2'-deoxyguanosine adduct levels in human lung. Carcinogenesis 17, 741-748.

Bohr, V.A., & Anson, R.M. (1995) DNA damage, mutation and fine structure DNA repair in aging. Mutat. Res. 338, 25-34.

Brash, D.E., & Hart, R.W. (1978) DNA damage and repair in vivo. J. Environ. Pathol. Toxicol. 2, 79-114.

Bucala, R., Model, P., & Cerami, A. (1984) Modification of DNA by reducing sugars: A possible mechanism of nucleic acid aging and age-related dysfunction in gene expression. Proc. Natl. Acad. Sci. USA 81, 105-109.

Bucala, R., Model, P. Russel, M., & Cerami, A. (1985) Modification of DNA by glucose-6-phosphate induces DNA rearrangements in an E. coli plasmid. Proc. Natl. Acad. Sci. USA 82, 8439-8442.

Bucala, R., Lee, A.T., Rourke, L., & Cerami, A. (1993) Transposition of an Alu-containing element induced by DNA-advanced glycosylation endproducts. Proc. Natl. Acad. Sci. USA 90, 2666-2670.

Cadet, J., Douki, T., & Ravanat, J.-L. (1997) Artifacts associated with the measurement of oxidized bases. Environ. Health Perspect. 105, 1034-1039.

Cai, Q., Tian, L., & Wei, H. (1996) Age-dependent increase of indigenous DNA adducts in rat brain is associated with a lipid peroxidation product. Exp. Gerontol. 31, 387-396.

Cedar, H. (1988) DNA methylation and gene activity. Cell, 53, 3-4.

Chen, Q., Fischer, A., Reagan, J.D., Yan, L.-J., & Ames, B.N. (1995) Oxidative DNA damage and senescence of human diploid fibroblasts. Proc. Natl. Acad. Sci. USA 92, 4337-4341.

Cheng, K.D., Cahill, D.S., Kassai, H., Nishimura, S., & Loeb, L.A. (1992) 8-Hydroxyguanine, an abundant form of oxidative DNA damage, caused G → T and A → C substitutions. J. Biol. Chem. 267, 166-172.

Chung, M.H., Kasai, H. Nishimuar, S., & Yu, B.P. (1992) Protection of DNA damage by dietary restriction. Free Rad. Biol. Med. 12, 523-525.

Claycamp, H.G. (1992) Phenol sensitization of DNA to subsequent oxidaitve damage in 8-hydroxyguanine assays. Carcinogenesis 13, 1289-1292.

Cooney, C. A., & Bradury, E. M. (1990) DNA Methylation and Chromosome Organization in Eukaryotes. In: The Eukaryotic Nucleus: Molecular Biochemistry and Macromolecular Assemblies (Strauss, P. & Wilson, S., Eds.) Vol.2, Telford Press, Caldwell, N.J., pp. 813-826.

Cooney, C. A. (1993) Are somatic cells inherently deficient in methylation metabolism? A proposed mechanism for DNA methylation loss, senescence and aging. Growth Dev. Aging 57, 261-273.

Cooper, D. N., & Krawczak, M. (1989) Cytosine methylation and the fate of CpG dinucleotides in vertebrate genomes. Hum. Genet. 83, 181-188.

Corral-Debrinski, M., Stepien, G., Shoffner, J.M., Lott. M.T., Kanter, K., & Wallace, D.C. (1991) Hypoxemia is associated with mitochondrial DNA damage and gene induction: implications for cardiac disease. J. Am. Med. Assoc. 266, 1812-1820.

Curtis, H.J. (1971) Genetic factors in aging. Adv. Genet. 16, 305-324.

Cundy, K.C., Kohen, R., & Ames B.N. (1989) Determination of 8-hydroxydeoxyguanosine in human urine: a possible assay for in vivo oxidative DNA damage. Basic Life Sci. 49, 479-482.

Cutler, R. G. (1976) Cross-linking Hypothesis of Aging: DNA Adducts in Chromatin as a Primary Aging Process. In: Aging, Carcinogenesis and Radiation Biology: The Role of Nucleic Acid Addition Reactions (Smith, K. C., Ed.) Plenum Press, New York, pp. 443-492.

Cutler, R. G. (1985a) Antioxidants and longevity of mammalian species. Basic Life Sci. 35. 15-73.

Cutler, R. G.(1985b) Dysdifferentation Hypothesis of Aging: A Review. In: Molecular Biology of Aging: Gene Stability and Gene Expression (Sohol, R. S., Bernbaum, L. S., & Cutler, R. G., Eds.) Raven Press, New York, pp. 307-340.

Dizdaroglu, M. (1991) Chemical determination of free radical-induced damage to DNA. Free   Rad. Biol. Med. 10, 225-242.

Doll, R., & Peto, R. (1981) The causes of cancer. J Natl. Cancer Inst. 66, 1191-1308.

Drinkwater, R. D., Blake, T. J., Morley A. A., & Turner, D. R. (1989) Human lymphocytes aged in vivo have reduced levels of methylation in transcriptionally active and inactive DNA. Mutat. Res. 219, 29-37.

Eloranta, T.O. (1977) Tissue distribution of S-adenosylmethionine in the rat: effects of age, sex, and methionine administration of the metabolism of S-adenosylmethionine, S-adenosylhomocysteine and polyamines. Biochem. J. 166, 521-529.

Engler, P., Weng, A., & Storb, U. (1993) Influence of CpG methylation and target spacing on V(D)J recombination in a transgenic substrate. Mol. Cell. Biol. 13, 571-577.

Finnegan, M.T.V., Herbert, K.E., Evans, M.D., Grifiths, H.R., & Lunec, J. (1996) Evidence for sensitization of DNA to oxidative damage during isolation. Free Rad. Biol. Med. 20, 93-98.

Fleming, J. E., Miguel, J., Cottrell, S. F., Yengoyan, L. S., & Economos, A. C. (1982) Is cell aging caused by respiration-dependent injury to the mitochondrial genome? Gerontology 28, 44-53.

Fraga, C. G., Shigenaga, M. K., Park, J. W., Degan, P., & Ames, B. N. (1990) Oxidative damage to DNA during aging: 8-hydroxy-2'-deoxyguanosine in rat organ DNA and urine. Proc. Natl. Acad. Sci. USA 87, 4533-4537.

Gallagher, J.E., Jackson, M.A., Geroge, M.H., Lewtas, J. & Robertson, I.G.C. (1989) Differences in detection of DNA adducts in the 32P-postlabeling assay after either 1-butanol extraction or nuclease P1 treatment. Cancer Lett. 45, 7-12.

Gaubatz, J. W. (1986) DNA damage during aging of mouse myocardium. J. Mol. Cell. Cardiol. 18, 1317-1320.

Gaubatz, J. W. (1989) Postlabeling analysis of indigenous aromatic DNA adducts in mouse myocardium during aging. Arch. Gerontol. Geriatr. 8, 47-54.

Gaubatz, J.W., & Tan, B.H. (1993) Age-related studies on the removal of 7-methylguanine from DNA of mouse kidney tissue following N-methyl-N-nitorsourea treatment. Mutat. Res. 295, 81-91.

Gaubatz, J.W. (1995) Genomic Instability During Aging of Postmitotic Mammalian Cells. In: Molecular Basis of Aging (Macieira-Coelhl, A., Ed.) CRC Press, Bocca Raton, FL, pp. 71-131.

Gaubatz, J.W., & Tan, B.H. (1997) Introduction, distribution, and removal of 7-methylguanine in different liver chromatin fractions of young and old mice. Mutat. Res. 375, 25-35.

Gensler, H.L., & Bernstein, H. (1981) DNA damage as the primary cause of aging. Q. Rev. Biol. 56, 279-303.

Gold, L.S., Slone, T.H., Stern, B.R., Manley, N.B., & Ames, B.N. (1992) Rodent carcinogens: setting priorities. Science 258, 261-265.

Gupta, R.C., & Earley, K. (1988) ^{32}P-adduct assay: comparative recoveries of structurally diverse DNA adducts in the various enhancement procedures. Carcinogenesis 9, 1687-1693.

Gupta, K.P., van Golen, K.L., Randerath, E., & Randerath, K. (1990) Age-dependent covalent alterations (I-compounds) in rat liver mitochondrial DNA. Mutat. Res. 237, 17-27.

Golbus, J., Palella, T. D., & Richardson, B. C. (1990) Quantitative changes in T cell DNA methylation occur during differentiation and ageing. Eur. J. Immunol. 20, 1869-1872.

Halle, J.P., Schmidt, C., & Adam, G. (1995) Changes in the methylation pattern of the c-myc gene during in vitro aging of IMR90 human embryonic fibroblasts. Mutat. Res. 316, 157-171.

Halliwell, B, & Dizdaroglu, M. (1992) The measurement of oxidaitve damage to DNA by HPLC and GC/MS techniques. Free Rad. Biol. Med. 16, 75-87.

Halliwell, B. (1996) Antioxidants in human health and disease. Annu. Rev. Nutr. 16, 33-50.

Harman, D. (1981) The aging process. Proc. Natl. Acad. Sci. USA 78, 7124-7128.

Harman, D. (1992) Free radical theory of aging. Mutat. Res. 275, 257-266.

Hart, R. W., & Setlow, R. B. Correlation between deoxyribonucleic acid excision-repair and life-span in a number of mammalian species. Proc. Natl. Acad. Sci. USA 71, 2169-2173.

Hayakawa, M., Torii, K., Sugiyama, S., Tanaka, M., & Ozawa, T. (1991) Age-associated accumulation of 8-hydroxydeoxyguanosine in mitochondria DNA of human diaphram. Biochem. Biophys. Res. Commun. 179, 1023-1029.

Hayakawa, M., Hattori, K., Sugiyama, S., & Ozawa, T. (1992) Age-associated oxygen damage and mutations in mitochondrial DNA in human hearts. Biochem. Biophys. Res. Commun. 189, 979-985.

Hayakawa, M., Sugiyama, S., Hattori, K., Takasawa, M., & Ozawa, T. (1993) Age-associated damage in mitochondrial DNA in human hearts. Mol. Cell. Biochem. 119, 95-103.

Hecht, S.S, & Hoffman, D. (1989) The relevance of nitrosamine to human cancer. Cancer Surv. 8, 273-294.

Helbock, H.J., Beckman, K.B., Shigenaga, M.K., Walter P.B., Woodall, A.A., Yeo, H.C., & Ames, B.N. (1998) DNA oxidation matters: The HPLC-electrochemical detection assay of 8-oxo-deoxyguanosine and 8-oxo-guanine. Proc. Natl. Acad. Sci. USA 95, 288-293.

Henle, E.S., Luo, Y., Glassmann, W., & Linn, S. (1996) Oxidative damage to DNA constitiuents by iron-mediated Fenton reactions. J. Biol. Chem. 271, 21177-21186.

Henner, W.D., Grunberg, S.M., & Haseltine, W.A. (1982) Sites and structures of (γ-irradiation-induced DNA strand breaks. J. Biol. Chem. 257, 11750-11754.

Hirano, T., Yamaguchi, Y., Hirano, H., & Kasai, H. (1995) Age-associated change of 8-hydroxyguanine repair activity in cultured human fibroblasts. Biochem. Biophys. Res. Commun. 214, 1157-1162.

Hirano, T., Yamaguchi, R., Asami, S., Iwamoto, N., & Kasai, H. (1996) 8-Hydroxyguanine levels in nuclear DNA and its repair activity in rat organs associated with age. J. Gerontol. 51A, B303-B307.

Holliday, R. (1984) The Significance of DNA Methylation in Cellular aging. In: The Molecular Biology of Aging: Basic Life Sciences (Woodhead, A. D., Blackett, A. D. & Hollaender, A., Eds.) Vol. 35, Plenum Press, New York, pp. 269-284.

Holliday, R. (1987) The inheritance of epigenetic defects. Science 238, 163-170.

Holmes, G.E., Bernstein, C., & Bernstein, H. (1993) Oxidative and other DNA damages as the basis of aging: a review. Mutat. Res. 275, 305-315.

Homma, Y., Tsunoda, M., & Kasai, H. (1994) Evidence for the accumulation of oxidative stress during cellular aging of human diploid fibroblasts. Biochem. Biophys. Res. Commun. 203, 1063-1068.

Howlett, D., Dalrymple, S., & Mays-Hoopes, L. L. (1989) Age-related demethylation of mouse satellite DNA is easily detectable by HPLC but not by restriction endonucleases. Mutat. Res. 219, 101-106.

Hutchinson, F. (1985) Chemical changes induced in DNA by ionizing radiation. Prog. Nucl. Acid Res. 32, 115-155.

Kaneko, T., Tahara, S., & Matsuo, M. (1996) Non-linear accumulation of 8-hydroxy-2'-deoxyguanosine, a marker of oxidized DNA-damage, during aging. Mutat. Res. 316, 277-285.

Kaneko, T., Tahara, S., & Matsuo, M. (1997) Retarding effect of dietary restriciton on the accumulation of 8-hydroxy-2'-dexoyguanosine in organs of Fischer 344 rats during aging. Free Rad. Biol. Med. 23, 76-81.

Karran, P., & Lindahl, T. (1980) Hypoxanthine in deoxyribonucleic acid: generation by heat-induced hydrolysis of adenine residues and release in free form by a deoxyribonucleic acid glycosylase from calf thymus. Biochemistry 19, 6005-6011.

Kirsh, M. E., Cutler, R. G., & Hartman, P. E. (1986) Absence of deoxyuridine and 5-hyroxymethyldeoxyuridine in the DNA from three tissues of mice of various ages. Mech. Ageing Dev. 35, 71-77.

Ku, H.-H., Brunk, U.T., & Sohal, R.S. (1993) Relationship between mitochondrial superoxide and hydrogen peroxide production and longevity of mammalian species. Free Rad. Biol. Med. 15, 621-627.

Lee, A.T., & Cerami, A. (1987a) The formation of reactive intermediates of glucose-6-phosphate and lysine capable of rapidly reacting with DNA. Mutat. Res. 179, 151-158.

Lee, A.T., & Cerami, A. (1987b) Elevated glucose-6-phophate levels are associated with plasmid mutations in vivo. Proc. Natl. Acad. Sci. USA 84, 8311-8314.

Lee, A.T., & Cerami, A. (1990) In vitro and in vivo reactions of nucleic acids with reducing sugars. Mutat. Res. 238, 185-191.

Lee, A.T., & Ceramin, A. (1991) Induction of transposition in response to elevated glucose-6-phosphate levels. Mutat. Res. 249, 125-133.

Li, D., & Randerath, K. (1990a) Strain differences of I-compounds in relation to organ sites of spontaneous tumorigenesis and non-neoplastic renal disease in mice. Carcinogenesis 11, 251-255.

Li, D., & Randerath, K. (1990b) Association between diet and age-related DNA modifications (I-compounds) in rat liver and kidney. Cancer Res. 50, 3991-3996.

Li, D., Xu, D., & Randerath, K. (1990a) Species and tissue specificities of I compounds as contrasted with carcinogen adducts in liver, kidney and skin of Sprague-Dawley rats, ICR mice and Syian hamsters. Carcinogenesis 11, 2227-2232.

Li, D., Xu, D., Chandar, N., Lombardi, B., & Randerath, K. (1990b) Persistent reduction of indigenous DNA modification (I-compound) levels in liver DNA from male Fischer rats fed choline-devoid diet and in DNA of resulting neoplasms. Cancer Res. 50, 7577-7580.

Li, D., Chen, S., & Randerath, K. (1992) Natural dietary ingredients (oats and alfalfa) induce covalent DNA modification (I-compounds) in rat liver and kidney. Nutr. Cancer 17, 205-216.

Li, D., & Randerath, K. (1992) Modulation of DNA modification (I-compound) levels in rat liver and kidney by dietary carbohydrate, protein, fat, vitamin, and mineral content. Mutat. Res. 275, 47-56.

Lindahl, T. & Nyberg, B. (1974) Heat-induced deamination of cytosine residues in deoxyribonucleic acid. Biochemistry 13, 3405-3410.

Loeb, L.A. (1985) Apurinic sites as mutagenic intermediates. Cell 40, 483-484.

Loeb, L.A., & Preston, B.D. (1986) Mutagenesis by apurinic/apyrimidinic sites. Ann. Rev. Genet. 20, 201-230.

Makita, Z., Vlassara, H., Cerami, A., & Bucala, R. (1992) Immunochemical detection of advanced glycosylation end products in vivo. J. Biol. Chem. 267, 5133-5138.

Mandal, P. C. & Yamamoto, O. (1986) Changes of fluorescence spectra of 2'-deoxyguanosine in aqueous solution by radiation. Biochem. Int. 12, 235-242.

Margison, G.P., & O'Connor, P.J. (1979) Nucleic Acid Modification by N-Nitroso Compounds. In: Chemical Carcinogens and DNA (Grover, L.P., Ed.) CRC Press, Boca Raton, FL, Vol.1, pp. 111-159.

Mays-Hoopes, L. L., Brown, A., & Huang, C. C. (1983) Methylation and rearrangement of mouse intracisternal A particle genes in development, aging and myeloma. Mol. Cell. Bio. 3, 1371-1380.

Mays-Hoopes, L. L., Chao, W., Butcher, H. C., & Huang, C. C. (1986) Decreased methylation of the major long interspersed repeated DNA during aging and in myeloma cells. Dev. Genet. 7, 65-73.

Mays-Hoopes, L. L. (1989) Age-related changes in DNA methylation: Do they represent continued developmental changes? Intl. Rev. Cytol. 114, 181-220.

Mecocci, P., MacGarvey, U., Kaufman, A.E., Koontz, D., Shoffner, J.M., Wallace, D.C., & Beal, M.F. (1993) Oxidative damage to DNA shows marked age-dependent increases in human brain. Ann. Neurol. 34, 609-616.

Miller, E.C., & Miller, J.A. (1981) Searches for ulitmate chemical carcinogens and their reactions with cellular macromolecules. Cancer 47, 2327-2345.

Miquel, J.G., & Fleming, J.E. (1984) A two-step hypothesis on the mechanism of in vitro cell aging: cell differentiation followed by intrinsic mitochondrial muatgenesis. Exp. Gerontol. 19, 31-36.

Mirvish, S.S. (1973) Formation of N-nitrosos compounds: chemistry, kinetics, and in vivo occurence. Toxicol. Appl. Pharmacol. 31, 325-351.

Mitra, S., & Kaina B. (1993) Regulation of repair of alkylation damage in mammalian genomes. Prog. Nucl. Acids Res. Mol. Biol. 44, 109-143.

Moller, L., & Hofer, T. (1997) [^{32}P]ATP mediates formation of 8-hydroxy-2'-deoxyguanosine from 2'-deoxyguanosine, a possible problem in the postlabeling assay. Carcinogenesis 18, 2415-2419.

Mullaart, E. Lohman, P.H.M., Berends, F., & Vijg, J (1990) DNA damage metabolism and aging. Mutat. Res. 237, 189-210.

Munscher, C., Muller-Hocker, J. and Kadenbach, B. (1993) Human aging is associated with various point mutations in tRNA genes of mitochondrial DNA. Biol. Chem. Hoppe Seyler 374, 1099-1104.

Mustonen, R., & Hemminki, K. (1992) 7-Methylguanine levels in DNA of smokers' and non-smokers' total white blood cells, grandulocytes, and lymphocytes. Carcinogenesis 13, 1951-1955.

Nath, R.C., Vulimiri, S.V., & Randerath, K. (1992) Circadian rhythm of covalent modifications in liver DNA. Biochem. Biophys. Res. Commun. 189,454-550.

Ono, T., Shinya, K., Uehara, Y., & Okada, S.(1989) Endogenous virus genomes become hypomethylated tissue-specifically during aging process in C57BL mice. Mech. Ageing Dev. 50, 27-36.

Orr, W. C., & Sohal, R. S. (1994) Extension of life-span by overexpression of superoxide dismutase and catalase in Drosophila melanogaster. Science 263, 1128-1130.

Pagani, F., Toniolo, D., & Vergani, C. (1990) Stability of DNA methylation of X-chromosome genes during aging. Somatic Cell Mol. Genet. 16, 79-84.

Park, J. W., & Ames, B. N. (1988a) 7-methylguanine adducts in DNA are normally present at high levels and increase on aging: analysis by HPLC with electrochemical detection. Proc. Natl. Acad. Sci. USA 85, 7467-7470.

Park, J. W., & Ames, B. N. (1988b) Correction. Proc. Natl. Acad. Sci. USA 85, 9508.

Park, J., Cundy, K. C., & Ames, B. N. (1989) Detection of DNA adducts by high-performance chromatography with electrochemical detection. Carcinogenesis 10, 827-832.

Papoulis, A., Al-Abed, Y., & Bucala, R. (1995) Identification of N2-(1-carboxyethyl)guanine (CEG) as a guanine advanced glycosylation end product. Biochemistry 34, 648-655.

Pegg, A.E. (1983) Alkylation and subsequent repair of DNA after exposure to dimethylnitrosamine and related carcinogens. Rev. Biochem. Toxicol., 5, 83-133.

Peterson, K., & Sapienza, C. (1993) Imprinting the genome: Imprinted genes, imprinting genes, and a hypothesis for their interaction. Ann. Rev. Genet. 27, 7-31.

Pogribny, I.P., Muskhelishvili, L., Miller, B.J., & James, S.J. (1997) Presence and consequence of uracil in preneoplastic DNA from folate/methyl-deficient rats. Carcinogenesis 18, 2071-2076.

Randerath, K., Reddy, M. V., & Disher, R. M. (1986) Age- and tissue-related DNA modifications in untreated rats: Detection by ^{32}P-postlabeling assay and possible significance for spontaneous tumor induction and aging. Carcinogenesis 7, 1615-1617.

Randerath, K., Putman, K. L., Randerath, E., Mason, G., Kelley, M., & Safe, S. (1988) Organ-specific effects of long term feeding of 2,3,7,8-tetrachlorodibenzo-p- dioxin and 1,2,3,7,8-pentachlorodibenzo-p-dioxin on I-compounds in hepatic and renal DNA of female Sprague-Dawley rats. Carcinogenesis 9, 2285-2289.

Randerath, K., Liehr, J. G., Gladek, A., & Randerath, E. (1989) Age-dependent covalent DNA alterations (I-compounds) in rodent tissues: species, tissue, and sex specificities. Mutat. Res. 219, 121-133.

Randerath, K., Li, D., & Randerath, E. (1990) Age-related DNA modifications (I-compounds): Modulation by physiological and pathological processes. Mutat. Res. 238, 245-253.

Randerath, K., Yang, P.F., Danna, T.F., Reddy, R., Watson, W.P., & Randerath, E. (1991) Bulky adducts detected by ^{32}P-postlabeling in DNA modified by oxidative damage in vitro. Comparison with rat lung I-compounds. Mutat. Res. 250, 135-144.

Randerath, K., Putman, K. L., Osterburg, H. H., Johnson, S. A., Morgan, D. G., & Finch, C. E. (1993a) Age-dependent increases of DNA adducts (I-compounds) in human and rat brain DNA. Mutat. Res. 295, 11-18.

Randerath, K., Hart, R. W., Zhou, G. D., Reddy, R., Danna, T. F., & Randerath, E. (1993b) Enhancement of age-related increases in DNA I-compound levels by calorie restriction: comparison of male B-N and F-344 rats. Mutat. Res. 295, 31-46.

Randerath, K. Zhou, G., Hart, R.W., Tuturro, A., & Randerath, K. (1993c) Biomarkers of aging: correlation of DNA I-compound levels with median lifespan of calorically restricted and ad libitum fed rats and mice. Mutat. Res. 295, 247-263.

Randerath, K., Gupta, K.P., & van Golen, K.L. (1993d) Altered fidelity of a nucleic acid modifying enzyme, T4 polynucleotide kinase, by safrole-induced DNA damage. Carciongenesis 14, 1523-1529.

Randerath, K., & Randerath, E.A. (1994) ^{32}P-postlabeling methods for DNA adduct detection: overview and critical evaluation. Drug Metabol. Rev. 261, 67-85.

Randerath, K., Randerath, E., Smith, C.V., & Chang, J. (1996) Structural origins of bulky oxidative DNA adducts (type II I-compounds) as deduced by oxidation of oligonucleotides of known sequence. Chem. Res. Toxicol. 9, 247-254.

Ravanat, J.L., Turesky, R.J., Gremaud, E., Trudel, I.J., & Stadler, R. (1995) Determination of 8-oxoguanine in DNA by gas chromatography-mass spectrometry and HPLC-electrochemical detection: overestimation of the backgound level of the oxidized base by the gas chromatography-mass spectrometry assay. Chem Res. Toxicol. 8, 1039-1045.

Razin, A. & Cedar, H. (1993) DNA methylation and embryogenesis, in DNA methylation: Molecular biology and biological significance, Jost, J. P. & Saluz, H. P.(eds.) Basel: Birkhauser Verlag, pp. 343-357.

Reddy, M.V., & Randerath, K. (1986) Nuclease P1-mediated enhancement of sensitivity of ^{32}P-postlabeling test for structurally diverse DNA adducts. Carcinogenesis 7, 1543-1551.

Reyes, A., Gissi, C., Pesole, G., & Saccone, C. (1998) Asymmetrical directional mutation pressue in the mitochondrial genome of mammals. Mol. Biol. Evol. 15, 957-966.

Reynolds, T.M. (1965) Chemistry of nonenzymatic browning. Adv. Food Res. 14, 165-283.

Richter, C., Park, J. W., & Ames, B. N. (1988) Normal oxidative damage to mitochondrial and nuclear DNA is extensive. Proc. Natl. Acad. Sci. USA 85, 6465-6467.

Rideout, W. M. III, Coetzee, G. A., Olumi, A. F., & Jones, P. A. (1990) 5-Methylcytosine as an endogenous mutagen in the human LDL receptor and p53 genes. Science 249, 1288-1290.

Ryberg, B., & Lindahl, T. (1982) Nonenzymatic methylation of DNA by the intracellular methyl group donor S-adenosyl-L-methionine is a potentially mutagenic reaction. EMBO J. 1, 211-216.

Sai, K., Takage, A., Umenura, T., Hasegawa, R., & Kurokawa, Y. (1992) Changes of 8-hydroxydeoxyguanosine levels in rat organ DNA during the aging process. J. Environ. Pathol. Toxicol. Oncol. 11, 139-143.

Saul, R. L., & Ames, B. N. (1986) Background levels of DNA damage in the population. in Mechanisms of DNA Damage and Repair, Simic, M. G., Grossman, L. & Upton, A. C., Eds., Plenum Press, New York, pp. 529-535.

Schmutte, C., & Jones, P.A. (1998) Involvement of DNA methylation in human carcinogenesis. Biol. Chem. 374, 377-388.

Segerback, D., & Vodicka, P. (1993) Recoveries of DNA adducts of polycyclic aromatic hydrocarbons in the ^{32}P-postlabeling assay. Carcinogenesis 14, 2463-2469.

Sharma, R. C., & Yamamoto, O. (1980) Base modification in adult animal liver DNA and similarity to radiation-induced base modification. Biochem. Biophys. Res. Commun. 96, 662-627.

Shibutani, S., Takeshita, M., & Grollman, A.P. (1991) Insertion of specific bases during DNA synthesis past the oxidation-damaged base 8-oxodG. Nature, 349, 431-434.

Shmookler Reis, R. J., & Goldstein, S. (1982) Variability of DNA methylation patterns during serial passage of human diploid fibroblasts. Proc. Natl. Acad. Sci. U.S.A. 79, 3949-3953.

Shmookler Reis, R. J., Finn, G. K., Smith, K., & Goldstein, S. (1990) Clonal variation in gene methylation: c-H-ras and (-hCG regions vary independently in human fibroblast linkages. Mutat. Res. 237, 45-57.

Sidorkina, O., Saparbaev, M., & Laval, J. (1997) Effects of nitrous acid treatment of the survival and mutagenesis of Escherichia coli cells lacking base excision repair (hypoxanthine-DNA glycosylase-ALK A protein) and/or nucleotide excision repair. Mutagenesis 12, 23-28.

Singer, S., & Kusmierek, J.T. (1982) Chemical mutagenesis. Annu. Rev. Biochem. 51, 665-693.

Singhal, R. P., Mays-Hoopes, L. L., & Eichhorn, G. L. (1987) DNA methylation in aging of mice. Mech. Ageing Dev. 41, 199-210.

Slagboom, P. E., & Vijg, J. (1989) Genetic instability and aging: theories, facts, and future perspectives. Genome 31, 373-385.

Slagboom, P. E., de Leeuw, W. J. F., & Vijg, J. (1990) Messenger RNA levels and methylation patterns of GAPDH and (-actin genes in rat liver, spleen and brain in relation to aging. Mech. Ageing Dev. 53, 243-257.

Sohal, R.S., & Dubey, A. (1994) Mitochondrial oxidative damage, hydrogen peroxide release, and aging. Free Rad. Biol. Med. 16, 621-626.

Sohal, R.S., Ku, H.-H., Agarwal, S., Forster, M.J., & Lal, H. (1994a) Oxidative damage, mitochondrial oxidant generation and antioxidant defences during aging and in response to food restriction in the mouse. Mech. Ageing Dev. 74, 121-133.

Sohal, R.S., Agarwal, S., Candas, M., Forster, M.J., & Lal, H. (1994b) Effect of age and caloric restriction on DNA oxidative damage in different tissues of C57BL/6 mice. Mech. Ageing Dev. 76, 215-224.

Stramentinole, G. Gualano, M., Catto, E. & Sergio, A. (1977) Tissue levels of S-adenosylmethionine in aging rats. J. Gerontol. 4, 392-394.

Tan, B. H., Bancsath, A., & Gaubatz, J. W. (1990) Steady-state levels of 7-methylguanine increase in nuclear DNA of postmitotic mouse tissues during aging. Mutat. Res. 237, 229-238.

Tannenbaum, S.R. (1987) Endogenous formation of N-nitroso compounds: a current perspective. IARC Sci. Publ. 292-296.

Tate, P. H., & Bird, A. P. (1993) Effects of DNA methylation on DNA-binding proteins and gene expression. Curr. Opin. Genet. Dev. 3, 226-231, 1993.

Tawa, R., Ono, T., Kurishita, A., Okada, S., & Hirose, S. (1990) Changes on DNA methylation level during pre- and postnatal periods in mice. Differentiation 45, 44-48.

Tawa, R., Ueno, S., Yamamoto, K., Yamamoto, Y., Sagisaka, K., Katakura, R., Kayama, T., Yoshimoto, T., Sakuri, H., & Ono, T. (1992) Methylated cytosine level in human liver DNA does not decline in aging process. Mech. Ageing Dev. 62, 255-261.

Tice, R. R., & Setlow, R. B. (1985) DNA Repair and Replication in Aging Organisms and Cells. In: Handbook of the Biology of Aging (Finch, C. E. & Schneider, E.L., Eds.) Van Nostrand Reinhold, New York, pp. 173-193.

Vlassara, H., Bucala, R., & Striker, L. (1994) Pathogenic effects of advanced glycosylation: Biochemical, biologic, and clinical implications for diabetes and aging. Lab. Invest. 70, 138-151.

Wang, Y.-J., Ho, Y.-S., Lo, M.-J., & Lin, J.-K. (1995) Oxidative modification of DNA bases in rat liver and lung during chemical carcinogenesis and aging. Chem. Biol. Interact. 94, 135-145.

Wang, D., Kreutzetr, D.A. & Essigmann, J.M. (1998) Mutagenicity and repair of oxidative DNA damage: insights from studies using defined lesions. Mutat. Res. 400, 99-115.

Weindruch, R., Warner, H.R., & Starke-Reed, P.E. (1993) Future Directions in Free Radical Research in Aging. In: Free Radicals in Aging (Yu, B.P., Ed.) CRC Press, Bocca Raton, FL, pp. 269-295.

Weiss, A., Keshet, I., Razin, A., & Cedar, H. (1996) DNA demethylation in vitro: involvement of RNA. Cell 86, 709-718.

Whong, W.Z., Stewart, J.D., & Ong, T. (1992) Comparison of DNA adduct detection between two enhancement methods of the ^{32}P-postlabeling assay in rat long cells. Mutat. Res. 283, 1-6.

Wilson, V. L., & Jones, P. A. (1983) DNA methylation decreases in aging but not immortal cells. Science 220, 1055-1057.

Wilson, V. L., Smith, R. A., Ma, S., & Cutler, R. G. (1987) Genomic 5-methyldeoxycytidine decreases with age. J. Biol. Chem. 262, 9948-9951.

Xiao, W., & Samson, L. (1993) In vivo evidence for endogenous DNA alkylation damage as a source of spontaneous mutations in eukaryotic cells. Proc. Natl. Acad. Sci. USA 90, 2117-2120.

Xiao, W., Chow, B.L., & Rathgeber, L. (1996) The repair of DNA methylation damage in Saccharomyces cerevisiae. Curr. Genet. 30, 461-468.

Yakes, F.M. & VanHouten, B. (1997) Mitochondrial DNA damage is more extensive and persists longer than nuclear DNA damage in human cells following oxidative stress. Proc. Natl. Acad. Sci. USA 94, 514-519.

Yamamoto, O., Fuji, I., Yoshida, T., Cox, A. B., & Lett, J. T. (1988) Age dependency of base modification in rabbit liver DNA. J. Gerontol. 43, B132-B136.

Yu, B.P. (1990) Food restriction research: past and present status. Rev. Biol. Res. Aging 4, 349-371.

Zhavoronkova, E. N., & Vanyushin, B. F. (1987) Methylation of DNA and its interaction with rat liver glucocorticoid-receptor complexes. Biochemistry 52, 870-877.

Zimmerman, C., Guhl, E., & Graessmann, A. (1997) Mouse DNA methyltransferase (MTase) deletion mutants that retain the catalytic domain display neither de novo nor maintenance methylation activity in vivo. Biol. Chem. 378, 393-405.

© 2001 Elsevier Science B.V. All rights reserved.
The Role of DNA Damage and Repair in Cell Aging
B.A. Gilchrest and V.A. Bohr, volume editors.

MODELS FOR STUYDING GENOMIC INSTABILITY DURING AGING

Jan Vijg, Heidi Giese and Martijn E.T. Dollé

*University of Texas Health Science Center and
CTRC Institute for Drug Development, San Antonio, TX*

1. Introduction

Historical Background

Aging can be defined as a series of time-related processes occurring in the adult individual that ultimately bring life to a close. To some extent, these processes can be understood as the coordinated action of the products of multiple genes. However, aging also has a major stochastic component. Random DNA alterations, induced by environmental and endogenous mutagens and carcinogens fall into this category. The consequence of such changes, i.e., a general loss of genome stability, has been considered as a most likely primary cause of aging, especially since the demonstration that late irradiation damage in animals, such as generalized atrophy and neoplasia, appeared to resemble premature aging (Henshaw et al., 1947). Subsequently, theories were formulated as to how spontaneous changes in DNA, i.e., induced by cosmic radiation, environmental mutagens and other DNA damaging agents, could cause aging (Failla, 1958; Szilard, 1959; Burnet, 1974). A major hindrance in gaining a full understanding of the potential impact of DNA alterations on a living organism was the initial lack of insight into the mechanisms by which DNA alterations arise and the factors that influence their rate of occurrence. Indeed, it was originally assumed that DNA as it occurs in the cell is intrinsically stable, even during its transmission from generation to generation. With the discovery of DNA repair (Boyce and Howard-Flanders, 1964; Pettijohn and Hanawalt, 1964; Setlow and Carrier, 1964; see also Hanawalt, 1994) it became clear that DNA alterations would be much more frequent, were it not for the continuous monitoring of the genome for changes in DNA chemical structure. From this concept was borne that the genes specifying DNA repair may belong to a category of genes called longevity assurance genes (Hart et al., 1979; Sacher, 1982). However, at the time, the enormous complexity of the various interconnected DNA repair and other DNA processing systems was still unknown (Friedberg et al., 1995).

Spontaneous DNA Damage

The ultimate source of genome instability is the continuous introduction of lesions in the DNA by a variety of exogenous and endogenous physical, chemical and biological agents. Breaks of the sugar-phosphate backbone are spontaneously induced, e.g., under the influence of body heat, apurinic and apyrimidinic sites arise as a consequence of DNA hydrolysis and naturally present methyl donors cause non-enzymatic methylation

(Lindahl, 1993). In addition, a variety of exogenous agents, such as ultraviolet light, ionizing radiation and a range of chemicals in food etc., can induce DNA damages in their target tissues, e.g., inter and intra-chromosomal cross links, DNA single and double strand breaks, bulky adducts and smaller adducts.

A major source of spontaneous DNA damage during aging could be free radicals, which have been considered as a most logical explanation for the various forms of cell and tissue damage observed to occur during aging (Harman, 1992; Ames et al., 1993; Martin et al., 1996a). As a by-product of oxidative phosphorylation and several other biological and physiological processes, oxygen radicals can induce a variety of damages into cellular DNA and other biological macromolecules like proteins and lipids. The lesions induced in DNA by free radicals are diverse and include a variety of adducts (Jaruga and Dizdaroglu, 1996) as well as abasic sites, cross links and DNA single and double strand breaks.

Attempts have been made to quantitate spontaneous DNA lesions in tissues and organs of animals and humans at different age levels. Interpretation of the results has been ambivalent due to the lack of sensitivity of most of the methods used (Mullaart et al., 1991). Even damage detected with highly specific and highly sensitive methods, such as high-performance liquid chromatography (with electrochemical detection), gas chromatography (with detection by mass spectrometry) and ^{32}P-postlabeling has been attributed to artifacts due to sample manipulation (Lindahl, 1993). This already suggests that the levels of spontaneous DNA damage are very low. This is in fact what has been found in most studies, although some forms of damage appeared to increase with age. For example, Te Koppele et al. (1996) reported about 3 copies of the specific oxidative DNA lesion 8-hydroxy-2'-deoxyguanosine (OH8dG) per 10^5 dG in the human brain, which is in the same range as results reported by others for rodents (Fraga et al., 1990). In a recent study of radiation-induced thymine glycol in human cell lines less than 4.3 thymine glycols were found per 10^9 bases of DNA for unirradiated samples (Le et al., 1998). At such low frequency, it would be unlikely for a lesion to occur within a gene or one of its regulatory regions, especially if it is realized that most of the lesions are not permanent but part of a steady state situation. That is, lesions are continuously removed through DNA repair pathways and re-introduced. Thus, it seems unlikely that under normal circumstances spontaneous DNA damage has any direct effect, in terms of impaired transcription. Somewhat higher steady state levels, however, could increase the level of spontaneous mutagenesis (see Fig. 1 and further below).

2. DNA Damage Processing: Genomic Instability

In order to become permanent, a DNA lesion must be fixed in the genome of a cell in the form of a mutation, that is, a heritable alteration that can be "corrected" only by elimination of the cell. Genetic stability and, hence, the limit to somatic mutation accumulation, is controlled by a number of sometimes interrelated and overlapping cellular functions, the most important of which are the various DNA replication, repair and recombination pathways. Within the structural constraints of the genome, these

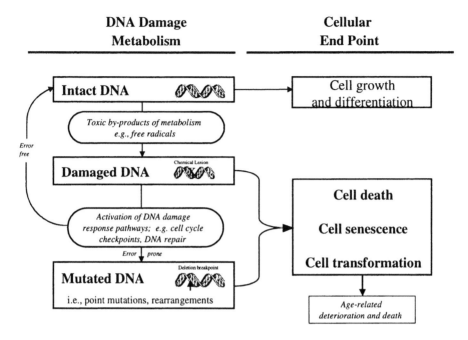

Fig. 1. Schematic depiction of the induction of DNA damage during aging, processing of the damage by, e.g, DNA repair systems, and the molecular end point in the form of irreversibly mutated genomic DNA. While the concept of cancer as a biological end point of mutations has now been fairly well established, a causal relationship between mutation accumulation and aging is still subject of investigation.

processes determine the molecular end point of DNA damage metabolism in each cell, tissue or organ, in the form of a mutation spectrum, i.e., a set of alterations in DNA sequence or DNA sequence organization (Fig. 1).

Mutations represent a two-edged sword. In unicellular organisms and in the germ line of metazoa they are needed as the substrate for evolution. However, at too high a rate mutations can lead to population extinction. In unicellular organisms sexual reproduction has been hypothesized to counterbalance mutation accumulation by rearranging the genetic material to re-create genomes with fewer or no deleterious mutations. According to the hypothesis known as Muller's ratchet (Muller, 1964), asexual lineages will tend to lose mutation-free genomes (due to genetic drift) and inevitably suffer from loss of viability. Senescence-like phenomena have been observed, for example, in asexually propagated protozoa lineages, and were attributed to the accumulation of deleterious mutations (Bell, 1988). More recent data on viruses and microbes also suggest that in the absence of sex deleterious mutations accumulate and lead to loss of fitness (Andersson and Hughes, 1996; Chao, 1997), which would be in agreement with Muller's ratchet.

In the somatic cells of multicellular organisms deleterious mutations are expected to accumulate, not unlike the situation in asexual unicellular lineages. In diploid organisms recessive mutations are masked, which limits adverse effects. Nevertheless, dominant and co-dominant mutations will also accumulate (Morley, 1982) and could eventually result in a senescent phenotype. Indeed, since no evolutionary advantage is gained from having better cellular maintenance and repair than strictly necessary to reach repro-ductive age, it follows that genomic mutation rate will never be low enough for indefinite survival. This fits in the "disposable soma theory", predicting that aging in general is caused by a lifelong accumulation of somatic damage (Kirkwood, 1989). Due to the lack of methods to measure genomic mutation rates in different organs and tissues of higher organisms, very little is known about spontaneous somatic mutagenesis and its possible relation with the degenerative aspects of the aging process.

3. Studying Genomic Instability in Vivo

To study genome stability in relation to aging two types of models are needed. First, one should be capable to quantify and characterize the molecular end point of DNA damage metabolism, that is, all types of mutations in various organs and tissues. Second, it will be necessary to evaluate the effect of deficiencies in various genome stability systems on both the aging phenotype and the in vivo somatic mutation spectrum over the life span. The resulting data set can then be used to derive conclusions as to which repair pathways are most relevant to both aging and mutation accumulation. This should lead to a rational approach to retard aging by manipulation of key repair pathways to maintain genome stability over longer periods of time.

A. *Models for Mutation Detection*

Since unlike DNA damages, mutations are not chemically distinct from the original situation they have been notoriously difficult to detect if occurring at low frequency. (Most methods for mutation detection are restricted to situations in which all or most of the cells carry the same mutation.) Current assays for detecting mutations at low frequency are based on selectable markers, represented by endogenous genes or transgenic reporter genes. Few endogenous genes lend themselves as mutational target genes. The best know assay system is the hypoxanthine phosphoribosyl transferase (HPRT) locus test. In this system mutation frequencies are scored as the frequency of cells capable of forming a clone in the presence of 6-thioguanine. Only cells with a mutation in the X-linked HPRT gene can survive this condition (Albertini et al., 1990). The results obtained with the assay suggest that mutant frequencies in humans increase with age from about 2×10^{-6} in young individuals to about 1×10^{-5} in middle aged and old individuals (Jones et al., 1995). In mice the mutant frequency appeared to be somewhat higher, that is, from about 5×10^{-6} in young animals to about 3×10^{-5} in middle aged mice (Dempsey et al., 1993). However, in both mice and humans these values are underestimates, due to the loss of HPRT mutants in vivo or in vitro. Indeed, results from

Grist et al. (1992), who assayed the HLA locus (using immunoselection for mutationally lost HLA antigen) in human lymphocytes, indicate 2–3 times higher mutant frequencies. Values higher than HPRT were also found with other assays involving selectable target genes and the discrepancy has been explained in terms of the inability of the HPRT test to detect mitotic recombination events (HPRT is X-linked) and a relatively strong in vivo selection against mutants (Grist et al., 1992). Indeed, HPRT mutant frequencies have been found to decrease with time following exposure to mutagenic agents, such as ethyl nitrosourea and radiation, and has been considered as less suitable for long term monitoring (da Cruz et al., 1996).

Virtually all studies on HPRT mutation frequencies have used human or mouse T lymphocytes. Martin et al. (1996b) applied the HPRT test to tubular epithelial cells of kidney tissue from 2 to 94 year-old human donors. Interestingly, the mutation frequencies found were much higher than the above mentioned values for blood lymphocytes and were also increasing with age, from about 5×10^{-5} to about 2.5×10^{-4}; the increase with age appeared to be exponential rather than linear. The higher mutant frequency in the kidney cells could reflect a relatively slow turnover as compared to T-cells.

With the development of transgenic mouse models, harboring chromosomally integrated reporter genes, the somatic mutation theory of aging could be directly tested by measuring mutations in every organ and tissue in a neutral gene (Gossen et al., 1989). The results thus far obtained with some of these models indicate an age-related increase in mutant frequency in spleen, from about 3×10^{-5} in mice of a few weeks old to 15×10^{-5} in 24-month old animals (Lee et al., 1994). These results were essentially confirmed by Ono et al. (1995). However, this first generation of models only allowed the detection of small mutations, such as basepair substitutions and very small deletions or insertions. To study a wider range of mutations, including large genomic rearrangements, in different somatic cell populations of higher organisms, we have generated a transgenic mouse model harboring multiple copies of a plasmid vector containing the lacZ reporter gene as a target (Gossen et al., 1995; Boerrigter et al., 1995). Several transgenic lines were generated, one with the vector cluster on chromosome 11 and another with two integration sites, on chromosome 3 and on chromosome 4. Copy numbers were roughly in the order of 10–20 copies per integration site per haploid genome. These two transgenic lines had a normal pattern of survival and did not differ from the parental C57Bl/6 strain (Dollé et al., 1997).

Procedures have been developed to recover the integrated plasmid vectors from genomic DNA with high efficiency (Gossen et al., 1993; Dollé et al., 1996; Fig. 2). Upon rescue, the plasmids are transferred into an E. coli host lacking the lacZ gene and harboring an inactivating mutation in the galE gene of the galactose operon. This latter mutation allows positive selection of plasmids with a mutated lacZ gene on medium containing the lactose analogue phenyl ß-D-galactoside (p-gal; Gossen et al., 1992). E. coli cells harboring wild type copies of the lacZ-plasmid metabolize p-gal to produce the toxic intermediate UDP-galactose. As a consequence of the galE mutation this cannot be converted into UDP-glucose and, hence, the cell will die (Dollé et al., 1996).

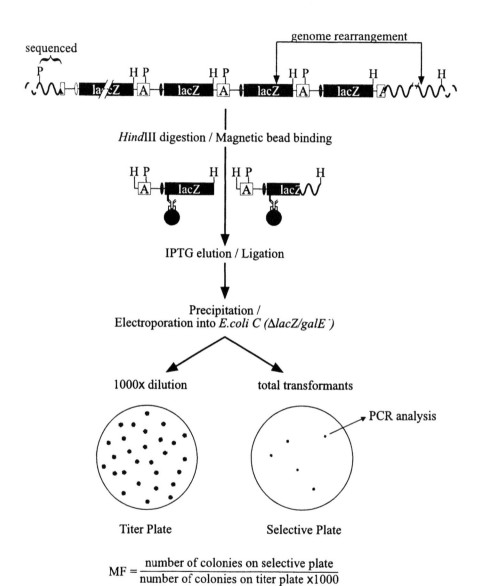

Fig. 2. Transgenic mouse model for studying mutation frequencies and spectra in different organs and tissues in relation to aging and cancer. Plasmids are integrated head to tail in the mouse genome. Several transgenic lines with the plasmid cluster at different chromosomal positions have been studied (Dollé et al., 1997). The plasmids are recovered from their integrated state by excision with a restriction enzyme, circularized by ligation and subsequently used to transform E. coli cells. A very small part is plated on the titer plate, which contains X-gal resulting in a blue halo round the colonies. The rest is plated on p-gal plates to select for the mutants, which appear as red colonies because of the presence of tetrazolium. The figure also indicates how genome rearrangement events with one breakpoint in a lacZ gene and the other in the 3'-mouse flanking sequence can be recovered and detected (for further details, see Boerrigter et al., 1995).

The plasmid-based model allows for the detection and characterization of all types of mutations, including large genomic rearrangements. As illustrated in Figure 2, deletions (or translocations) with one break point in a lacZ reporter gene and the other break point in the 3' mouse sequence flanking the plasmid cluster can be detected in the form of a recovered plasmid harboring a small piece of mouse DNA. The mouse DNA fragment is the part between the break point and the first upstream HindIII site in the mouse DNA. The occurrence of such mutations was empirically confirmed by physical mapping, using the cloned integration sites as one anchor point and the mouse sequence fragment in the mutant as the other. Some of these "mouse sequence mutants" are relatively small as indicated by their physical map position within a 70–80 kb P1 clone, which was obtained by screening a library with the cloned integration site as probe (Dollé et al., submitted). Others, however, proved to be much larger. Indeed, genetic mapping of the break point of one mouse sequence mutant (obtained from brain DNA), using a M. spretus/C57Bl/6 backcross panel, indicated a distance of 13 cM from the integration site (break point in the lacZ reporter) on chromosome 4 (Dollé et al., submitted).

Using the plasmid model, we demonstrated that mutant frequencies in vivo increase with age in some but not in all tissues. For example, an age-related increase in mutant frequency (from about 4×10^{-5} in the young adults to about 15×10^{-5} in old animals) was only observed in the liver, but not in the brain (Fig. 3A; Dollé et al., 1997). Moreover, although at young age the mutation frequencies in liver and brain were not significantly different in that study, in the liver the frequency of large genomic rearrangements in the mutation spectrum was significantly greater than in the brain (at all age levels). The frequency of these genomic rearrangements in the liver was found to increase rapidly at late age (Fig. 3B). The increased susceptibility to spontaneous mutagenesis of liver versus brain corresponds with observed higher frequency of focal pathological lesions in the mouse liver as compared to brain (Bronson and Lipman, 1991).

B. *Influence of Inactivated Genome Stability Systems on in vivo Mutation Accumulation*

To obtain information as to the functional implications of background patterns of organ-specific mutation accumulation, crosses can be made between the reporter gene mice and mouse mutants in which either completely or partially inactivating mutations have been selectively introduced in specific genes involved in genome stability processes. Since genome stability systems are considered as major longevity assurance systems it is reasonable to assume that inactivation of one or more of such pathways accelerates in parallel both age-related deterioration and death and mutation accumulation. Such a finding would be in keeping with the observation that in humans heritable mutations in various genes involved in DNA repair, recombination and transcription, are responsible for human syndromes mimicking accelerated aging, e.g., ataxia telangiectasia, xeroderma pigmentosum, Cockayne syndrome, Werner syndrome, Bloom syndrome, Li-Fraumeni syndrome (Martin and Turker, 1994). Several of these syndromes, such as

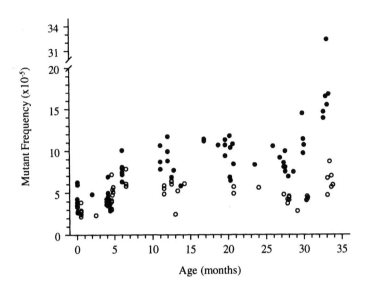

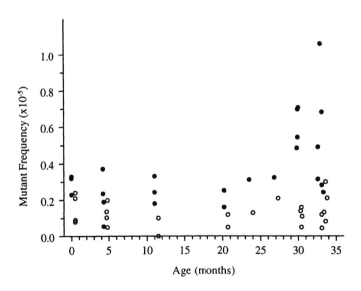

Fig. 3. Spontaneous frequencies of all mutants (top) and frequencies of genome rearrangements (bottom) in liver and brain of lacZ-plasmid transgenic mice (line 60) as a function of age. Closed circles: liver; open circles: brain. To prevent data overlap, all data points for brain were shifted 0.5 month to the right.

Werner syndrome, are further characterized by a high level of spontaneous mutations, most of them DNA rearrangements, in their T lymphocytes (Fukuchi et al., 1989). Based on the existence of these diseases, it is conceivable that allelic variation at loci harboring genes involved in DNA repair are responsible, at least partly, for the observed variation in life span among normal human individuals.

Several crosses of reporter gene mice and genome stability mutant mice have now been made and the results thus far obtained will be discussed. The first involved the p53 tumor suppressor gene, which exerts DNA damage control by shutting down cell cycle advance, facilitating removal of the lesions by DNA repair enzymes (Jacks and Weinberg, 1996). In addition to the prevention of DNA damage to be converted into mutations, p53 is also thought to act in an apoptosis pathway to eliminate cells with high loads of DNA damage and/or mutations (Ko and Prives, 1996; Levine, 1997). Mice with inactivated p53 genes develop tumors and die generally within six months of age (Donehower et al., 1992). It is not unreasonable to assume that the high tumor frequency in these mice at such early age is due, at least in part, to an accelerated mutation accumulation in both the normal and tumor tissues. This would increase the chance of the initiating mutation(s) to occur and greatly facilitate later mutations providing the tumor cells with increasingly effective growth attributes.

The possible influence of p53 on spontaneous mutation frequencies has been directly tested by Sands et al. (1995) and Nishino et al. (1995), by crossing a mouse model harboring a reporter gene as part of a bacteriophage lambda vector with a p53-knockout mouse. The hybrids permitted the analysis of mutation frequencies and spectra in different organs and tissues at a neutral locus in a p53-deficient background. Surprisingly, no elevated mutation frequencies were found, either in the normal tissues or in the tumors. The authors provided several possible explanations for a normal mutation frequency in an animal lacking an important genome protection mechanism. These explanations included the limited sensitivity of the chromosomally integrated bacteriophage lambda vectors used, which is mainly restricted to point mutations (see also Gossen et al., 1995). Possibly more importantly, the animals were only tested at an early age, i.e., at 2–3 months.

We have tested the sensitivity of p53-defective mice for spontaneous mutations by crossing p53 knock-out mice with the plasmid-based lacZ mutation model, which is more sensitive for DNA rearrangement types of mutations. We then analyzed mutation frequencies and spectra in spleen of these hybrid p53-deficient mice over the entire life span of about 6 months. The results confirmed the normal mutation frequencies at early age also found by Sands et al. (1995) and Nishino et al (1995), but also indicated an accelerated mutation accumulation at later ages, once development has been completed (Giese et al., unpublished; Fig. 4). Many of these mutations were size-change mutations including large deletions. This accelerated mutation accumulation was found not to be due to a higher sensitivity to mutation induction, which might be expected as a consequence of defective cell cycle arrest. Indeed, similar numbers of mutations were found after treatment with the powerful mutagen ENU of p53-deficient and control mice.

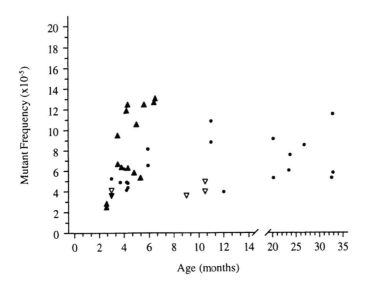

Fig. 4. *Spontaneous frequencies of all mutants in spleen of Trp53⁻⁻, lacZ hybrids (closed triangles) compared to Trp53⁻ ⁻, lacZ mice (open triangles) and the lacZ parental strain (closed circles) as a function of age.*

One step further in DNA damage control is DNA repair, the many different pathways of which can act in both mitotic and postmitotic cells. A major repair pathway thought to act predominantly in mitotic tissues is DNA mismatch repair (MMR). The function of MMR is to monitor the newly synthesized DNA strand for incorrect or mismatched bases, which are removed and replaced by the correct equivalents (Modrich, 1994). MMR overlaps with other repair pathways, in that it is also involved in the repair of chemical damage to DNA and the processing of recombination intermediates (Modrich, 1994). Kolodner and co-workers were the first to identify and clone human homologs of bacterial and yeast DNA mismatch repair genes (Fishel et al., 1993). As it turned out, germ line mutations in a number of these genes, e.g., MLH1 and MSH2, have been associated with hereditary nonpolyposis colorectal cancer in humans (Fishel et al., 1993; Nicolaides et al., 1994). Tumor cells from these patients (homozygously mutated) demonstrate accelerated mutation of simple repeat sequences (microsatellites) in the genome (Perucho, 1996).

Several MMR-deficient knock-out mice have been made, some of which were crossed with bacteriophage lambda-based reporter mice. The results indicated greatly elevated (up to 100-fold) spontaneous mutation frequencies in several organs and tissues these hybrid mice (Narayanan et al., 1997; Andrew et al., 1997). The type of mutations (frequent 1-bp deletions) suggested replication slippage errors as the mechanism of induction. Interestingly, these data immediately refute the possibility that such high levels of mutation are incompatible with normal development; however, at this stage nothing is known about the situation at later ages since complete life span studies on these mice have not yet been performed. It is possible that high levels of point mutations only increase the rate of cancer, but not of aging. Indeed, the mutations relevant for aging could be predominantly large rearrangements, such as those that have been shown to occur at a high rate in lymphocytes of Werner's syndrome patients (Fukuchi et al., 1989). An increase in the rate of large deletion mutations can be expected to have more of an effect on genome functioning than a comparable increase in point mutations (see above).

A second major DNA repair mechanism in mammalian cells is nucleotide excision repair (NER). This system entails multiple steps that employ a number of proteins to eliminate a broad spectrum of structurally unrelated lesions such as UV-induced photoproducts, chemical adducts, as well as intra-strand crosslinks and some forms of oxidative damage. Deficiency in NER has been shown to be associated with human inheritable disorders, such as xeroderma pigmentosum (XP), Cockayne's syndrome (CS) and trichothiodystrophy (TTD). These disorders are characterized by UV-sensitivity, genomic instability and various signs of premature ageing (Bootsma, 1998). XP patients can be classified into at least 7 complementation groups (XPA-XPG), XPA being the most frequent. XPA deficiency inactivates both global and transcription-coupled NER and causes severe symptoms, the most prominent of which are a > 2000-fold increased frequency of UVB-induced skin cancer and accelerated neurodegeneration (Bootsma, 1998). The XPA-protein, in combination with replication protein A, has been proposed to be involved in DNA damage recognition, i.e., the pre-incision step of NER.

Mice deficient in the NER gene XPA have been generated and appeared to mimic the phenotype of humans with xeroderma pigmentosum, that is, increased sensitivity to UVB-induced skin cancer. This was found to be associated with an almost complete lack of UV-induced excision repair, measured by unscheduled DNA synthesis in cultured fibroblasts, and a much lower survival of fibroblasts after UV-irradiation or treatment with DMBA (de Vries et al., 1995). However, at early age NER-deficient mice do not show spontaneous abnormalities. These mice develop normally, indistinguishable from wild type mice. At older age, i.e., from about 15 months onwards, XPA$^{-/-}$ mice show an increased frequency of hepatocellular adenomas (de Vries et al., 1997), which in normal mice is only a common pathological lesion at old age (Bronson and Lipman, 1991). The lack of spontaneous abnormalities in young mice might be due to the fact, that under normal conditions mice have only limited exposure to NER-mediated DNA damage. Hence, at early age NER appears to be dispensable. (This is in contrast to base excision repair, the inactivation of which is lethal; Sobol et al., 1996.)

In order to test the hypothesis that loss of NER causes accelerated mutation accumulation, preceding the onset of accelerated tumor formation and ageing, XPA-deficient mice were crossed with the lacZ transgenic mice previously used to monitor mutation accumulation in liver and brain during ageing (Fig. 3). In the hybrid XPA$^{-/-}$, lacZ mice, mutant frequencies were analysed in liver and brain at early ages: 2 months,

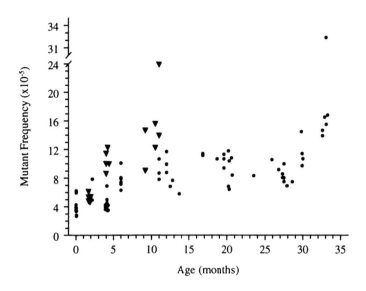

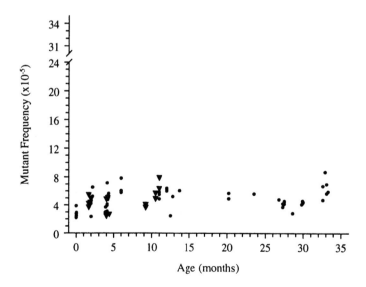

Fig. 5. Spontaneous frequencies of all mutants in liver (top) and brain (bottom) of XPA   , lacZ hybrids (closed triangles) compared to the lacZ parental strain (closed circles) as a function of age.

4 months and 9-12 months. Fig. 5 shows the effect of the XPA deficiency on the mutant frequency as compared to the normal lacZ control background (Giese et al., 1999). In liver of 2-month old mice, mutant frequencies were still comparable to those for the lacZ control animals. In 4-month old XPA$^{-/-}$, lacZ mice, mutant frequencies were significantly increased by a factor of 2. A further, albeit smaller increase was observed between 4 and 9-12 months. At the latter age level mutant frequencies were in the range of the maximum level reported earlier for 25-34-month old lacZ control mice, with a similar high individual variation (Fig. 5A). In brain, mutant frequencies were not found to increase over the age levels studied, which is in keeping with the lack of an age-related increase reported previously for the lacZ control mice (Fig. 5B). Mutant spectra were analysed from liver and brain of 2-month and 9-12-month old XPA$^{-/-}$, lacZ mice (data not shown). The results were quite similar to those reported earlier (Dollé et al., 1997), i.e., an about equal fraction of size change versus point mutations and a higher level of genome rearrangements in liver than in brain.

The results of this study indicate that a deficiency in the NER gene XPA causes an early, accelerated accumulation of somatic mutations in liver, but not in brain. The increase in mutant frequency in liver is in keeping with the higher incidence of spontaneous liver tumors several months later (de Vries et al., 1997). Therefore it is tempting to conclude that the higher mutant frequencies in liver predict an organ-specific predisposition to cancer. In the corresponding human syndrome, xeroderma pigmentosum, internal tumor development is rare, but neurological abnormalities are frequently observed (Cleaver and Kraemer, 1995). In XPA-deficient mice such abnormalities have thus far not been found. Phenotypical differences between XPA-deficient mice and human XPA might be due to tissue-dependent variation in the levels of endogenous DNA damaging species. This may lead in humans to deleterious effects in brain, whereas in mice it may cause a higher frequency of spontaneous liver tumors. Humans with XP rarely survive beyond the third decade of life, a consequence of the dramatic increase in sunlight induced skin cancer (Cleaver and Kraemer, 1995). Skin cancer does not occur in rodents, which have a fur that cannot be penetrated by UV and are also kept under conditions not permitting exposure to sunlight. This explains the lack of such a phenotype of the XPA mutation at early ages. In the young animals NER could be essentially redundant, but become increasingly important at later ages. Indeed, it has been repeatedly argued that loss of redundancy in, e.g., cell number, gene copy number, functional pathways, could be responsible for the gradual loss of individual stability and increased incidence of disease associated with ageing (Strehler and Freeman, 1980).

4. General Discussion and Future Perspectives

Control of life span can be considered as a balance between somatic maintenance and reproductive effort (Kirkwood, 1989). Therefore, aging can be explained in terms of an accumulation of somatic damage and it is argued in this paper that somatic mutations, i.e., irreversible alterations in DNA sequence, are a major component of the spectrum.

Are mutation frequencies in mammalian cells in vivo high enough to explain the functional and pathological endpoints of aging?

At present, there is surprisingly little information on the frequencies of spontaneous mutations in mammalian cells in vivo, and most of the discussion has been focused on cancer, not aging. Loeb (1991) adopted the very low frequency of 1.8×10^{-7} per cell generation per locus (HPRT) in the context of his proposal of a mutator phenotype. A mutator phenotype, i.e., an acquired somatic mutation rendering a tumor cell lineage more susceptible to spontaneous mutagenesis, was considered necessary to account for the multistep nature of cancer pathogenesis in humans (Loeb, 1991). A mutator phenotype would not only explain the multiple mutations necessary to give rise to a tumor, but also the large number of apparently random mutations so characteristic of the tumor phenotype. Although Tomlinson et al. (1996) provided a model of tumorigenesis in which selection appeared to be more important than a high mutation rate, mutator phenotypes have actually been demonstrated in a number of cancers and cancer cell lines (Perucho, 1996). Indeed, in a subset of hereditary nonpolyposis colorectal cancer patients, a form of cancer caused by germline mutations in mismatch repair genes, a high level of somatic mutations (i.e., microsatellite instability) was found in their tumors and sometimes also in their non-neoplastic cells (Parsons et al., 1995).

Without neglecting the occurrence of mutator phenotypes in some tumors, it seems prudent to consider at this stage the possibility that the somatic mutation frequency in vivo in normal cells is much higher than the very low figures thus far adopted. As mentioned above, the mutation rate at HPRT is likely to be higher than most data suggest, due to selection against cells with a mutated HPRT gene and the lack of somatic recombination of this X-linked gene. Results from Grist et al. (1992), using the HLA-A locus as the selectable target, indicate mutation frequencies between 0.71×10^{-5} in neonates to 6.53×10^{-5} in elderly individuals. This is not dramatically different from the aforementioned results obtained with the lacZ reporter mice (Fig. 3A). It should also be noted that mutations do not necessarily depend on cell proliferation but rather on chronological time (Strauss, 1992; MacPhee, 1995; Morley, 1996). Indeed, mutations in postmitotic cells could be the result of errors during DNA repair processes (see, for example, MacPhee, 1995). In fact, it is conceivable that mutations accumulate more readily in postmitotic tissues than in actively proliferating ones, due to increased possibilities of negative selection in the latter. Therefore, rather than mutations per cell generation, it seems more useful to investigate the total number of mutations per cell or diploid genome at a given point during the life span.

How could somatic mutation loads of the magnitude given above be responsible for the adverse effects associated with the aging process in mammals? First, they could to some extent explain the well known age-related increase in the incidence of cancer (Ames et al., 1993). Second, somatic mutations could be responsible for the loss of cells in various organs, such as the brain. In this respect, empirically determined in vivo mutation frequencies are always underestimates due to the elimination of cells with a high mutation load. Finally, random somatic mutations, especially large genome rearrangements, could lead to impaired cell functioning rather than cell death or cell transformation. In this view, the accumulation of random mutations would result in a mosaic of cells at various levels of deficiency. It is conceivable that, especially large structural alterations at mutational hotspots, could gradually impair genome functioning. Indeed, rather than a catalog of useful genes interspersed with functionless DNA, each chromosome is now viewed as a complex information organelle with sophisticated maintenance and control systems. Destabilization of these structures by DNA mutations may lead to changes in gene expression, for example, by influencing patterns of DNA methylation and conformation (Van Holde and Zlatanova, 1993).

Finally, at the current state of technology a proposed role of somatic DNA alterations in aging is a testable hypothesis. As described above, defects and partial defects in every possible genome stability system can now be modeled in transgenic mice. Crosses between these models and the lacZ-plasmid model will allow to identify the functional pathways most relevant for the preservation of the genome, especially at the later stages of life. Unraveling the different pathways of genome stabilization and a better understanding of their relative importance and long-term impact on health and disease at old age will provide new targets for prevention and therapeutic intervention based on reducing mutation loads (Vijg and Wei, 1995).

Acknowledgments

This work was supported by NIH grants PO1 1801 AG10829-01, 1 P30 AG13314-01 and 1 RO1 ES/CA 08797-01.

References

Albertini, R.J., Nicklas, J.A., O'Neill, J.P., and Robison, S.H. (1990). In vivo somatic mutations in humans: mesurement and analysis. Annu. Rev. Genet. 24, 305-326.

Ames, B.N., Shigenaga, M.K., and Hagen, T.M. (1993). Oxidants, antioxidants, and the degenerative diseases of aging. Proc. Natl. Acad. Sci. USA 90, 7915-7922.

Andersson, D.I., and Hughes, D. (1996). Muller's ratchet decreases fitness of a DNA-based microbe. Proc. Natl. Acad. Sci. USA 93, 906-907.

Andrew, S.E., Reitmair, A.H., Fox, J., Hsiao, L., Francis, A., McKinnon, M., Mak, T.W., and Jirik, F.R. (1997). Base transitions dominate the mutational spectrum of a transgenic reporter gene in MSH2 deficient mice. Oncogene 15, 123-129.

Bell, G. (1988). Sex and Death in Protozoa, Cambridge University Press, Cambridge, U.K.

Boerrigter, M.E.T.I., Dollé, M.E.T., Martus, H-J, Gossen, J.A., and Vijg, J. (1995). Plasmid-based transgenic mouse model for studying in vivo mutations. Nature 377, 657-659.

Bootsma, D. (1998). Genetic Basis of Human Cancer. Vogelstein B, Kinzler KW (eds). Chapter 13, McGraw-Hill: New York, pp. 245-274.

Boyce, R., and Howard-Flanders, P. (1964). Release of ultraviolet light-induced thymine dimers from DNA in E. coli K-12. Proc Natl Acad Sci USA 51, 293-300.

Burnet, F.M. (1974). Intrinsic mutagenesis: a genetic approach to aging. Wiley, New York.

Bronson, R.T., and Lipman, R.D. (1991). Reduction in rate of occurrence of age-related lesions in dietary restricted laboratory mice. Growth, Development and Aging 55, 169-184.

Chao, L. (1997). Evolution of sex and the molecular clock in RNA viruses. Gene 205, 301-308.

Da Cruz, A.A., Curry, J., Curado, M.P., and Glickman, B. (1996). Monitoring hprt mutant frequency over time in T-lymphocytes of people accidentally exposed to high doses of ionizing radiation. Environ. Mol. Mutagen. 27, 165-175.

De Vries, A., van Oostrom, C.T.M., Hofhuis, F.M.A., Dortant, P.M., Berg, R.J.W., de Gruijl, F.R., Wester, P.W., van Kreijl CF, Capel PJA, van Steeg H, and Verbeek SJ (1995). Increased susceptibility to ultraviolet-B and carcinogens of mice lacking the DNA excision repair gene XPA. Nature 377, 169-173.

De Vries, A., Van Oostrom, C.Th.M., Dortant, P.M., Beems, R.B., Van Kreijl, C.F., Capel, P.J.A., and Van Steeg, H. (1997). Spontaneous liver tumours and Benzo[a]pyrene-induced lymphomas in XPA-deficient mice. Mol. Carcinogen. 19, 46-53.

Dempsey, J.L., Pfeiffer, M., and Morley, A.A. (1993). Effect of dietary restriction on in vivo somatic mutation in mice. Mutat. Res. 291, 141-145.

Dollé, M.E.T., Martus, H-J., Gossen, J.A., Boerrigter, M.E.T.I., and Vijg, J. (1996). Evaluation of a plasmid-based transgenic mouse model for detecting in vivo mutations. Mutagenesis 11, 111-118.

Dollé, M.E.T., Giese, H., Hopkins, C.L., Martus, H-J., Hausdorff, J.M., and Vijg, J. (1997). Rapid accumulation of genome rearrangements in liver but not in brain of old mice. Nature Genet. 17, 431-434.

Donehower, L.A., Harvey, M., Slagle, B.L., McArthur, M.J., Montgomery, Jr. C.A., Butel, J.S., and Bradley, A. (1992). Mice deficient for p53 are developmentally normal but susceptible to spontaneous tumours. Nature 356, 215-221.

Failla, G. (1958). The aging process and carcinogenesis. Ann. N.Y. Acad. Sci. 71, 1124-1135.

Fishel, R., Lescoe, M.K., Rao, M.R.S., Copeland, N.G., Jenkins, N.A., Garber, J., Kane, M., and Kolodner, R. (1993). The human mutator gene homolog MSH2 and its association with hereditary nonpolyposis colon cancer. Cell 75, 1027-1038.

Fraga, C.G., Shigenaga, M.K., Park, J.W., Degan, P., and Ames, B.N. (1990). Oxidative damage to DNA during aging: 8-Hydroxy 2'-deoxyguanosine in rat organ DNA and urine. Proc Natl Acad Sci USA 87, 4533-4537.

Friedberg, E.C., Walker, G.C., Siede, W. (1995). DNA repair and mutagenesis. ASM Press, Washington DC.

Fukuchi, K., Martin, G.M., and Monnat, R.J. Jr. (1989). Mutator phenotype of Werner syndrome is characterized by extensive deletions. Proc. Natl. Acad. Sci. USA 86, 5893-5897.

Giese, H., Dollé, M.E.T., Hezel, A., van Steeg, H., and Vijg, J. (1999). Accelerated accumulation of somatic mutations in mice deficient in the nucleotide excision repair gene XPA. Oncogene 18, 1257-1260.

Gossen, J.A., de Leeuw, W.J.F., Tan, C.H.T., Lohman, P.H.M., Berends, F., Knook, D.L., Zwarthoff, E.C., and Vijg J (1989). Efficient rescue of integrated shuttle vectors from transgenic mice: a model for studying gene mutations in vivo. Proc. Natl. Acad. Sci. USA 86, 7971-7975.

Gossen, J.A., Molijn, A.C., Douglas, G.R., and Vijg, J. (1992). Application of galactose-sensitive E.coli strains as selective hosts for LacZ⁻ plasmids. Nucleic Acids Res. 20, 3254.

Gossen, J.A., W.J.F. de Leeuw, A.C. Molijn, and Vijg, J. (1993). Plasmid rescue from transgenic mouse DNA using LacI repressor protein conjugated to magnetic beads. BioTechniques 14, 624-629.

Gossen, J.A., Martus, H-J., Wei, J.Y., and J. Vijg. (1995). Spontaneous and X-ray-induced deletion mutations in a lacZ plasmid-based transgenic mouse model. Mutat. Res. 331, 89-97.

Grist, S.A., McCarron, M., Kutlac, A., Turner, D.R., and Morley AA (1992). In vivo human somatic mutation: frequency and spectrum with age. Mutat. Res. 266, 189-196.

Hanawalt, P.C. (1994). Evolution of concepts in DNA repair. Env. Mol. Mutagenesis 23, 78-85.

Harman, D. (1992). Role of free radicals in aging and disease. Ann. N.Y. Acad. Sci. 673, 126-137.

Hart, R.W., D'Ambrosio, S.M., Ng., K.J., and Modak, S.P. (1979). Longevity, stability and DNA repair. Mech. Ageing Dev. 9, 203-223.

Henshaw, P.S., Riley, E.F., and Stapleton, G.E. (1947). The biologic effects of pile radiations. Radiology 49, 349-364.

Jacks, T., and Weinberg, R.A. (1996). Cell-cycle control and its watchman. Nature 381, 643-644.

Jaruga, P., and Dizdaroglu, M. (1996). Repair of products of oxidative DNA base damage in human cells. Nucleic Acids Res. 24, 1389-1394.

Jones, I.M., Thomas, C.B., Tucker, B., Thompson, C.L., Pleshanov, P., Vorobtsova, I., and Moore II 2ⁿᵈ. (1995). Impact of age and environment on somatic mutation at the hprt gene of T lymphocytes in humans. Mutation Res. 338, 129-139.

Kirkwood, T.B.L. (1989). DNA, mutations and aging. Mutat Res., 219, 1-7.

Ko, L. J. and Prives, C. (1996). p53: puzzle and paradigm. Genes and Development 10, 1054-1072

Le, X.C., Xing, J.Z., Lee, J., Leadon, S.A., Weinfeld, M. (1998). Inducible repair of thymine glycol detected by an ultrasensitive assay for DNA damage. Science 280, 1066-1069.

Lee, A.T., DeSimone, C., Cerami, A. and Bucala, R. (1994). Comparative analysis of DNA mutations in lacI transgenic mice with age. FASEB J. 8, 545-550.

Levine, A.J. (1997). p53, the cellular gatekeeper for growth and division. Cell 88, 323-331.

Lindahl, T. (1993). Instability and decay of the primary structure of DNA. Nature 362, 709-715.

Loeb, L.A. (1991). Mutator phenotype may be required for multistage carcinogenesis. Cancer Res. 51, 3075-3079.

Lowe, S.W., Schmitt, S.W., Smith, S.W., Osborne, B.A., and Jacks, T. (1993). p53 is required for radiation-induced apoptosis in mouse thymocytes. Nature 362, 847-849.

MacPhee, D.G. (1995). Mismatch repair, somatic mutations, and the origins of cancer. Cancer Res. 55, 5489-5492.

Martin, G.M., Austad, S.N., Johnson, T.E. (1996a). Genetic analysis of ageing: role of oxidative damage and environmental stresses. Nature Genet. 13, 25-34.

Martin, G.M., Ogburn, C.E., Colgin, L.M., Gown, A.M., Edland, S.D., and Monnat, Jr. R.J. (1996b). Somatic mutations are frequent and increase with age in human kidney epithelial cells. Hum. Mol. Genet. 5, 215-221.

Morley, A.A. (1982). Is ageing the result of dominant and co-dominant mutations? J. Theor. Biol. 98, 469-474.

Morley, A.A. (1996). The estimation of in vivo mutation rate and frequency from samples of human lymphocytes. Mutation Res. 357, 167-176.

Muller, H.J. (1964). The relation of recombination to mutational advance. Mutation Res. 1, 2-9.

Narayanan, L., Fritzell, J.A., Baker, S.M., Liskay, R.M., Glazer, P.M. (1997). Elevated levels of mutation in multiple tissues of mice deficient in the DNA mismatch repair gene Pms2. Proc. Natl. Acad. Sci. USA 94, 3122-3127.

Nicolaides, N.C., Papadopoulos, N., Liu, B., Wei, Y-F., Carter, K.C., Ruben, S.M., Rosen, C.H., Haseltine, W.A., Gleischmann, R.D., Fraser, C.M., Adams, M.D., Venter, J.C., Dunlop, M.G., Hamilton, S.R., Petersen, G.M., de la Chapelle, A., Vogelstein, B., and Kinzler, K.W. (1994). Mutations of two PMS homologues in hereditary nonpolyposis colon cancer. Nature 371, 75-80.

Nishino, H., Knöll, A., Buettner, V.L., Frisk, C.S., Maruta, Y., Haavik, J. and Sommer, S.S. (1995). p53 wild-type and p53 nullizygous Big Blue transgenic mice have similar frequencies and patterns of observed mutation in liver, spleen and brain. Oncogene 11, 263-270

Ono, T., Miyamura, Y., Ikehata, H., Yamanaka, H., Kurishita, K., Yamamoto, T., Suzuki, T., Nohmi, T., Hayashi, M., and Sofuni, T. (1995). Spontaneous mutant frequency of lacZ gene in spleen of transgenic mouse increases with age. Mutation Res. 338, 183-188.

Parsons, R., Li, G-M., Longley, M., Modrich, P., Liu, B., Berk, T., Hamilton, S.R., Kinzler, K.W., and Vogelstein, B. (1995). Mismatch repair deficiency in phenotypically normal human cells. Science 268, 738-740.

Perucho, M. (1996). Microsatellite instability: The mutator that mutates the other mutator. Nature Med. 2, 676-681.

Pettijohn, D., and Hanawalt, P.C. (1964). Evidence for repair-replication of ultraviolet damaged DNA in bacteria. J. Mol. Biol. 9, 395-410.

Sands, A.T., Suraokar, M.B., Sanchez, A., Marth, J.E., Donehower, L.A., and Bradley, A. (1995). p53 deficiency does not affect the accumulation of point mutations in a transgene target. Proc. Natl. Acad. Sci. USA 92, 8517-8521.

Sacher, G.A. (1982). Evolutionary theory in gerontology. Persp. Biol. Med. 25, 339-353.

Setlow, R.B., and Carrier, W.L.L. (1964). The disappearance of thymine dimers from DNA: An error-correcting mechanism. Proc. Natl. Acad. Sci. USA 81, 7397-7401.

Sobol, R.W., Horton, J.K., Kuhn, R., Gu, H., Singhal, R.K., Prasad, R., Rajewsky, K., Wilson, S.H. (1996). Requirement of mammalian DNA polymerase-beta in base-excision repair. Nature 379, 183-186

Strauss, B.S. (1992). The origin of point mutations in human tumor cells. Cancer Res. 52, 249-253.

Szilard, L. (1959). On the nature of the aging process. Proc. Natl. Acad. Sci. USA 45, 30-45.

Te Koppele, J.M., Lucassen, P.J., Sakkee, A.N., Van Asten, J.G., Ravid, R., Swaab, D.F., Van Bezooijen, C.F.A. (1996). 8OHdG levels in brain do not indicate oxidative DNA damage in Alzheimer's Disease. Neurobiology of Aging 17, 819-826.

Tomlinson, I.P.M., Novelli, M.R., and Bodmer, W.F. (1996). The mutation rate and cancer. Proc. Natl. Acad. Sci. USA 93, 14800-14803.

Van Holde, K.. and Zlatanova, J. (1993). Unusual DNA structures, chromatin and transcription. BioEssays 16, 59-68.

Vijg, J., and Wei, J.Y. (1995). Understanding the biology of aging: the key to prevention and therapy? J. Am. Ger. Soc. 43, 426-434.

© 2001 Elsevier Science B.V. All rights reserved.
The Role of DNA Damage and Repair in Cell Aging
B.A. Gilchrest and V.A. Bohr, volume editors.

EFFECTS OF AGING ON GENE SPECIFIC REPAIR

Arlan Richardson[1,2] and *ZhongMao Guo*[2]

[1]Geriatric Research, Education and Clinical Center, South Texas Veterans Health Care System, San Antonio;
[2]Department of Physiology, University of Texas Health Science Center at San Antonio, TX

1. The general features of DNA repair pathways

DNA in cells is constantly exposed to insults from both endogenous and exogenous factors that may cause breaks in the phosphodiester backbone, cleavage of N-glycosylic bonds, alteration of structure of the purine or pyrimidine bases, intrastrand or interstrand crosslinks, and distortion of helix by intercalation (Friedberg et al., 1995). Cells have evolved multiple pathways to repair these DNA lesions (Carr and Hoekstra, 1995). In general, these pathways include direct reversal, mismatch repair, recombinational repair, base excision repair and nucleotide excision repair. The simplest DNA repair pathways only involve single enzymes that catalyze a direct reversal of a specific damage, e.g. alkyltransferase, which removes the methyl group from O^6-methylguanine (Harris et al., 1983) and DNA ligase, which rejoins single-strand breaks in DNA with nucleotide termini containing 3'-OH and 5' phosphate groups (Lehman, 1974). Mismatch repair corrects mispaired bases, which often occur as a consequence of an error in replication. For example, the *E. coli* methyl-directed mismatch repair involves four proteins (MutH, MutL, MutS and MutU) for damage recognition and incision (Modrich, 1991). The resulting gap is filled by DNA polymerase III and sealed by DNA ligase. The current understanding of mismatch repair in mammalian cells is poor; however, the importance of this system has been dramatically illustrated by defects in the human homologues of MutS and MutL, which may be the primary cause of hereditary nonpolyposis colorectal cancer in humans (Prolla, 1998). Recombinational repair is the process that occurs when both strands of the DNA helix are damaged and there is no intact template for the DNA polymerase to copy, e.g., when double-strand breaks and interstrand cross-links occur. This pathway has been relatively well characterized in bacteria (Van Houten, 1990); however, it is poorly understood in mammalian cells. Base excision repair and nucleotide excision repair (NER) are involved in the repair of much of the DNA damage that occurs in cells. Through base excision repair, simple base modifications, e.g., oxidation and monofunctional alkylation, are repaired (Wallace, 1994). Typically, this pathway is characterized by a series of glycosylases that specifically cleave the glycosylic bond between the damaged base and the deoxyribose. The resulting apurinic/apyrimidinic site is removed by AP endonuclease and deoxyribophosphodiesterase, and the gap is filled by DNA polymerase ß and sealed by DNA ligase (Demple and Harrison, 1994). Through NER, bulky helix-distorting lesions such as cyclobutane primidine dimers (CPDs) and large alkylating adducts are repaired. In addition, some oxidative DNA lesions are also repaired by NER.

An important recent development in the area of DNA repair is the observation that NER is heterogeneous in the genome. Bohr et al. (1985) initially studied the repair of UV-induced DNA damage in the dihydrofolate reductase (DHFR) gene and the DNA fragment upstream of the DHFR gene as well as the genome overall in Chinese ovary cells. They found that the DHFR gene was more rapidly repaired than the upstream DNA fragments and the genome overall. For example, more than 60% of the CPDs were removed from the DHFR gene 26 hours after UV-irradiation compared to 15% from the genome overall. Little damage was removed from the upstream fragment of the DHFR gene. Subsequently, other studies with a variety of organisms ranging from *E. coli* (Mellon and Hanawalt, 1989) and yeast (Sweder and Hanawalt, 1992) to cells from rodents (Bohr et al., 1985; Guo et al., 1998b; Mellon et al., 1987) and humans (Mellon et al., 1986; Venema et al., 1991) have shown that the removal of CPDs from transcriptionally active genes/DNA fragments is more efficient than from non-transcribed DNA. In addition, Mellon et al. (1987) observed that the transcribed strand of the DHFR gene in cells from Chinese hamsters and humans was preferentially repaired compared to the non-transcribed strand, e.g., approximately 70 to 80% of CPDs were removed from the transcribed strand compared to 10 to 30% from the non-transcribed strand. Subsequently, the selective removal of DNA damage from the transcribed strand of transcriptionally active genes also has been found in *E. coli* (Mellon and Hanawalt, 1989), yeast (Leadon and Hanawalt, 1992; Sweder and Hanawalt, 1992) and a variety of mammalian cells (Bohr et al., 1986; Fiala et al., 1995; Guo et al., 1998b; Ho and Hanawalt, 1991; Lan et al., 1994; Mullenders et al., 1993; Petersen et al., 1995; Vrieling et al., 1991). The rapid repair of the transcribed strand of transcriptionally active genes is defined as strand-specific repair, transcription-coupled repair, or preferential repair (Bohr, 1995a; Glazer and Rohlff, 1994). Transcriptionally independent DNA repair is referred to as global gemone repair or bulk genome repair (Bohr, 1995a). The current understanding of both subpathways of NER in mammalian cells has been reviewed from different perspectives by other authors (Sancar, 1996; Wood, 1997). This review presents a brief survey of this subject in order to better understand the studies of NER with respected to aging.

2. Mechanism of the global NER

Global genome repair is a complex process, which involves multiple proteins, including seven proteins (A through G) that are associated with the genetic disease xeroderma pigmentosum (XP). Global genome repair involves four steps: damage recognition, dual incision, repair synthesis and ligation.

A. *Damage recognition*

Although it has not been demonstrated experimentally, damage recognition is generally believed to be the rate-limiting step of NER (Sancar, 1996). Recently, investigators in Hoeijmakers laboratory showed that a complex containing xeroderma pigmentosum group C protein (XPC) and HR23B protein is the earliest detector of damage that initiates global genome repair (Sugasawa et al., 1998). Their observation suggests that the XPC-HR23B complex, after binding to a lesion, induces a specific conformational change (including local opening of DNA), which then recruits other factors, such as XPA and RPA. XPA and RPA are DNA binding proteins, and both proteins have a higher affinity for damaged DNA than for undamaged DNA. The functional form of XPA and RPA appears to be a complex because these two proteins associate tightly, and the complex has higher affinity for damaged DNA than either component alone. XPC does not appear to be required for transcription-coupled repair because cells obtained from patients with XP complementary group C normally remove CPDs from the transcribed strand of transcriptionally active genes despite a deficiency in repair of the non-transcribed strand of these genes (Venema et al., 1991). In addition to XPC-HR23B and XPA-RPA, XPE is also believed to be involved in damage recognition (Henricksen and Wold, 1994; Wang et al., 1991). The role of XPE in the repair process remains unclear at present time. Cells obtained from patients with XP complementary group E are only slightly UV sensitive and moderately defective in NER (Cleaver and Kraemer, 1989); therefore, the XPE protein is assumed to play an auxiliary role in NER (Wood, 1997).

B. *Incision*

The incision process of NER involves the XPF-ERCC1 complex and a complex known as TFIIH. These proteins may be recruited to the site of DNA damage by XPC and XPA. The TFIIH complex consists of nine polypeptides and plays a role in both transcription and NER (Akiyama et al., 1994). In the initiation step of transcription, TFIIH and other basal transcription factors are assembled with RNA polymerase II on the promoter. Once transcription starts, TFIIH and the other transcription factors are released from the DNA. TFIIH has two functions in NER. First, it cooperates with RAP to recruit XPG to the lesion (Park et al., 1995; Sancar, 1996). Second, two subunits of TFIIH (XPB and XPD) have helicase activity (Drapkin et al., 1994), which is believed to open the duplex at the lesion site (Sancar, 1996). At this point, DNA topoisomerase(s) may be involved in relieving the strain on the superstructure of DNA (Stevnsner and Bohr, 1993). These conformational changes expose the damaged site to the endonuclease action of the XPF-ERCC1 complex and XPG. XPF-ERCC1 makes an incision at the 22nd to 24th phosphodiester bond 5' to the damaged site, and XPG makes an incision at the 5th phosphodiester bond 3' to the damage site. The 27 to 29-oligomer and the incision complex are then released from DNA possibly through the action of RFC (Sancar, 1994).

C. *Repair synthesis and ligation*

Removal of damage from DNA through the above two steps creates a gap that must be filled to generate a functional duplex. This step is initiated by RFC, which recruits proliferating cell nuclear antigen (PCNA) and DNA polymerase δ or ε onto the resulting gap. DNA polymerase δ or ε associates with PCNA and initiates repair synthesis (Sancar, 1996). Recently, it has been shown that mutations in either DNA polymerase δ or ε make the yeast cells sensitive to UV-irradiation (Budd and Campbell, 1995). Thus, both DNA polymerase δ and ε appear to be necessary for repair synthesis. The importance of PCNA has also been shown by the fact that depletion of PCNA from the cell-free extracts completely eliminates repair synthesis (Shivji et al., 1992). The gap is finally filled with a nascent DNA patch and sealed by a DNA ligase (Arlett, 1986).

3. Mechanism of Transcription-Coupled NER

Preferential repair of the transcribed strand of transcriptionally active genes has been shown to be related to transcription. For example, Mellon and Hanawalt (1989) found that induction of lactose operon of *E. coli* by isopropyl-β-D-thiogalactoside increased the removal of CPDs from the transcribed strand of the lactose operon but not the non-transcribed strand. In mammalian cells, preferential repair is confined only to genes transcribed by RNA polymerase II (Friedberg, 1996), and repression of RNA polymerase II abolishes the preferential repair of the transcribed strand of the transcriptionally active genes. For example, Christians and Hanawalt (1992) observed that inhibition of RNA polymerase II by α-amanitin eliminated the preferential repair of the transcribed strand of the DHFR gene in Chinese hamster ovary cells. Similarly, Leadon and Lawrence (1991) found that the removal of UV-induced CPDs and aflatoxin-induced damage from the transcribed strand of the transcriptionally active metallothionein II_A gene in human fibrosarcoma cells was reduced by α-amanitin treatment to a level similar to the non-transcribed strand of this gene and other transcriptionally inactive genes.

In addition to CPDs, other DNA lesions have been reported to show preferential repair, and these DNA lesions are listed in Table 1. It appears that DNA lesions that block the elongating RNA polymerase complex show preferential repair even if it is not a bulky, helix-distorting type of lesion. For example, thymine glycol is not a helix-distorting type of damage, and it is believed to be repaired primarily by the base excision repair pathway. Preferential removal of this type of DNA damage from transcriptionally active genes suggested that this lesion elicits a 'crossover' repair that occurs through the transcription-coupled repair pathway by arresting transcription elongation (Hanawalt, 1995). In contrast, 7-methylguanine, which does not block transcription, does not show preferential repair (Scicchitano and Hanawalt, 1989).

Several factors may contribute to the rapid repair of transcriptionally active genes. For example, the transcribed genes may attach to the nuclear matrix, thereby providing a higher accessibility to the repair enzymes. This concept is supported by the study of Venema et al., (1992) using a human cell line containing two alleles of an inactivated adenosine deaminase (ADA) gene in which the complete promoter was deleted. They found that despite the decrease in the initial rate of repair for the DNA fragment

Table 1. Substrates for gene-specific repair and preferential repair in eukaryotic cells

DNA-damage agents	Adducts	References
UV-irradiation	CPDs	Bohr et al., (1985), Mollon et al. (1987)
	6-4 dimers	van Hoffen et al. (1995), Vreeswijk et al., (1994)
Psoralen	interstrand crosslinks	Islas et al., (1994)
	Adducts of pyrimidine	Meniel et al., (1993)
Nitrogen mustard	interstrand crosslinks	Larminant et al., (1993)
	alkylated purines	Wassermann, Damgaard (1994)
Cisplatin	interstrand crosslinks	Larminant et al., (1993)
	intrastrand crosslinks	May et al., (1993)
N-methyl-N'-nitro-N-nitrosoguanidine	alkylated purines and pyrimidines	Chary et al., (1991)
1-nitrosopyrene	adducts of Guanine	McGregor et al., (1998)
Benzo[a]pyrenediolepoxide	adducts of Guanine	McGregor et al., (1998), Chen et al., (1992)
N-ethyl-N-nitrosourea	alkylated purines	Thomale et al., (1994), Sitaram et al., (1997)
N-acetoxy-2-acetylaminofluorene	adducts of Guanine	van Oosterwijk et al., (1996)
Hydrogen peroxide	thymine glycols	Leadon et al., (1995)
γ-rays	single strand breaks	Igusheva et al., (1993)
Acridine orange + Visible light	Formamidopyrimidines 8-hydroxydeoxyguanosines	Taffe et al., (1996)

containing the inactivated ADA gene, the extent of repair of this DNA fragment was similar to the transcriptionally active gene but higher than the non-transcribed DNA. Another factor that could increase access of repair proteins to sites of DNA damage is the locally opened chromatin structure of the transcriptionally active genes (Venema et al., 1992). The accessibility of enzymes to DNA lesions undoubtedly effects the rate of repair for specific genes/DNA regions; however, this does not appear to explain the preferential repair of the transcribed strand of transcriptionally active genes.

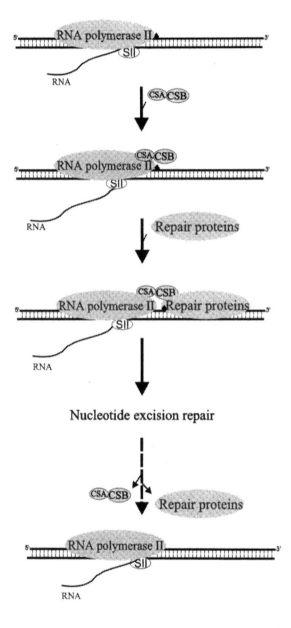

Fig. 1. **Model for transcription-coupled repair in human cells.** *The pathway shown is adapted from Hanawalt (1994): 1) RNA polymerase II stalls at the site of DNA damage. A complex containing the CSA and CSB proteins recognizes and binds to the arrested transcription complex. 2) The incomplete transcript is transferred to the transcription elongation factor SII. 3) CSA/CSB recruits repair proteins to the damage site to remove the damage. 4) CSA/CSB and repair proteins are released after repair and transcription resumes, thereby completing the RNA transcript.*

Recently, studies have shown that the major factor modulating preferential repair of transcribed genes is the interaction of transcription-repair coupling factor (TRCF) with the stalled RNA polymerase at the damage site. TRCF was Initially isolated from *E. coli.* by Selby and Sancar (1993). TRCF binds to and releases the RNA polymerase blocked at the lesion. This factor may then interact with the excision repair complex to remove the lesion (Hanawalt, 1994). In mammalian cells, the mechanism of transcription-repair coupling may be more complicated than that in bacteria. A complex containing at least two proteins, which are defective in patients with Cockayne's syndrome (CSA and CSB) has been proposed to perform a function analogous to that of TRCF in bacteria. Protein CSB has been shown to bind to the XPA protein and the TFIIH complex (Selby and Sancar, 1997). The CSA protein also has a motif that is involved in protein-protein interactions (Henning et al., 1995). Genetic studies have shown that cells defective in either of these two proteins are proficient in repairing UV-induced damage in the inactive regions of the genome. However, they are unable to repair damage in active genes at an accelerated rate, i.e., all damage is repaired at the slower rate for DNA that is not actively transcribed (Venema et al., 1990). Based on the chemical properties of CSA and CSB and the phenotype of cells defective in them, the model shown in Fig. 1 has been proposed for transcription-repair coupling in human cells (Hanawalt, 1994; Sancar, 1996). RNA polymerase II stalls at a lesion and the arrested transcription complex is recognized by a CSA/CSB heterodimer. The CSA/CSB heterodimer then recruits the repair proteins to the damaged site, thereby accelerating the rate of DNA repair. The main difference between this model and the prokaryotic model is that the nascent transcript and RNA polymerase II do not dissociate from the lesion during repair in human cells (Hanawalt, 1994). The transcription elongation factor SII may provide a provisional residence for the incomplete transcript during repair. Following excision and repair synthesis, RNA polymerase II resumes transcription at the repaired template, which avoids the initiation of transcription every time that RNA polymerase II encounters a lesion.

In addition to CSB and CSA, proteins required for mismatch repair may also be involved in transcription-repair coupling because mutations in genes required for mismatch repair, e.g., MutS, MutL in bacteria and their homologs in mammalian cells, have been found to abolish transcription-coupled repair. In addition, the defect in transcription-coupled repair can be corrected in a mutant cell line in which mismatch repair deficiency has been corrected by chromosome transfer (Mellon et al., 1996).

An argument also has been made that strand specific repair may be simply due to a special affinity of TFIIH to the stalled transcription complex at the lesion (Friedberg, 1996). The requirement for the resumption of transcription after stalling of RNA polymerase II may recapitulate some of the requirements for transcription initiation at the natural promoter. Hence, TFIIH may be recruited to sites of arrested transcription as the core complex around which a repairosome is then assembled (Friedberg, 1996). Currently there is no experimental evidence to support this view.

Table 2. Effect of age on UV-induced unscheduled DNA synthesis

Species/Cells	Ages Studied	Reference
Decrease with age		
Mouse		
Hepatocytes	1-18 months	Srivastava, Busbee (1992)
Neurons	4-24 months	de Sousa et al. (1986)
Lymphocytes	3-30 months	Licastro, Walford (1988)
Fibroblasts	2-30 months	Kempf et al. (1998)
Rat		
Fibroblasts	3-31 months	Tilley et al. (1992)
Kidney cells	5-30 months	Weraachakul et al. (1989)
Hepatocytes	5-30 months	Weraachakul et al. (1989)
Hepatocytes	3-20 months	Kennah et al. (1985)
Hepatocytes	6-32 months	Plesko, Richardson (1984)
Fibroblasts	6-40 months	Vijg et al. (1985)
Hamster		
Brain cells	1-18 months	Gensler (1981b)
Rabbit		
Chondrocytes	3-18 months	Lipman, Sokoloof (1985)
Human		
Keratinocytes	17-77 years	Nette et al., (1984)
Leukocytes	13-94 years	Lambert et al., (1979)
Lymphocytes	20-90 years	Lezhava et al., (1979)
No change with age		
Mouse		
Lymphocytes	4-24 months	de Sousa et al. (1986)
Rat		
Hepatocytes	6 –24 months	Sawada et al. (1988)
Hamster		
Lung & kidney cells	1-18 months	Gensler (1981a)
Rabbit		
Chondrocytes	18-52 months	Lipman, Sokoloof (1985)
Chrondrocytes	3-24 months	Krystal et al., (1983)
Human		
Keratinocytes	17-67 years	Liu et al. (1985)
Lymphocytes	17-74 years	Kovacs et al. (1984)
Chrondrocytes	23-63 years	Krystal et al. (1983)
Fibroblasts	1-95 years	Hall et al. (1982)
Keratinocytes	1-70 years	Liu et al. (1982)

4. Previous Studies of the Effect of Aging on NER

Over the past three decades, a number of studies have been conducted to determine the relationship between DNA repair and aging. Recent reviews have provided a thorough discussion of these studies (Bohr and Anson, 1995b; Walter et al., 1997). In this chapter, we focus on the repair of DNA damage by NER pathway in relation to aging, specially the repair of UV-induced DNA damage. Two approaches have been taken in these studies: experiments that have measured the ability of cells from animals of various life spans to repair DNA damage by the NER pathway and experiments that have compared the ability of cells from animals of different ages to repair DNA damage by the NER pathway. The correlation between DNA repair capacity and species life span is discussed in another chapter in this book.

There have been a large number of studies comparing the repair of UV-induced DNA damage by cells/tissues from young and old organisms. In these studies, which are listed in Table 2, DNA repair was measured as unscheduled DNA synthesis (UDS) after the cells were exposed to UV radiation. UDS has been observed to decrease approximately 30 to 50% in cells from mice, hamsters, rats, rabbits and humans. However, there are also studies with cells from these animal models that show no change in UDS with age.

The major problem with the previous studies that have measured UV-induced DNA repair by UDS is the assay. UDS is a relatively crude assay, which neither directly nor accurately measures the removal of a specific type of damage. Rather UDS measures UV-induced non-replicative DNA synthesis. Thus, changes in the specific activity of the thymidine precursor pool or replicative DNA synthesis could affect the level of DNA synthesis measured as repair in the UDS assay. A second concern is the limited correlation between UV-induced UDS and the sensitivity of cells to UV-irradiation, which brings into question how accurately UDS reflects true DNA repair capacity of a cell. For example, no correlation was found between UV-induced UDS in mammalian cells and the viability of cells to UV-irradiation (Bohr et al., 1986). In addition, keratinocytes from old human subjects showed greater sensitivity to UV-irradiation even though no age-related difference was observed in UV-induced UDS (Liu et al., 1985). A third important limitation of the UDS assay for measuring DNA repair is that it measures overall genome repair. This is a problem when measuring repair of DNA damage, e.g., UV-induced DNA damage, by the NER pathway because NER is heterogeneous.

In addition to measuring NER by UDS, other techniques have been used to study more accurately the effect of age on NER. For example, Boerrigter et al. (1995) injected mice with benzo(a)pyrene and then measured the removal of benzo(a)pyrene-DNA adducts from the genome using the ^{32}P-postlabeling assay. In this study, the investigators observed a significant decrease in DNA repair in multiple organs of the old mice. The radioactive postlabeling assay directly measures the presence and removal of the damaged bases; therefore, it is a more accurate measure of NER than the UDS assay. However, this assay does not measure the repair of the specific genes; therefore this study was only a measure of global genome repair by the NER pathway.

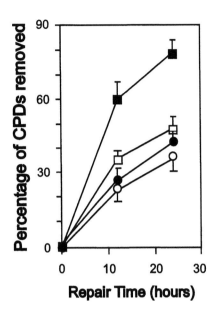

*Fig. 2. **Removal of CPDs from the genome overall and from the MHC^{emb} and albumin genes.** Primary cultures of hepatocytes were irradiated at 10 J/m² of UV-irradiation. Graph shows the percentage of CPDs removed from the genome overall (O), MHC^{emb} gene (●), and the transcribed (■) and non-transcribed (□) strands of the albumin gene. The data are taken from Guo et al. (1998b).*

Moriwaki et al. (1996) developed a system to measure the repair of UV-induced DNA damage by transfecting a UV-irradiated plasmid into human primary skin fibroblasts or lymphocytes and then measuring the expression of the reporter gene (chloramphenicol acetyltransferase) contained in the plasmid as an assay of the NER by the cells. They found an age-related decline in repair of DNA damage at a rate of 0.6% per year from the first to the tenth decade of life. However, the repair of the plasmid DNA may not be an accurate reflection of the ability of the cell to carry out NER because higher-order structure of DNA and DNA-protein interactions in chromatin play a role in the access of repair proteins to the lesion (Wang et al., 1991). In addition, it is well documented that the transcription of cells changes with age (Van Remmen et al., 1995); therefore, changes in the transcription of the reporter gene may be due to factors other than NER.

In summary, the data generated by the previous studies that have mueasured NER of UV-indiced DNA damage by UDS, postlabeling, and plasmids strongly suggest that global genome repair by NER pathway is reduced in tissues of old organisms.

5. The Effect of Aging on the Repair of Specific Genes by the NER Pathway

The methods used in the studies described above give no information on the repair of specific genes or the differential DNA repair of the genome by the NER pathway. However, the biological consequences of the unrepaired DNA damage (i.e., the accumulation of mutations) and the ability of cells to repair DNA damage by the NER pathway certainly depends on the location of the lesion in the genome. Therefore, it is important to measure the repair of the specific genes or regions of DNA in order to understand how aging affects NER. Recently, we utilized the assay developed by Bohr et al. (1985) to measure gene-specific repair of UV-induced DNA damage (cyclobutane pyrimidine dimers, CPDs) in primary cultures of rat hepatocytes (Guo et al., 1998b). As shown in Fig. 2, CPDs were more rapidly removed from transcriptionally active genes in cultured hepatocytes than either transcriptionally inactive genes/fragments or the genome overall. In addition, the transcribed strand of transcriptionally active genes were preferentially repaired as compared to the non-transcribed strand. These data demonstrate that primary cultures of non-dividing rat hepatocytes show differential repair of UV-induced DNA damage that is comparable to what has been reported for transformed, proliferating mammalian cell lines (Bohr et al., 1985; Mellon et al., 1987). Therefore, using this system to measure DNA repair, it was possible for us to compare NER in transcriptionally active genes as well as inactive genes/DNA fragments for hepatocytes isolated from rats of different ages.

To study the effect of age on gene specific repair by the NER pathway, we have measured the ability of hepatocytes isolated from young/adult (6 months) and old (24 months) rats to repair CPDs induced by UV-irradiation in three specific gene fragments and the genome overall. The three gene fragments studied were: the albumin gene, the PEP-carboxykinase (PEPCK) gene and the embryonic myosin heavy chain (MHC^{emb}) gene. The albumin gene is highly transcribed in hepatocytes. The PEPCK gene is also expressed in hepatoyctes; however, the transcription of the PEPCK gene is about one-fifth that of the albumin gene. The transcription of PEPCK in hepatocytes is inducible by cAMP to a level similar to the albumin gene (Salavert and Iynedjian, 1982). Therefore, by studying the repair of CPDs in the albumin and PEPCK gene fragments in the presence and absence of cAMP, we were able to measure the effect of NER on genes that are transcribed at various levels. The MHC^{emb} gene is transcriptionally inactive in the adult liver; therefore, this gene allowed us to study the effect of age on a gene that is not expressed.

The data in Figs. 3 and 4 show that the effect of age on NER is heterogeneous in actively transcribed genes; it differs between the transcribed and non-transcribed strands. For example, Fig. 3 shows that the rate (12 hours after UV-irradiation) of removal of CPDs from the transcribed strand of the albumin gene was 40% lower for hepatocytes isolated from old rats than for hepatocytes isolated from young/adult rats. In contrast, no significant difference was observed in the ability of hepatocytes isolated from young/adult and old rats to repair the non-transcribed strand of the albumin gene.

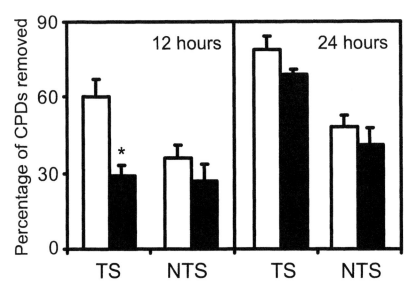

*Fig. 3. **Effect of age on the removal of CPDs from the albumin gene.** Hepatocytes were isolated from 6- (open bars) and 24-month-old rats (solid bars). DNA samples were extracted from the cultured hepatocytes 0, 12 and 24 h after 10 J/m² of UV irradiation. The removal of CPDs from the transcribed strand (TS) and non-transcribed strand (NTS) of the albumin fragment was determined. The values are expressed as mean ±S.E. of six to eight separate experiments in which hepatocytes were pooled from two rats for each experiment, and the data were statistically analyzed (*P<0.05 as compared to the 6-month-old rats). The data are taken from Guo et al., (1998a).*

However, when hepatocytes are given 24 hours to repair the CPDs, no age-related difference in the repair of the transcribed strand was observed.

Fig. 4 shows the ability of hepatocytes to remove CPDs from the DNA fragment containing the PEPCK gene. In the absence of cAMP (Fig. 4a and 4b), no difference was observed in the ability of hepatocytes from young/adult and old rats to remove CPDs from either the transcribed or non-transcribed strands of the PEPCK gene. However, when cAMP was added to the cultures of hepatocytes and PEPCK transcription was induced 4- to 5-fold, the rate of removal of CPDs from the transcribed strand of the PEPCK gene was significantly induced in hepatocytes from young/adult rats and was greater than that observed for hepatocytes from the old rats (Fig. 4c). The rate of removal of CPDs from the non-transcribed strand of PEPCK in the presence of cAMP was similar for hepatocytes from young/adult and old rats (Fig. 4c). The removal of CPDs from the transcribed strand did not show any age-related difference 24 hours after UV-irradiation (Fig. 4d). Thus, our studies with albumin and PEPCK suggest that age only affects the ability of cells to rapidly repair the transcribed strand of highly

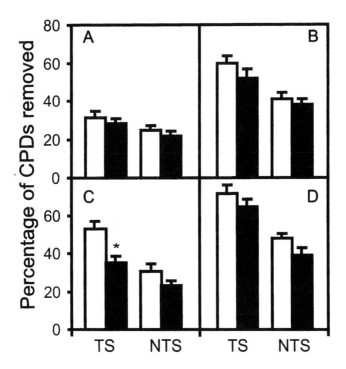

Fig. 4. **Effect of age on the removal of CPDs from the PEPCK gene.** *Hepatocytes were isolated from 6- (open bars) and 24-month-old (solid bars) rats. The cultured hepatocytes were incubated in absence (A, B) and presence (C, D) of cAMP. DNA was extracted from cultured hepatocytes 0, 12 and 24 hours after 10 J/m^2 of UV irradiation, and the removal of CPDs from the PEPCK gene at 12 (A, C) and 24 hours (B, D) was measured. Values are expressed as mean ± S.E. of six to eight separate experiments in which hepatocytes were pooled from two rats for each experiment, and the data were statistically analyzed (*P<0.05 as compared to the 6-month-old rats). The data are taken from Guo et al., (1998c).*

expressed genes. No age-related differences in repair of the non-transcribed strand of either albumin or PEPCK was observed. Our studies with PEPCK also suggest that transcribed genes that are not highly transcribed, e.g., in the absence of cAMP, may not show any age-related change in the rate of repair of the transcribed strand.

Our laboratory has begun to study the mechanism responsible for the age-related decrease in the rate of repair of the transcribed strand of highly transcribed genes. We have found that the age-related decrease in removal of CPDs from the transcribed strand of the albumin fragment is not due to changes in transcription because the basal level of albumin transcription in hepatocytes does not change significantly with increasing age (Guo et al., 1998a). Thus, the decreased repair of the transcribed strand of the albumin fragment without a change in albumin transcription suggests that the coupling of transcription and DNA repair is compromised in cultured hepatocytes isolated from old

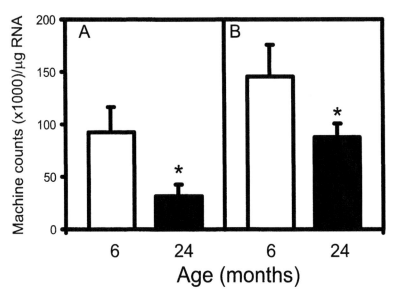

*Fig. 5. **Effect of age on CSA and CSB mRNA levels in rat liver.** Total cellular RNA was extracted from liver tissues of 6-month- (open bars) and 24-month-old (solid bars) rats and the mRNA levels were detected using dot blotting hybridization as described by Guo et al., (1998a). The CSA and CSB probes were obtained from Dr. Hoeijmakers (Medical Genetics Center, Erasmus University Rotterdam, The Netherlands) and were the 1.7-kb of mouse CSA cDNA and the 2.5-kb of mouse CSB cDNA, respectively. The mRNA levels of CSA (A) and CSB (B) are expressed as mean ± standard error of five separate experiments. * The values for the 24-month-old rats were significantly different from 6-month-old rats at P<0.05.*

rats. The results from our study on the PEPCK fragment also support this view. cAMP increased both PEPCK transcription and the rate of repair of the transcribed strand of the PEPCK fragment in cultured hepatocytes isolated from young rats. However, cAMP did not increase the rate of repair of the transcribed strand of the PEPCK fragment in cultured hepatocytes isolated from old rats fed *ad libitum* even though PEPCK transcription was induced 4- to 5-fold (Guo et al., 1998c). Thus, it is possible that the effect of age on the repair of the transcribed strand of the transcriptionally active genes arises at the level of coupling repair to transcription. As shown in Fig. 1, CSA and CSB are believed to be involved in coupling transcription and DNA repair (Venema et al., 1990). Therefore, an age-related decrease in the levels or activities of these proteins might give rise to reduced repair of the transcribed strand of the transcriptionally active genes when the gene is highly transcribed. We have recently measured the levels of the mRNA transcripts for Cockayne's syndrome group A and B genes in rat liver. The data in Fig. 5 demonstrate that the levels of these two transcripts decrease significantly with age in rat liver. For example, CSA mRNA levels decreased over 65% between 6 and 24 months of age while CSB mRNA levels decreased 40%. Therefore, we propose that the age-related decrease in the repair of the transcribed strand of highly expressed genes

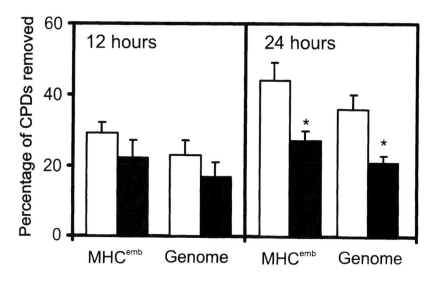

Fig. 6. **Effect of age on removal of CPDs from the MHC^emb gene and the genome overall.**
Hepatocytes were isolated from 6- (open bars) and 24-month-old rats (solid bars). DNA was extracted from the cultured hepatocytes 0, 12 and 24 hrs after 10 J/m² of UV irradiation. The removal of CPDs from the MHC^emb gene and the genome overall was measured. The values are expressed as mean ± S.E. of six to eight separate experiments in which hepatocytes were pooled from two rats for each experiment, and the data were statistically analyzed (* P<0.05 as compared to 6-month-old rats). The data are taken from Guo et al. (1998a).

(e.g., albumin and PEPCK in the presence of cAMP) might be due to changes in the activities of proteins involved in coupling of transcription to NER. We do not believe that the age-related decline in the repair of the transcribed strand of the highly expressed albumin and PEPCK genes is due to changes in general DNA repair proteins that are used in damage recognition, incision, repair synthesis and ligation because the rate of removal of CPDs from the non-transcribed strand of albumin and PEPCK did not change significantly with age.

We have also studied the effect of age on NER of transcriptionally inactive genes in cultured hepatocytes. Fig. 6 shows data for the removal of CPDs from the DNA fragment containing the MHC^emb gene, which is not expressed in adult liver. Although no age-related difference in repair was observed 12 hours after UV-irradiation, a 40% decrease in the repair of the MHC^emb gene was observed 24 hours after UV-irradiation for hepatocytes isolated from old rats. We have observed a similar decrease in the repair of other non-transcribed DNA fragments (Guo et al., 1998a). The data in Fig. 6 also show that the age-related decrease in the removal of CPDs from the MHC^emb gene is similar to the age-related decrease in removal of CPDs from the genome overall. At the present time, there is one published report in which the effect of age on the repair of transcriptionally inactive regions of DNA has been studied. Kruk et al., (1995) measured the repair of telomeres in the primary cultures of human fibroblasts obtained from young and old individuals after exposure to UV-light. They found the removal of CPDs from the telomeres was approximately 40% less for fibroblasts from a 75-year-old subject compared to a 9-year-old subject. The age-related decrease in the ability of the cells to repair non-transcribed, silent regions of the DNA could play a role in the age-related accumulation of DNA damage and mutations, which consistently have been observed to occur with increasing age (for reviews, see Bohr and Anson, 1995b; Kruk et al., 1995).

The results from our study also suggest that the age-related decrease in the repair of transcriptionally inactive genes/regions and the genome overall is not due to an age-related deficiency in general repair proteins because the removal of CPDs from the non-transcribed strand of the albumin and PEPCK fragments was not altered with age (Guo et al., 1998a; Guo et al., 1998c). It is most likely that the age-related changes in the structure of the chromosomes are responsible for the age-related decrease in the repair of transcriptionally inactive genes/regions. For example, several studies suggest that chromatin becomes more condensed with age because of an increases in protein-DNA cross-links and internucleosome interaction (for a review, see Medvedev, 1984). In addition, Mozzhukhina et al., (1991) showed that the chromatin template activity for α and β DNA polymerases declined with age. They suggested that this age-related change was due to an increased chromosomal condensation. Increased chromosomal condensation could also reduce the access of DNA repair proteins to the DNA and thereby decrease the repair of the lesions.

Are the age-related changes in DNA repair that we have observed physiologically important in aging? Our data suggest that the changes we have observed in DNA repair may play a role in aging. For example, the transcription of the albumin gene after UV-irradiation support the hypothesis advanced by Alexander (1967) that the accumulation of DNA damage could be important in aging by interfering with gene expression. We have shown that the ability of primary cultures of hepatocytes from old rats to recover from UV-irradiation and express albumin is delayed compared to hepatocytes isolated from young/adult rats (Guo et al., 1998a). The potential importance of the age-related decrease in preferential repair (i.e., the repair of the transcribed strand) in aging is also illustrated by a genetic disease, Cockayne's syndrome. This disease is associated with a loss of preferential repair of the transcribed strand of transcriptionally active genes (Venema et al., 1990). Interestingly, patients with Cockayne's syndrome display many characteristics of normal aging, such as typical "old face", osteoporosis, atheroscleorosis and calcification of cerebral blood vessels (Hanawalt, 1994; Warner and Price, 1989). Our preliminary data demonstrate that the levels of the mRNA transcripts for the two proteins involved in Cockayne's syndrome, CSA and CSB, decrease significantly with age in rat liver. In general, age-related changes in mRNA levels are paralleled by a similar change in the levels/activities of the protein encoded by the mRNA species (Van Remmen et al., 1995). Therefore, it is exciting to speculate that age-related decline in the preferential repair of DNA by the NER pathway might be similar to Cockayne's syndrome and therefore, could contribute to the aging process.

The decline in the repair of non-transcribed DNA also could be important physiologically in aging because it may be an important factor in the age-related increase in cancer. For example, cells from the patients with XPC, which show a decline in repair of non-transcribed DNA but no change in repair of the transcribed strand of transcriptionally active genes (Venema et al., 1991), have an increased incidence of skin cancers (Kraemer et al., 1994). In contrast, patients with Cockayne's syndrome, which shows a defect in preferential repair but not the repair of non-transcribed DNA, do not show an increased incidence of cancer (Warner and Price, 1989). From this limited information, it would appear that the age-related decline in the ability of cells to repair regions of the genome that are not expressed could lead to an increased vulnerability to developing cancer, which is a hallmark of aging.

Acknowledgment

This work was supported in part by Grants AG15134 and AG01548 from the National Institute on Aging and the Office of Research and Development, Department of Veterans Affairs.

References

Akiyama, K., Kagawa, S., Tamura, T., Shimbara, N., Takashina, M., Kristensen, P., Hendil, K.B., Tanaka, K., and Ichihara, A. (1994). Replacement of proteasome subunits X and Y by LMP7 and LMP2 induced by interferon-gamma for acquirement of the functional diversity responsible for antigen processing. FEBS Lett. *343*, 85-88.

Alexander, P. (1967). The role of DNA lesions in processes leading to aging in mice. Sym.Soc.Exp.Biol. *21*, 29-51.

Arlett, C.F. (1986). Human DNA repair defects. J.Inherited Metab. *1*, 69-84.

Boerrigter, M.E.T.I., wei, J.Y., and Vijg, J. (1995). Induction and repair of benzo(a)pyrene-DNA adducts in C57BL/6 and BALB/c mice: association with aging and longevity. Mech.Ageing Dev. *82*, 31-50.

Bohr, V.A. (1995a). DNA repair fine structure and its relations to genomic instability. Carcinogenesis *16*, 2885-2892.

Bohr, V.A. and Anson, R.M. (1995b). DNA damage, mutation and fine structure DNA repair in aging. Mutation Res *338*, 25-34.

Bohr, V.A., Okumoto, D.S., and Hanawalt, P.C. (1986). Survival of UV-irradiated mammalian cells correlates with efficient DNA repair in an essential gene. Proc.Natl.Acad.Sci.USA *83*, 3830-3833.

Bohr, V.A., Smith, C.A., Okumoto, D.S., and Hanawalt, P.C. (1985). DNA repair in an active gene: Removal of pyrimidine dimers from the DHFR gene of CHO cells is much more efficient than in the genome overall. Cell *40*, 359-369.

Budd, M.E. and Campbell, J.L. (1995). DNA polymerases required for repair of UV-induced damage in *Saccharomyces cerevisiae*. Mol.Cell.Biol. *15*, 2173-2179.

Carr, A.M. and Hoekstra, M.F. (1995). The cellular responses to DNA damage. Trends Cell Biol. *5*, 32-40.

Chary, P., Mitchell, C.E., and Kelly, G. (1991). Chemical carcinogen-induced damage to the c-*neu* and c-*myc* protooncogenes in rat lung epithelial cells: Possible mechanisms for differential repair in transcriptionally active genes. Cancer Lett. *58*, 57-63.

Chen, R.H., Maher, V.M., Brouwer, J., van de Putte, P., and McCormick, J.J. (1992). Preferential repair and strand-specific repair of benzo[a]pyrene diol epoxide adducts in the HPRT gene of diploid human fibroblasts. Proc.Natl.Acad.Sci.USA *89*, 5413-5417.

Christians, F.C. and Hanawalt, P.C. (1992). Inhibition of transcription and strand-specific DNA repair by alpha-amanitin in Chinese hamster ovary cells. Mutation Res. *274*, 93-101.

Cleaver, J.E. and Kraemer, K.H. (1989). Xeroderma Pigmentosum. In The metabolic basis of inherited disease. C.R. Scriver, A.L. Beaudet, W.S. Sly, and D. Valle, eds. (New York: Mcgraw-Hill Book Co.), pp. 2949-2971.

de Sousa, J., de Boni, U., and Cinader, B. (1986). Age-related decrease in ultraviolet induced DNA repair in neurons but not in lymph node cells of inbred mice. Mech.Aeing Dev. 36, 1-12.

Demple, B. and Harrison, L. (1994). Repair of oxidative damage to DNA: enzymology and biology. Ann.Rev.Biochem. 63, 915-948.

Drapkin, R., Reardon, J.T., Ansari, A., Huang, J.-C., Zawel, L., Sancar, A., and Reinberg, D. (1994). Dual role of TFIIH in DNA excision repair and intranscription by RNA polymerase II. Nature (London) 368, 769-772.

Fiala, E.S., Sodum, R.S., Hussain, N.S., Rivenson, A., and Dolan, L. (1995). Secondary nitroalkanes: Induction of DNA repair in rat hepatocytes, activation by aryl sulfotransferase and hepatocarcinogenicity of 2-nitrobutane and 3-nitropentane in male F344 rats. Toxicology 99, 89-97.

Friedberg, E.C. (1996). Relationships between DNA repair and transcription. Ann.Rev.Biochem. 65, 15-42.

Friedberg, E.C., Walker, G.C., and Siede, W. (1995). DNA damage. In DNA repair and mutagenesis. Friedberg E.C., Walker G.C., and Siede W., eds. (Washington, D.C.: ASM press), pp. 1-58.

Gensler, H.L. (1981a). Low level of UV-induced unscheduled NA synthesis in postmitotic brain cells of hamsters: possible relevance to aging. Exp.Gerontol. 16, 199-207.

Gensler, H.L. (1981b). The effect of hamster age on UV-induced unscheduled DNA synthesis in freshly isolated lung and kidney cells. Exp.Gerontol. 16, 59-68.

Glazer, R.I. and Rohlff, C. (1994). Transcriptional regulation of multidrug resistance in breast cancer. Breast Cancer Res.Treat. 31, 263-271.

Guo, Z., Heydari, A.R., and Richardson, A. (1998a). Nucleotide excision repair of actively transcribed versus non-transcribed DNA in rat hepatocytes. Exp Cell Res 245, 228-238.

Guo, Z., Heydari, A.R., Wu, W.-T., Yang, H., Sabia, M.R., and Richardson, A. (1998b). Thecharacterization of gene specific DNA repair by primary cultures of rat hepatocytes. J.Cell Physiol. 176, 314-322.

Guo, Z., Van Remmen, H., Wu, W.T., and Richardson, A. (1998c). Effect of cAMP-induced transcription on the repair of the phosphoenolpyruvate carboxykinase gene by hepatocytes isolated from young and old rats. Mutation Res 409, 37-48.

Hall, J.D., Almy, R.E., and Scherer, K.L. (1982). DNA repair in cultured human fibroblasts does not decline with donor age. Exp.Cell Res. 139, 351-359.

Hanawalt, P.C. (1994). Transcription-coupled repair and human disease. Science 266, 1957-1958.

Hanawalt, P.C. (1995). DNA repair comes of age. Mutation Res. 336, 101-113.

Harris, A.L., Karran, D., and Lindahl, T. (1983). O^6-Methylguanine DNA methyltransferase of human lymphoid cells: structural and kinetic properties and absence in repair-deficient cells. Cancer Res. 43, 3247-3525.

Henning, K.A., Li, L., Iyer, N., McDaniel, L.D., Reagan, M.S., Legerski, R., Schultz, R.A., Stefanini, M., Lehmann, A.R., and Mayne, L.V. (1995). The Cockayne syndrome group A gene encodes a WD repeat protein that interacts with CSB protein and a subunit of RNA polymerase II TFIIH. Cell 82, 555-564.

Henricksen, L.A. and Wold, M.S. (1994). Replication protein A mutants lacking phosphorylation sites for p34cdc2 kinase support DNA replication. J.Biol.Chem. 269, 24203-24208.

Ho, L. and Hanawalt, P.C. (1991). Gene-specific DNA repair in terminally differentiating rat myoblasts. Mutation Res. 255, 123-141.

Igusheva, O.A., Bil'din, V.N., and Zhestianikov, V.D. (1993). The repair of gamma-induced single-strand beaks in the transcribed and notranscribed DNA of Hela cells. Tsitologiia 35, 54-63.

Islas, A.L., Baker, F.J., and Hanawalt, P.C. (1994). Transcription-coupled repair of psoralen cross-links but not monoadducts in Chinese hamster ovary cells. Biochemistry 33, 10794-10799.

Kempf, C., Schmitt, M., Danse, J.M., and Kempf, J. (1998). Correlation of DNA repair synthesis with ageing in mice, evidenced by quantitative autoradiography. Mech.Aeing Dev. 26, 183-194.

Kennah, H.E.2d., Coetzee, M.L., and Ove, P. (1985). A comparison of DNA repair synthesis in primary hepatocytes from young and old rats. Mech.Aeing Dev. 29, 283-298.

Kovacs, E., Weber, W., and Muller, H. (1984). Age-related variation in the DNA repair synthesis after UV-C irradiation in unstimulated lymphocytes of healthy blood donors. Mutation Res. 131, 231-237.

Kraemer, K.H., Levy, D.D., Parris, C.N., Gozukara, E.M., Moriwaki, S., Adelberg, S., and Seidman, M.M. (1994). Xeroderma pigmentosum and related disorders: Examining the linkage between defective DNA repair and cancer. J.Invest.Dermatol. *103 Suppl.*, 96S-101S.

Kruk, P.A., Rampino, N.J., and Bohr, V.A. (1995). DNA damage and repair in telomeres: Relation to aging. Proc.Natl.Acad.Sci.USA *92*, 258-262.

Krystal, G., Morris, G.M., Lipman, J.M., and Sokoloff, L. (1983). DNA repair by articular chondrocytes. I. Unscheduled DNA synthesis following ultraviolet irradiation in monolayer culture. Mechanisms of Ageing & Development *21*, 83-96.

Lambert, B., Ringborg, U., and Skoog, L. (1979). Age-related decrease of ultraviolet-induced DNA repair synthesis in human peripheral leukocytes. Cancer Res. *39*, 2792-2795.

Lan, R., Greenoak, G.E., and Moran, C. (1994). Pyrimidine dimer induction and removal in the epidermis of hairless mice: Inefficient repair in the genome overall and rapid repair in the H-*ras* sequence. Photochem.Photobiol. *59*, 356-361.

Larminant, F., Zhen, W., and Bohr, V.A. (1993). Gene-specific DNA repair of interstrand cross-links induced by chemotherapeutic agents can be preferential. J.Biol.Chem. *268*, 2649-2654.

Leadon, S.A., Barbee, S.L., and Dunn, A.B. (1995). The yeast RAD2, but not RAD1, gene is involved in the transcription-coupled repair of thymine glycols. Mut.Res. *337*, 169-178.

Leadon, S.A. and Hanawalt, P.C. (1992). Strand-selective repair of DNA damage in the yeast GAL7 gene requires RNA polymerase II. Proc.Natl.Acad.Sci.USA *89*, 10692-10700.

Leadon, S.A. and Lawrence, D.A. (1991). Preferential repair of DNA damage on the transcribed strand of the human metallothionein genes requires RNA polymerase II. Mutation Res. *255*, 67-78.

Lehman, I.R. (1974). DNA ligase: structure, mechanism, and function. Science *186*, 790-797.

Lezhava, T.A., prokofeva, V.V., and Mikhelson, V.M. (1979). Weakening of the ultraviolet ray-induced unscheduled DNA synthesis in human lymphocytes in extreme old age. Tsitologiia *21*, 1360-1363.

Licastro, F., Weindruch, R., Davis, L.J., and Walford, R.L. (1988). Effect of dietary restriction upon the age-associated decline of lymphocyte DNA repair activity in mice. Age *11*, 48-52.

Lipman, J.M. and Sokoloff, L. (1985). DNA repair by articular chondrocytes. III. Unscheduled DNA synthesis following ultraviolet light irradiation of resting cartilage. Mech.Ageing Dev. *32*, 39-55.

Liu, S.C., Meagher, K., and Hanawalt, P.C. (1985). Role of solar conditioning in DNA repair response and survival of human epidermal keratinocytes following UV irradiation. J.Invest.Dermatol. *85*, 93-97.

Liu, S.C., Parsons, C.S., and Hanawalt, P.C. (1982). DNA repair response in human epidermal keratinocytes from donors of different age. J.Invest.Dermatol. *79*, 330-335.

May, A., Nairn, R.S., Okumoto, D.S., Wassermann, K., Stevnsner, T., Jones, J.C., and Bohr, V.A. (1993). Repair of individual DNA strands in the hamster dihydrofolate reductase gene after treatment with ultraviolet light, alkylating agents, and cisplatin. J.Biol.Chem. *268*, 1650-1657.

McGregor, W.G., Mah, M.C., Chen, R.W., Maher, V.M., and McCormick, J.J. (1998). Lack of correlation between degree of interference with transcription and rate of strand specific repair in the HPRT gene of diploid human fibroblasts. J.Biol.Chem. *270*, 27222-27227.

Medvedev, Z.A. (1984). Age changes of chromatin. Mech.Ageing Dev. *28*, 139-154.

Mellon, I., Bohr, V.A., Smith, C.A., and Hanawalt, P.C. (1986). Preferential DNA repair of an active gene in human cells. Proc.Natl.Acad.Sci.USA *83*, 8878-8882.

Mellon, I. and Hanawalt, P.C. (1989). Induction of the *Escherichia coli* lactose operon selectively increases repair of its transcribed DNA strand. Nature *342*, 95-98.

Mellon, I., Rajpal, D.K., Koi, M., Boland, C.R., and Champe, G.N. (1996). Transcription-coupled repair deficiency and mutations in human mismatch repair genes. Science *272*, 557-560.

Mellon, I., Spivak, G., and Hanawalt, P.C. (1987). Selective removal of transcription-blocking DNA damage from the transcribed strand of the mammalian DHFR gene. Cell *51*, 241-249.

Meniel, V., Brouwer, J., and Averbeck, D. (1993). Evidence for preferential repair of 3-carbethoxypsoralen plus UVA induced DNA lesions in the active MAT alpha locus in Saccharomyces cerevisiae using the UvrABC assay. Mutagenesis *8*, 467-471.

Modrich, P. (1991). Mechanisms and biological effects of mismatch repair. Ann.Rev.Genet. *25*, 229-253.

Moriwaki, S., Ray, S., Tarone, R.E., Kraemer, K.H., and Grossman, L. (1996). The effect of donor age on the processing of UV-damaged DNA by cultured human cells: reduced DNA repair capacity and increased DNA mutability. Mutation Res. *364*, 117-123.

Mozzhukhina, T.G., Chabanny, V.N., Levitsky, E.L., and Litoshenko, A.Y. (1991). Age-related changes of supranucleosomal structures and DNA- synthesizing properties of rat liver chromatin. Gerontology *37*, 181-186.

Mullenders, L.H.F., Hazekamp-van Dokkum, A., Kalle, W.H.J., Vrieling, H., Zdzienicka, M.Z., and van Zeeland, A.A. (1993). UV-induced photolesions, their repair and mutations. Mut.Res. *299*, 271-276.

Nette, E.G., Xi, Y.P., Sun, Y.K., Andrews, A.D., and King, D.W. (1984). A correlation between aging and DNA repair in human epidermal cells. Mech.Ageing Dev. *24*, 283-292.

Park, C.-H., Mu, D., Reardon, J.T., and Sancar, A. (1995). The general transcription-repair factor TFIIH is recruited to the excision repair complex by the XPA protein independent of the TFIIE transcription factor. J.Biol.Chem. *270*, 4896-4902.

Petersen, L.N., Orren, D.K., and Bohr, V.A. (1995). Gene-specific and strand-specific DNA repair in the G_1 and G_2 phases of the cell cycle. Mol.Cell.Biol. *15*, 3731-3737.

Plesko, M.M. and Richardson, A. (1984). Age-related changes in unscheduled DNA synthesis by rat hepatocytes. Biochem.Biophys.Res.Commun. *118*, 730-735.

Prolla, T.A. (1998). DNA mismatch repair and cancer. Curr.Opin.Cell Biol. *10*, 311-316.

Salavert, A. and Iynedjian, P.B. (1982). Regulation of phosphoenolpyruvate carboxykinase(GTP) synthesis in rat liver cells. Rapid induction of specific mRNA by glucagon or cyclic AMP and permissive effect of dexamethasone. J.Biol.Chem. *257*, 13404-13412.

Sancar, A. (1994). Mechanisms of DNA excision repair. Science *266*, 1954-1956.

Sancar, A. (1996). DNA excision repair. Ann.Rev.Biochem. *65*, 43-81.

Sawada, N. and Ishikawa, T. (1988). Reduction of potential for replication but not unscheduled DNA synthesis in hepatocytes isolated from ages as compared to young rats. Cancer Res. *48*, 1618-1622.

Scicchitano, D.A. and Hanawalt, P.C. (1989). Repair of N-methylpurines in specific DNA sequences in Chinese hamster ovary cells: Absence of strand specificity in the dihydrofolate reductase gene. Proc.Natl.Acad.Sci.USA *86*, 3050-3054.

Selby, C.P. and Sancar, A. (1993). Molecular mechanism of transcription-repair coupling. Science *260*, 53-58.

Selby, C.P. and Sancar, A. (1997). Human transcription-repair coupling factor CSB/ERCC6 is a DNA-stimulated ATPase but is not a helicase and does not disrupt the ternary transcription complex of stalled RNA polymerase II. J.Biol.Chem. *272*, 1885-1890.

Shivji, M.K.K., Kenny, M.K., and Wood, R.D. (1992). Proliferating cell nuclear antigen is required for DNA excision repair. Cell *69*, 367-374.

Sitaram, A., Plitas, G., Wang, W., and Scicchitano, D.A. (1997). Functional nucleotide excision repair is required for the preferential removal of N-ethylpurines from the transcribed strand of the dihydrofolate reductase gene of Chinese hamster overy cells. Mol.Cell.Biol. *17*, 567-570.

Srivastava, V.K. and Busbee, D.L. (1992). Decreased fidelity of DNA polymerases and decreased DNA excision repair in aging mice: Effects of caloric restriction. Biochem.Biophys.Res.Commun. *182*, 712-721.

Stevnsner, T. and Bohr, V.A. (1993). Studies on the role of topoisomerases in general, gene- and strand-specific DNA repair. Carcinogenesis *14*, 1841-1850.

Sugasawa, K., Ng, J.M.Y., Masutani, C., Iwai, S., van der Spek, P., Eker, A.P.M., Hanaoka, F., Bootsma, D., and Hoeijmakers, J.H.J. (1998). Xeroderma Pigmentosum Group C protein Complex is the Initiator of Global Genome Nucleotide Excision Repair. Mol.Cell *2*, 223-232.

Sweder, K.S. and Hanawalt, P.C. (1992). Preferential repair of cyclobutane pyrimidine dimers in the transcribed strand of a gene in yeast chromosomes and plasmids is dependent on transcription. Proc.Natl.Acad.Sci.USA *89*, 10696-10700.

Taffe, B.G., Larminant, F., Laval, J., Croteau, D.L., Anson, R.M., and Bohr, V.A. (1996). Gene-specific nuclear and mitochondrial repair of formamidopyrimidine DNA glycosylase-sensitive sites in Chinese hamster overy cells. Mut.Res. *364*, 183-192.

Thomale, J., Hochleitner, K., and Rajewsky, M.F. (1994). Differential formation and repair of the mutagenic DNA alkylation product O^6-ethylguanine in transcribed and nontranscribed genes of the rat. J.Biol.Chem. *269*, 1681-1686.

Tilley, R., Miller, S., Srivastava, V., and Busbee, D. (1992). Enhanced unscheduled DNA synthesis by secondary cultures of lung cells established from calorically restricted aged rats. Mech.Ageing Dev. *63*, 165-176.

Van Hoffen, A., Venema, J., Meschini, R., van Zeeland, A.A., and Mullenders, L.H.F. (1995). Transcription-coupled repair removes both cyclobutane pyrimidine dimers and 6-4 photoproducts with equal efficiency and in a sequential way from transcribed DNA in xeroderma pigmentosum group C fibroblasts. EMBO J. *14*, 360-367.

Van Houten, B. (1990). Nucleotide excision repair in Escherichia coli. Microbiol.Rev. *54*, 18-51.

van Oosterwijk, M.F., Filon, R., Kalle, W.H., Mullender, L.H., and van Zeeland, A.A. (1996). The sensitivity of human fibroblasts of N-acetoxy-2-acetylaminofluorene is determined by the extent of transcription-coupled repair, and /or their capacity to counteract RNA synthesis inhibition. Nucleic Acids Res. *24*, 4653-4659.

Van Remmen, H., Ward, W., Sabia, R.V., and Richardson, A. (1995). Effect of age on gene expression and protein degradation. In Handbook of physiology volume on aging. E.J. Masoro, ed. (New York: Oxford University Press), pp. 171-234.

Venema, J., Bartosová, Z., Natarajan, A.T., van Zeeland, A.A., and Mullenders, L.H.F. (1992). Transcription affects the rate but not the extent of repair of cyclobutane pyrimidine dimers in the human adenosine deaminase gene. J.Biol.Chem. *267*, 8852-8856.

Venema, J., Mullenders, L.H.F., Natarajan, A.T., van Zeeland, A.A., and Mayne, L.V. (1990). The genetic defect in Cockayne syndrome is associated with a defect in repair of UV-induced DNA damage in transcriptionally active DNA. Proc.Natl.Acad.Sci.USA *87*, 4707-4711.

Venema, J., Van Hoffen, A., Karkagi, V., Natarajan, A.T., van Zeeland, A.A., and Mullender, L.H.F. (1991). Xeroderma pigmentosum complementation group C cells remove pyrimidine dimers selectively from the transcribed strand of active genes. Mol.Cell Biol. *11*, 4128-4134.

Vijg, J., Mullaart, E., Lohman, P.H., and Knook, D.L. (1985). UV-induced unscheduled DNA synthesis in fibroblasts of aging inbred rats. Mutation Res. *146*, 197-204.

Vreeswijk, M.P., Van Hoffen, A., Westland, B.E., Vrieling, H., van Zeeland, A.A., and Mullender, L.H.F. (1994). Analysis of repair of cyclobutane pyrimidine dimers and pyrimidone 6-4 pyrimidine photoproducts in transcriptionally active and inactive genes in Chinese hamster cells. J.Biol.Chem. *269*, 31858-31863.

Vrieling, H., Venema, J., Van Rooyen, M.L., Van Hoffen, A., Menichini, P., Zdzienicka, M.Z., Simons, J.W.I.M., Mullenders, L.H.F., and van Zeeland, A.A. (1991). Strand specificity for UV-induced DNA repair and mutations in Chinese hamster HPRT gene. Nucleic Acids Res. *19*, 2411-2415.

Wallace, S.S. (1994). DNA damages processed by base excision repair: Biological consequences. Int.J.Radiat.Biol. *66*, 579-589.

Walter, C.A., Grabowski, D.T., Street, K.A., Conrad, C.C., and Richardson, A. (1997). Analysis and modulation of DNA repair in aging. Mech.Ageing Dev. *98*, 203-222.

Wang, Z., Wu, X., and Friedberg E.C. (1991). Nucleotide excision repair of DNA by human cell extracts is supressed in reconstituted nucleosomes. J.Biol.Chem. *266*, 22472-22478.

Warner, H.R. and Price, A.R. (1989). Involvement of DNA repair in cancer and aging. J.Gerontol. *44*, 45-54.

Wassermann, K. and Damgaard, J. (1994). Ongoing activity of RNA polymerase II confers preferential repair of nitrogen mustard-induced N-alkylpurines in the hamster dihydrofolate reductase gene. Cancer Res. *54*, 175-181.

Weraarchakul, N., Strong, R., Wood, W.G., and Richardson, A. (1989). The effect of aging and dietary restriction on DNA repair. Exp.Cell Res. *181*, 197-204.

Wood, R.D. (1997). Nucleotide excision repair in mammalian cells. J.Biol.Chem. *272*, 23465-23468.

© 2001 Elsevier Science B.V. All rights reserved.
The Role of DNA Damage and Repair in Cell Aging
B.A. Gilchrest and V.A. Bohr, volume editors.

POLY(ADP-RIBOSE) POLYMERASE AND AGING

Dean S. Rosenthal, Cynthia M. Simbulan-Rosenthal,
Wen Fang Liu, and Mark E. Smulson

Department of Biochemistry and Molecular Biology, Georgetown University
School of Medicine, Washington, DC

1. Introduction

Poly(ADP-ribose) polymerase (PARP) is a major nuclear protein associated with chromatin that contains zinc fingers and binds to either double- or single-strand DNA breaks. Upon binding to DNA, PARP is activated, and forms covalent homopolymers of poly(ADP-ribose) (PAR) attached to a number of nuclear proteins, including itself, and proteins involved in DNA replication, DNA repair and apoptosis. Nuclear NAD, which comprises 95% of the total cellular NAD, is the substrate for polymer formation. PARP has been implicated in numerous biological functions, involved with the breaking and rejoining of DNA (Berger et al., 1979; Jacobson and Jacobson, 1989; Satoh and Lindahl, 1992; Satoh et al., 1993; Ding et al., 1994; Smulson et al., 1994). In addition, other functions have been ascribed for PARP in which the role of DNA strand breaks is not clear. For example PARP has been demonstrated to play a role as a coactivator of gene transcription (Meisterernst et al., 1997; Kannan et al., 1999). In addition, the binding of PARP to specific nuclear proteins has been shown to alter their activity, in the absence of DNA breaks or NAD (Simbulan et al., 1993).

An attempt to demonstrate a direct role for PARP in longevity was made in several studies (Burkle et al., 1992; Grube and Burkle, 1992) in which the investigators showed that poly(ADP-ribosyl)ation inhibited DNA amplification induced by carcinogens. Further, a strong positive correlation was observed between inducible PARP activities in permeablized mononuclear leukocytes of 13 mammalian species and their life spans. By Western blot analysis, no correlation between the amount of PARP protein and life span was found, suggesting higher PARP specific activity in longer-lived species (Grube and Burkle, 1992). Burkle et al. also investigated the formation of PAR *in vivo* in mononuclear blood cells derived from rats and humans by immunofluorescence following γ-irradiation (Burkle et al., 1994). A significantly higher percentage of PAR-positive cells was found in humans, even though the number of DNA strand breaks were not significantly different in the two species, suggesting that the higher poly(ADP-ribosyl)ation capacity of long-lived species might more efficiently help to reduce the accumulation of DNA damage and of genetic alterations, as compared with short-lived species. We originally mapped the PARP gene to chromosome 1q41-q42 and PARP-like sequences to chromosomes 14q13-q32 and 13q34 (Cherney et al., 1987); the latter pseudogene interrupts a *pol*-like element (Lyn et al., 1993) and exhibits two-allele polymorphism (Lyn et al., 1993) associated with predisposition to several cancers

(Bhatia et al., 1990). However, an attempt to correlate inter-individual variation in longevity with PARP found no association between case/control genotypic frequencies at the PARP chromosome 1 locus, when comparing the genotypic pools of subjects older than 100 years with those of younger subjects matched for sex and geographic area (De Benedictis et al., 1998). A correlation was noted for a different (THO) locus, indicating that this PARP locus did not affect inter-individual variability in life expectancy. The mechanism by which PARP might play a role in the aging process is unclear. However, its roles in DNA repair, genomic stability, and cell death could conceivably contribute to its putative association with longevity.

2. PARP, DNA Repair and Genomic Stability

A number of studies have employed chemical inhibitors (Morgan and Cleaver, 1982; Burkle et al., 1990; Waldman and Waldman, 1991), dominant negative mutants (Schreiber et al., 1995; Kupper et al., 1996), and PARP antisense RNA (Ding et al., 1992; Ding and Smulson, 1994) to examine the function of PARP. These studies have demonstrated that PARP plays a role in reducing the frequency of DNA strand breaks, recombination, gene amplification, micronuclei formation, and sister chromatid exchanges (SCE), all of which are markers of genomic instability, in cells exposed to DNA-damaging agents. PARP-deficient cell lines are hypersensitive to carcinogenic agents and also display increased SCE, implicating PARP as a guardian of the genome that facilitates DNA repair and protects against DNA recombination (Chatterjee et al., 1999).

PARP knockout mice have been independently generated from the interruption of exon 2 (Wang et al., 1995), exon 4 (de Murcia et al., 1997), and most recently, exon 1 (Masutani et al., 1999) of the PARP gene on chromosome 1. PARP knockout mice with a disrupted PARP gene neither express intact PARP nor exhibit significant poly(ADP-ribosyl)ation (Wang et al., 1995; de Murcia et al., 1997; Masutani et al., 1999). Primary fibroblasts derived from exon 2 PARP knockout mice show an elevated frequency of spontaneous SCE and micronuclei formation in response to treatment with genotoxic agents (Wang et al., 1995; Wang et al., 1997) providing further support for a role of PARP in the maintenance of genomic integrity. Exon 4 PARP knockout mice exhibit extreme sensitivity to γ-irradiation and methylnitrosourea and also show increased genomic instability as revealed by a high level of SCE (de Murcia et al., 1997). Immortalized cells derived from these animals are characterized by retarded cell growth, G_2/M block, and chromosomal instability on exposure to DNA-alkylating agents, presumably because of a severe defect in DNA repair (Trucco et al., 1998).

We recently utilized immortalized fibroblasts derived from exon 2 PARP knockout mice (PARP$^{-/-}$), as well as from control animals of the same strain (PARP$^{+/+}$) to study the role of PARP in genomic stability. FACS analysis initially revealed that these cells exhibit mixed ploidy, including a tetraploid cell population, indicative of genomic instability (Andreassen et al., 1996). The tetraploid population was not observed in PARP$^{+/+}$ cells. Further, this tetraploid cell population was no longer apparent in PARP$^{-/-}$

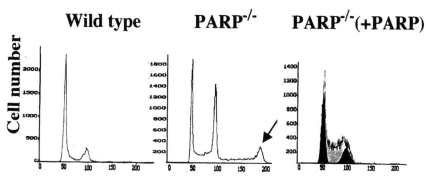

Relative fluorescence

Figure 1. Flow cytometric analysis of immortalized wild-type, PARP⁻, and PARP⁻(+PARP) fibroblasts. Cells were harvested 18 h after release from serum deprivation. Nuclei were then prepared and stained with propidium iodide for flow cytometric analysis. In addition to the two major peaks of nuclei at G_0-G_1 and G_2-M apparent in the DNA histograms of wild-type and PARP⁻ (+PARP) cells, the DNA histograms of PARP⁻ cells exhibit a third peak corresponding to the G_2-M peak of an unstable tetraploid cell population (arrow).

cells retransfected with PARP cDNA, suggesting that the reintroduction of PARP into PARP⁻/⁻ cells may have resulted in selection against this genomically unstable population (Fig. 1).

We characterized the genetic alterations associated with PARP depletion by comparative genomic hybridization (CGH) analysis, a cytogenetic technique that detects chromosomal gains and losses in the test DNA as a measure of genetic instability (Kallioniemi et al., 1992; du Manoir et al., 1993). Although CGH is now commonly used for mapping DNA copy number changes in human tumor genomes, few studies to date have utilized this technique to evaluate genetic instability in transgenic mouse models (Shi et al., 1997; Weaver et al., 1999). CGH analysis revealed that PARP⁻/⁻ mice or immortalized PARP⁻/⁻ cells exhibited gains in regions of chromosomes 4, 5, and 14, as well as a deletion in chromosome 14 (Fig. 2). We further investigated the effect of stable transfection of immortalized PARP⁻/⁻ fibroblasts with PARP cDNA on the genetic instability of these cells. Reintroduction of PARP cDNA into PARP⁻/⁻ cells appeared to confer stability because these chromosomal gains were no longer detected in these cells, further supporting an essential role for PARP in the maintenance of genomic stability (Fig. 2).

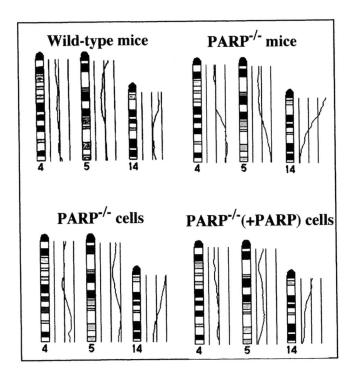

Figure 2. Comparison of the CGH profiles of chromosomes 4, 5, and 14 among wild-type mice, PARP⁻ mice, immortalized PARP⁻ fibroblasts, and PARP⁻ (+PARP) fibroblasts. Average ratio profiles were computed for all chromosomes and used for the mapping of changes in copy number, with only the results for chromosomes 4, 5, and 14 shown. The three vertical lines to the right of the chromosome ideograms represent values of 0.75, 1, and 1.25 (left to right, respectively) for the fluorescence ratio between the test DNA and the normal control DNA. The ratio profile (curve) was computed as a mean value of at least eight metaphase spreads. A ratio of ≥1.25 was regarded as a gain and a ratio of ≤0.75 as a loss. PARP⁻ (+PARP) fibroblasts did not show the gains at 4C5-ter, 5F-ter, or 14A1-C2 that were apparent in both PARP⁻ mice and immortalized PARP⁻ cells, although they retained the partial loss at 14D3-ter.

In our study, we noted the absence of immunoreactive p53 from these cells as revealed by immunoblot analysis. A previous report indicated that primary fibroblasts from exon 2 PARP⁻/⁻ mice (Agarwal et al., 1997) also show reduced basal levels of p53 as well as a defective induction of p53 in response to DNA damage, suggesting that PARP-dependent signaling may influence this response. Cells that are unable to synthesize PAR because of unavailability of NAD (Whitacre et al., 1995) also show a reduced p53 response. Thus, PARP may regulate genomic stability, at least in part, via p53. Given that the loss of p53 from diploid cells promotes the survival of cells with severe DNA damage and the development of tetraploidy (Cross et al., 1995; Ramel et al., 1995; Yin et al., 1999), the presence of a tetraploid population among the immortalized PARP$^{-/-}$ cells was consistent with the apparent absence of p53. p53 is involved in the maintenance of diploidy as a component of the spindle checkpoint (Cross et al., 1995) and by regulating centrosome duplication (Fukasawa et al., 1996). A functional association of PARP and p53 has been suggested by immunoprecipitation experiments (see below). Thus, the increased genetic instability of PARP$^{-/-}$ mice and cells are consistent with their deficiencies in PARP and p53.

3. PARP and Cell Death

How does PARP confer genomic stability? One possibility is that in addition to its role in DNA repair, PARP is involved in the elimination of cells that have accumulated an unacceptable level of DNA damage. Because poly(ADP-ribosyl)ation is stimulated by DNA fragmentation, the potential role for PARP in cell death via NAD and ATP depletion has been proposed previously (Berger et al., 1983). This idea has been supported by recent studies in which both exon 1 (Masutani et al., 1999) and exon 2 (Burkart et al., 1999; Pieper et al., 1999) PARP$^{-/-}$ animals have been shown to be resistant to steptozotocin-induced pancreatic islet cell death, associated with NAD depletion in PARP$^{+/+}$ animals. We have also collaborated in a study that demonstrated that exon 2 PARP$^{-/-}$ animals are resistant to the neurotoxin MPTP-induced parkinsonism (Mandir et al., 1999). Exon 2 PARP$^{-/-}$ animals are also more resistant to ischemic injury (Eliasson et al., 1997; Endres et al., 1997; Szabo et al., 1997; Zingarelli et al., 1999).

To investigated whether PARP might play an active role in programmed cell death, we first used a human osteosarcoma cell line that undergoes a "slow," spontaneous apoptotic death (Rosenthal et al., 1997a). On reaching confluency, approximately 6 days under our culture conditions, these cells undergo the morphological and biochemical changes characteristic of apoptosis. Internucleosomal DNA cleavage was apparent at day 7 and increased until day 10, at which time virtually all of the cells have undergone apoptosis. Cells from duplicate cultures were incubated for up to 10 days and fixed at daily intervals for examination of nuclear poly(ADP-ribosyl)ation with antiserum to PAR. After 3 days, the nuclei of all attached cells stained intensely for the PAR. The *in vivo* synthesis of PAR was markedly reduced afterwards (Fig. 3A). Our results support the idea that nuclear disruption involving strand breaks may be present in the earliest stages of apoptosis, before morphological changes and the appearance of the characteristic nucleosome ladder.

Kaufmann et al. (1993) first demonstrated that PARP undergoes proteolytic cleavage during chemotherapy-induced apoptosis. By immunoblot analysis with epitope-specific antibodies, it was demonstrated that programmed cell death was accompanied by early cleavage of PARP into 85- and 24- kDa fragments that contain the active site and the DNA-binding domain (DBD) of the enzyme, respectively. This latter domain is required for PARP activity. The purification and characterization of caspase-3, responsible for the cleavage of PARP during apoptosis was performed by Nicholson et al. (1995). This enzyme is composed of two subunits of 17 and 12 kDa that are derived from a common

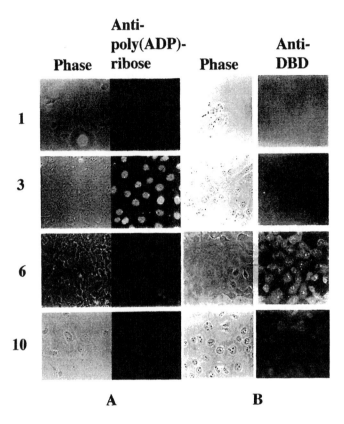

Figure 3. Antibodies that specifically recognize poly (ADP-ribose) and the caspase-3 cleavage product of PARP (DBD) detect early and late stage functions for PARP. Osteosarcoma cells undergoing 10-day spontaneous apoptosis were fixed at the indicated times and subjected to immunofluorescent analysis utilizing antibodies specific for poly(ADP-ribose) (A), or the caspase-3 cleavage product of PARP (DBD; B).

proenzyme, which is related to interleukin-1β-converting enzyme and to CED-3, the product of a gene required for programmed cell death in *Caenorhabditis elegans* (Yuan et al., 1993). The identity of this protease was also demonstrated by Tewari et al. (1995). To measure PARP cleavage in intact cells, we subjected human osteosarcoma cells to immunofluorescence analysis with antibodies that recognize the DBD but not intact PARP (Rosenthal et al., 1997a,b). As with the other markers, samples were analyzed each day throughout the total 10-day period; samples from immediate (day 1), early (day 3), mid- (day 6), and late (10) stages of apoptosis are shown in the Fig. 3B. Immunofluorescence analysis detected the PARP DBD in human osteosarcoma cells only after 6 to 7 days in culture, a time at which the abundance of both PARP and PAR is decreasing, PARP-cleavage activity is increasing, and internucleosomal DNA cleavage is present (Nicholson et al., 1995). The pattern of staining for the DBD also differed markedly from that of full-length PARP. Whereas PARP staining was present throughout the nucleus, the DBD showed a more localized punctate pattern in the region of the nucleolus and throughout the nucleus-disrupted cytoplasm.

Therefore, catalytic activation of PARP occurs early in osteosarcoma cell death, while the cleavage of PARP and the accumulation of a large number of DNA strand breaks occur later in the apoptotic process. The concomitant loss of poly(ADP-ribosyl)ation of target proteins appears to be characteristic of later stages of apoptosis during which cells become irreversibly committed to death. This may conserve NAD and ATP during the later stages of apoptosis. Recently, the requirement for PARP cleavage to prevent necrosis associated with depletion of NAD has been confirmed using PARP$^{-/-}$ cells that express a caspase-resistant mutant of PARP (Herceg and Wang, 1999).

The generality of an early burst of poly(ADP-ribosyl)ation was confirmed with human HL-60 cells, mouse 3T3-L1, and immortalized fibroblasts derived from wild-type mice (Simbulan-Rosenthal et al., 1998). The effects of eliminating this early transient modification of nuclear proteins by depletion of PARP protein either by antisense RNA expression or by gene disruption on various morphological and biochemical markers of apoptosis were then examined.

In 3T3 L1 cells stably transfected with a construct expressing dexamethasone (Dex)-inducible PARP antisense RNA, Dex induced a time-dependent depletion of PARP, with only ~5% of the protein remaining after 72 h. A combination of anti-Fas and cycloheximide induced a marked increase in caspase-3-like activity in control 3T3-L1 cells that had been preincubated in the absence or presence of Dex. This effect was maximal 24 h after induction of apoptosis, as indicated by the generation of the 89- and 24-kDa cleavage fragments of PARP in an *in vitro* assay (Fig. 4A, top). No caspase-3 activity was apparent in PARP-antisense cells that had been depleted of PARP by preincubation with Dex before exposure to anti-Fas and cycloheximide (Fig. 4A, bottom).

To confirm that procaspase-3 is proteolytically processed to an active p17 subunit during apoptosis in control 3T3-L1 cells, and to determine whether the transient early poly(ADP-ribosyl)ation is necessary for this activation, control and antisense cells were preincubated with Dex, exposed to anti-Fas and cycloheximide for indicated times, and cell extracts were subjected to immunoblot analysis with antibodies to the p17 subunit of caspase-3. Whereas procaspase-3 was proteolytically processed to p17 by 24 h, coinciding with the peak of *in vitro* caspase-3-like PARP-cleavage activity, in control cells, proteolytic processing of procaspase-3 was not apparent in the PARP-depleted antisense cells (Fig. 4B). Furthermore, using DNA fragmentation analysis as another assay for apoptosis, control 3T3-L1 cells exposed to anti-Fas and cycloheximide for 24 h exhibited marked internucleosomal DNA fragmentation (DNA ladders), but not the PARP-depleted antisense cells exposed to these inducers for the same time (Fig. 4C). Similar to our previous studies, we noted that the earliest stages of apoptosis were

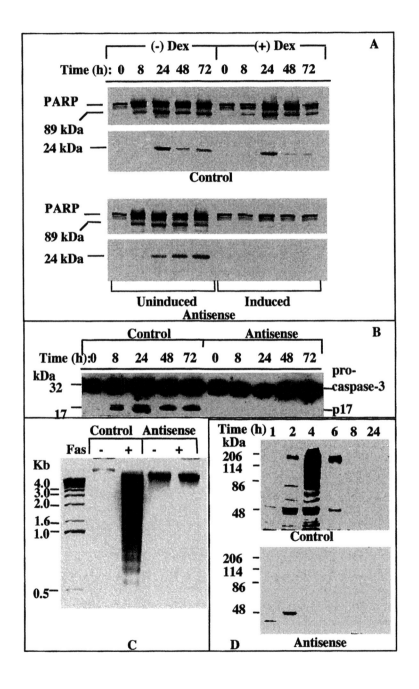

Figure 4 (left page). Effects of PARP depletion by antisense RNA expression on the increase in caspase-3-like activity (A), proteolytic processing of procaspase-3 (B), internucleosomal DNA fragmentation (C) and synthesis of PAR during Fas-mediated apoptosis in 3T3-L1 cells. Mock-transfected (A, top) and PARP-antisense (A, bottom) 3T3-L1 cells were preincubated in the absence or presence of 1 μM Dex for 72 h and then incubated with anti-Fas (50 ng/ml) and cycloheximide (10 μg/ml) for the indicated times. Cytosolic extracts were prepared and assayed for in vitro PARP-cleavage activity with [³⁵S]PARP as substrate. (B) 3T3-L1 control and antisense cells were preincubated with Dex for 72 h and then exposed to anti-Fas and cycloheximide for the indicated times as in (A). Cell extracts were subjected to immunoblot analysis with a monoclonal antibody to the p17 subunit of caspase-3 (B) or to PAR (D). The positions of procaspase-3 and p17 are indicated. (C) Total genomic DNA was extracted and internucleosomal DNA ladders characteristic of apoptosis was detected by agarose gel electrophoresis and ethidium bromide staining.

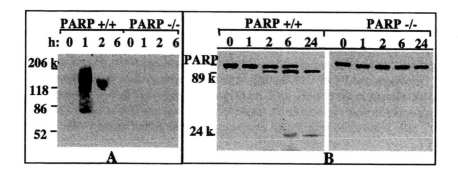

Figure 5. Effects of anti-Fas and cycloheximide on poly(ADP-ribosyl)ation of nuclear proteins and caspase-3-like activity in immortalized fibroblasts from wild-type and PARP knockout mice. (A) PARP⁻ and PARP⁻⁻ fibroblasts were exposed to anti-Fas (100 ng/ml) and cycloheximide (10 μg/ml) for the indicated times, after which extracts were subjected to immunoblot analysis with antibodies to PAR. (B) Cytosolic extracts from cells treated as in (A) were assayed for in vitro PARP-cleavage activity with [³⁵S]PARP as substrate.

associated with a burst of PAR synthesis. This early synthesis of PAR was eliminated by the expression of PARP antisense RNA (Fig. 4D).

Cells derived from animals depleted of PARP were also unable to undergo Fas-mediated apoptosis. Anti-Fas and cycloheximide induced a rapid synthesis of PAR in PARP$^{+/+}$ cells, which was not observed in PARP$^{-/-}$ cells (Fig. 5A). At later times, a marked increase in caspase-3-like activity is observed in PARP$^{+/+}$ cells; this effect was maximal 24 h after induction of apoptosis, as indicated by the *in vitro* cleavage of PARP into 89-kDa and 24-kDa fragments. In contrast, no such increase in caspase-3-like activity (Fig. 5B) nor processing of procaspase-3 (Simbulan-Rosenthal et al., 1998) was evident in PARP$^{-/-}$ cells after exposure to anti-Fas and cycloheximide for up to 24 h.

PARP$^{+/+}$ cells showed substantial nuclear fragmentation and chromatin condensation 24 h after induction of Fas-mediated apoptosis; ~97% of nuclei exhibited apoptotic morphology by this time. In contrast, no substantial changes in nuclear morphology were apparent in the PARP$^{-/-}$ fibroblasts even after exposure to anti-Fas and cycloheximide for 24 h or 48 h (Simbulan-Rosenthal et al., 1998).

PARP$^{-/-}$ fibroblasts were stably transfected with a plasmid expressing wild-type PARP (Alkhatib et al., 1987). Three different cell clones, as well as pooled clones expressed PARP protein similar to the PARP$^{+/+}$ cells. These cells were induced to undergo apoptosis by exposure to anti-Fas and cycloheximide for up to 48 h. PARP$^{+/+}$ cells as well as PARP$^{-/-}$ clones stably transfected with PARP exhibited significant caspase-3-like activity after 48 h; ~95% of the PARP protein was cleaved *in vivo* to the 24 kDa cleavage fragment by 48 h. As expected, PARP was not expressed in the PARP$^{-/-}$ fibroblasts nor in $^{-/-}$ cells transfected with vector alone. Consistently, whereas exposure to anti-Fas and cycloheximide induced marked internucleosomal DNA fragmentation in PARP$^{+/+}$ fibroblasts and PARP$^{-/-}$ cells stably transfected with PARP, no apoptotic DNA ladders were evident in the PARP$^{-/-}$ cells when similarly treated (Fig. 6A). Furthermore, exposure to anti-Fas plus cycloheximide for 48 h induced apoptotic nuclear morphology in PARP$^{-/-}$ cells transfected with PARP almost to the same extent as the PARP$^{+/+}$ cells (Fig. 6B).

Thus depletion of PARP by antisense in 3T3L1, or by knockout of PARP attenuates Fas plus cycloheximide-mediated apoptosis. In addition, the reintroduction of PARP in independent clones of PARP$^{-/-}$ cells reestablished the response. We interpret these results to indicate that PARP plays an active role early in apoptosis either by depletion of NAD and ATP, or via the modification of nuclear proteins involved in apoptosis. Furthermore, these studies are consistent with earlier results using chemical inhibitors (which require a degree of caution in their interpretation), indicating that the activation of PARP is required for apoptosis to occur in some systems (Agarwal et al., 1988; Monti et al., 1994; Nosseri et al., 1994; Kuo et al., 1996). It should be pointed out that another study indicated that PARP$^{-/-}$ cells underwent similar apoptosis compared to PARP$^{+/+}$ cells (Wang et al., 1997). Whether these differing observations are due to the use of different culture conditions or reagents (e.g. primary vs. immortalized cells) remains to be determined. However, it has recently been shown that expression of caspase-3-resistant PARP (Boulares, 1999; Herceg and Wang, 1999), as well as wild-type PARP (Boulares, 1999) in exon 2 PARP$^{-/-}$ cells results in an earlier onset of the apoptotic blank page is to keep page count right

4. Poly(ADP-ribosyl)ation of p53

In addition to undergoing automodification, PARP catalyzes the poly(ADP-ribosyl)ation of such nuclear proteins as histones, topoisomerases I and II (Kasid et al., 1989; Simbulan-Rosenthal et al., 1996), SV40 large T antigen (Baksi et al., 1987), DNA polymerase α, proliferating cell nuclear antigen (PCNA), and approximately 15 protein components of the DNA synthesome (Simbulan-Rosenthal et al., 1996). The modification of nucleosomal proteins also alters the nucleosomal structure of the DNA containing strand breaks and promotes access of various replicative and repair enzymes to these sites (Butt et al., 1980; Poirier et al., 1982). We have obtained some potentially relevant targets for poly(ADP-ribosyl)ation during the burst of PAR synthesis at the early stages of apoptosis, including p53. p53, a tumor suppressor nuclear phosphoprotein, reduces the occurrence of mutations by mediating cell cycle arrest in G1 or G2/M or inducing apoptosis in cells that have accumulated substantial DNA damage, thus, preventing progression of cells through S phase before DNA repair is complete (Kastan et al., 1992; O'Connor et al., 1993; Levine, 1997). p53 is induced by a variety of apoptotic stimuli and is required for apoptosis in many cell systems (Fisher, 1994). Overexpression of p53 is sufficient to induce apoptosis in various cell types (Yonish-Rouach et al., 1991). Interestingly, p53 can utilize transcription activation of target genes and/or direct protein-protein interaction to initiate p53-dependent apoptosis.

Both PARP activity and p53 accumulation are induced by DNA damage, and both proteins have been implicated in the normal cellular responses to such damage. Whereas PAR synthesis increases within seconds after induction of DNA strand breaks (Berger and Petzold, 1985), the amount of wild-type p53, which is usually low because of the short half-life (20 min) of the protein, increases several hours after DNA damage as a result of reduced degradation (Kastan et al., 1991; Fritsche et al., 1993). A functional association of PARP and p53 has recently been suggested by coimmunoprecipitation of each protein *in vitro* by antibodies to the other (Wesierska-Gadek et al., 1996; Vaziri et al., 1997). It was recently shown that p53 is poly(ADP-ribosyl)ated *in vitro* by purified PARP, and that binding of p53 to a specific p53 consensus sequence prevents its covalent modification (Wesierska-Gadek et al., 1996). We recently showed that modification of p53 by poly(ADP-ribosyl)ation also occurs *in vivo*, and that it represents one of the early acceptors of poly(ADP-ribosyl)ation during apoptosis in human osteosarcoma cells (Simbulan-Rosenthal et al., 1999). Given that the *in vivo* half-life of PAR chains on an acceptor has been estimated to be about 1 to 2 min, we additionally explored how this postranslational modification of p53 is altered at the onset of caspase-3 mediated cleavage and inactivation of PARP during the later stages of the death program.

Human osteosarcoma cells were plated under conditions that result in spontaneous apoptosis over a 10-day period (Nicholson et al., 1995; Rosenthal et al., 1997a). Biochemical markers of apoptosis were initially observed at day 5 and maximized around days 7 to 9, including caspase-3-mediated *in vitro* PARP-cleavage activity (Fig. 7A), proteolytic processing of the caspase 3 proenzyme (CPP32) to its active form (p17) (Fig. 7B), and internucleosomal DNA fragmentation.

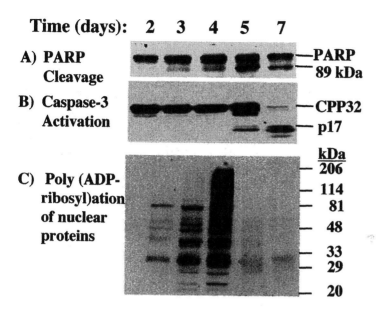

Figure 7. Time courses of in vitro PARP cleavage (A), proteolytic activation of procaspase-3 to its active form (p17) (B) and poly(ADP-ribosyl)ation of nuclear proteins (C) during apoptosis in osteosarcoma cells (C). (A) At the indicated times of confluence-associated spontaneous apoptosis, cell extracts were prepared and caspase-3-like PARP-cleavage activity in cytosolic extracts was assayed with [³⁵S]PARP as substrate. (B) Extracts were derived, and equal amounts of protein (30 µg) were subjected to immunoblot analysis with mAb to the p17 subunit of caspase-3 (B), or to PAR (C). The positions of full-length PARP and of its 89- and 24-kDa cleavage products as well as procaspase-3 and p17 are indicated in (A and B). The positions of the molecular size standards (in kilodaltons) are indicated in (C).

Consistent with previous studies showing p53 accumulation during early apoptosis in different cell lines, immunoblot analysis with anti-p53 mAbs of extracts of osteosarcoma cells at various stages of spontaneous apoptosis revealed that endogenous levels of p53 protein were significantly increased as early as days 2 to 3, maximized at day 4, and declined thereafter. Immunoblot analysis with antibodies to PARP to monitor *in vivo* PARP cleavage during the same time frame showed that ~ 50% of endogenous PARP was cleaved to its 89 kDa fragment by day 7 and complete cleavage of PARP was noted by day 9 (Simbulan-Rosenthal et al., 1999).

When the same extracts were subjected to immunoblot analysis with antibodies to PAR, low levels of polymer were observed at day 2 of apoptosis (Fig. 7C), indicating the absence of DNA strand breaks, PARP activity, or both. However, poly(ADP-ribosyl)ation of nuclear proteins was markedly increased at day 3 and was maximal at day 4, a stage at which all the cells were still viable and could be replated, prior to any evidence of internucleosomal DNA fragmentation. Subsequently, a marked decline in

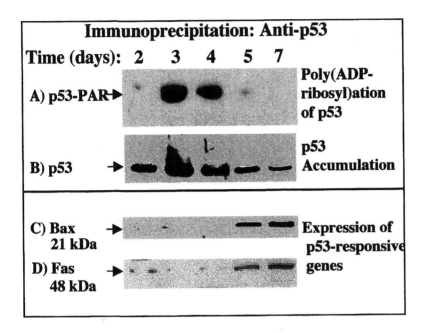

Figure 8. Poly(ADP-ribosyl)ation of p53 and time courses of Bax and Fas expression during spontaneous apoptosis in human osteosarcoma cells (A) At the indicated times during spontaneous apoptosis, cell extracts were prepared and equal amounts of total protein (100 µg) were subjected to immunoprecipitation with a mAb to p53. The immunoprecipitated proteins were then subjected to immunoblot analysis with a mAb to PAR. The immunoblot shown in A was stripped of antibodies by incubation for 30 min at 50°C with a solution containing 100 mM 2-mercaptoethanol, 2% SDS, and 62.5 mM Tris-HCl pH (6.7), and reprobed with antibodies to p53 (B). Bax (C), and Fas (D) antibodies were used to reprobe immunoblots shown in Fig. 7.

poly(ADP-ribosyl)ation of nuclear proteins was observed at later time points (days 7 to 9), concomitant with the onset of substantial DNA fragmentation, proteolytic activation of caspase 3, and caspase-3 mediated *in vitro* and *in vivo* cleavage of PARP.

To confirm if p53 undergoes poly(ADP-ribosyl)ation *in vivo* during apoptosis in human osteosarcoma cells, cell extracts were derived at various times during spontaneous apoptosis and subjected to immunoprecipitation with an anti-p53 mAb. The immunoprecipitated proteins were then subjected to immunoblot analysis with mAb to PAR. This approach revealed marked poly(ADP-ribosyl)ation of p53 at the early stages of apoptosis (days 3 to 4) (Fig. 8A), coincident with the burst of PAR synthesis during this stage. The extent of poly(ADP-ribosyl)ation of p53 declined concomitant with the onset of both *in vitro* and *in vivo* caspase-3 mediated PARP cleavage. Reprobing of the blot with polyclonal antibodies to p53 confirmed that the modified protein was in fact p53

(Fig. 8B). The observation that p53 is specifically poly(ADP-ribosylated) during the early stages of spontaneous apoptosis in human osteosarcoma cells suggests that this postranslational modification may play a role in regulating its function during the early phases of the cell death cascade.

PARP can modulate the catalytic activity of a number of DNA-binding nuclear enzymes by catalyzing their poly(ADP-ribosyl)ation. In most instances, poly(ADP-ribosyl)ation inhibits the activity of the modified protein, presumably because of a marked decrease in DNA-binding affinity caused by electrostatic repulsion between DNA and PAR. Thus, postranslational modification of p53 may also alter DNA binding to specific DNA sequences in the promoters of target genes associated with the induction of p53-mediated apoptosis, such as those encoding Bax, IGF-BP3 (Chinnaiyan et al., 1996), or Fas (Muller et al., 1998). The time course of accumulation and poly(ADP-ribosyl)ation of p53 during the early stages of apoptosis was thus correlated with the induction of expression of the p53-responsive genes *bax* and *Fas*. Immunoblot analysis of extracts of cells at various stages of apoptosis in osteosarcoma cells with antibodies to either Bax or Fas revealed that expression of both Bax and Fas (8C, D) were negligible before and at the peak of p53 accumulation and poly(ADP-ribosyl)ation (days 3 and 4). Although p53 accumulation was already significantly elevated by day 2, expression of Bax and Fas was markedly induced only at day 5, concomitant with a decline in PAR attached to p53 and the onset of caspase-3 mediated PARP cleavage and inactivation. The coincident decrease in PAR covalently bound to p53 and induction of Bax and Fas expression suggests that poly(ADP-ribosyl)ation may regulate p53 function early in apoptosis; caspase-3 mediated cleavage of PARP may release p53 from poly(ADP-ribosyl)ation–induced inhibition at the later stages of the apoptotic cascade.

Accordingly, p53 may represent a potentially relevant target for poly(ADP-ribosyl)ation during the burst of PAR synthesis at the early periods of apoptosis. Colocalization of PARP and p53 in the vicinity of large DNA breaks and their physical association (Wesierska-Gadek et al., 1996; Vaziri et al., 1997), suggest that poly(ADP-ribosyl)ation may regulate the DNA binding ability and, consequently, the function of p53. The accumulation of p53 may be due to induced expression of the protein by the apoptotic stimuli or stabilization by inhibition of p53 degradation via modification of the protein. These results suggest a negative regulatory role for PARP and/or PAR early in apoptosis, since subsequent degradation of PAR attached to p53 coincided with the increase in caspase-3 (PARP-cleavage) activity as well as the induction of expression of the p53-responsive genes *bax* and *Fas* at a stage when cells are irreversibly committed to death. Although the mechanism(s) of action of the Bax/Bcl-2 family of gene products during apoptosis remain to be clarified, induction of Bax expression may influence the decision to commit to apoptosis since homodimerization of Bax promotes cell death and heterodimerization of Bax with Bcl-2 inhibits the anti-apoptotic function of Bcl-2 (Chinnaiyan et al., 1996). Wild-type p53, but not mutant p53, also upregulates Fas expression during chemotherapy-induced apoptosis, and p53-responsive elements were recently identified within the first intron and the promoter of the *Fas* gene (Muller et al., 1998). Binding of Fas to Fas ligand recruits the adapter molecule FADD via shared protein motifs (death domains), resulting in subsequent activation or amplification of the caspase cascade leading to apoptosis.

5. PARP and Cellular Senescence

In addition to oxidative damage that occurs from environmental sources as well as metabolic processes, aging in most normal cells is accompanied by the progressive loss of telomeric DNA. This persistent loss of telomeres leads to a critically short telomere length, leading to genomic instability, cell cycle arrest, and senescence. Activation of p53 has been proposed to play a role in the senescent response to telomere shortening (Vaziri and Benchimol, 1996). PARP may also be responsible for initiating cellular senescence in response to telomere shortening, since PARP is involved in the post-translational activation of p53 protein in aging cells in response to telomere loss from replication, hyperoxia, or depletion of ATM (Vaziri et al., 1997). These authors demonstrated that p53 protein can associate with PARP and inhibition of PARP activity leads to abrogation of p21 and mdm2 expression in response to DNA damage. Further, inhibition of PARP activity leads to an extension of cellular lifespan. As the authors themselves point out, however, chemical inhibitor studies need to be interpreted with caution in light of their potential lack of specificity. Another study found that down-regulation of PARP protein expression was associated with senescence, which the authors proposed as a causative factor in increased DNA damage associated with cell aging (Salminen et al., 1997), although PARP down-regulation is associated with other forms of G1 arrest. Recently, tankyrase, a protein with PARP catalytic activity, has been shown to play a role in the poly(ADP-ribosyl)ation of telomeric repeat binding factor-1 (TRF-1) (Smith et al., 1998), a protein that appears to be involved in progressive telomere shortening (van Steensel and de Lange, 1997). Tankyrase has been shown to bind to and poly (ADP-ribosyl)ate TRF-1 and prevents its binding to telomeres in vitro. As yet, no *in vivo* studies on tankyrase have been reported in the literature.

6. Conclusions

PARP has been shown to play active roles in the response to diverse forms of cellular damage resulting from normal metabolic processes associated with aging, as well as environmental factors. The response may depend upon the level and type of damage, as well as the cell type. In severely damaged cells, PARP activation induces poly(ADP-ribosyl)ation of key nuclear proteins, including p53, and a concomitant lowering of NAD and ATP levels, resulting ultimately in cell death, the form of which (apoptosis vs. necrosis) may depend upon the time of onset of caspase-mediated PARP cleavage. In mildly damaged cells, PARP may signal a repair response. Finally, PARP may regulate p53 and other gene products important for senescence in response to telomere shortening.

Acknowledgements

This work was supported in part by grants CA25344 and 1PO1 CA74175 from the National Cancer Institute, by the United States Air Force Office of Scientific Research (grant AFOSR-89-0053), and by the United States Army Medical Research and Development Command (contract DAMD17-90-C-0053; to M.E.S.) and DAMD 17-96-C-6065 (to D.S.R).

References

Agarwal, M., Agarwal, A., Taylor, W., Wang, Z. Q., and Wagner, E. (1997). Defective induction but normal activation and function of p53 in mouse cells lacking PARP. Oncogene 15, 1035-1041.

Agarwal, S., Drysdale, B. E., and Shin, H. S. (1988). Tumor necrosis factor-mediated cytotoxicity involves ADP-ribosylation. J Immunol 140, 4187-92.

Alkhatib, H. M., Chen, D. F., Cherney, B., Bhatia, K., Notario, V., Giri, C., Stein, G., Slattery, E., Roeder, R. G., and Smulson, M. E. (1987). Cloning and expression of cDNA for human poly(ADP-ribose) polymerase. Proc. Natl. Acad. Sci. 84, 1224-8.

Andreassen, P., Martineau, S., and Margolis, R. (1996). Chemical induction of mitotic checkpoint override in mammalian cells results in aneuploidy following a transient tetraploid state. Mutat. Res. 372, 181-194.

Baksi, K., Alkhatib, H., and Smulson, M. E. (1987). In vivo characterization of the poly ADP-ribosylation of SV40 chromatin and large T antigen by immunofractionation. Exptl. Cell Res. 172, 110-123.

Berger, N., A., Petzold, S. J., and Berger, S. J. (1979). Association of poly (ADP-rib) synthesis with cessation of DNA synthesis and DNA fragmentation. Biochim. Biophys. Acta 564, 90-104.

Berger, N. A., and Petzold, S. J. (1985). Identification of the requirements of DNA for activation of poly (ADP-ribose) polymerase. Biochemistry 24, 4352-5.

Berger, N. A., Sims, J. L., Catino, D. M., and Berger, S. J. (1983). Poly(ADP-ribose)polymerase mediates the suicide response to massive DNA damage: Studies in normal and DNA-repair defective cells. In ADP-ribosylation, DNA repair and cancer, M. Miwa, O. Hayaishi, S. Shall, M. Smulson and T. Sugimura, eds. (Tokyo: Japan Scientific Societies Press), pp. 219-226.

Bhatia, K. G., Cherney, B. W., Huppi, K., Magrath, I. T., Cossman, J., Sausville, E., Barriga, F., Johnson, B., Gause, B., Bonney, G., Neequayi, J., DeBernardi, M., and Smulson, M. (1990). A Deletion Linked to a Poly (ADP-Ribose) polymerase Gene on Chromosome 13q33-qter occurs Frequently in the Normal Black Population as well as in Multiple Tumor DNA. Cancer Research 50, 5406-13.

Boulares, A. H., Yakovlev, A. G., Ivanova, V., Stoica, B. A., Wang, G., Iyer, S., and Smulson, M. (1999). Role of poly(ADP-ribose) polymerase cleavage in apoptosis. J Biol Chem 274, 22932-40.

Burkart, V., Wang, Z. Q., Radons, J., Heller, B., Herceg, Z., Stingl, L., Wagner, E. F., and Kolb, H. (1999). Mice lacking the poly(ADP-ribose) polymerase gene are resistant to pancreatic beta-cell destruction and diabetes development induced by streptozocin. Nat Med 5, 314-9.

Burkle, A., Grube, K., and Kupper, J. H. (1992). Poly(ADP-ribosyl)ation: its role in inducible DNA amplification, and its correlation with the longevity of mammalian species. Exp Clin Immunogenet 9, 230-40.

Burkle, A., Heilbronn, R., and zur, H. H. (1990). Potentiation of carcinogen-induced methotrexate resistance and dihydrofolate reductase gene amplification by inhibitors of poly(adenosinediphosphate-ribose) polymerase. Cancer Res 50, 5756-60.

Burkle, A., Muller, M., Wolf, I., and Kupper, J. H. (1994). Poly(ADP-ribose) polymerase activity in intact or permeabilized leukocytes from mammalian species of different longevity. Mol Cell Biochem 138, 85-90.

Butt, T. R., DeCoste, B., Jump, D., Nolan, N., and Smulson, M. E. (1980). Characterization of a putative poly adenosine disphosphate ribose chromatin complex. Biochemistry 19, 5243-5249.

Chatterjee, S., Berger, S., and Berger, N. (1999). Poly(ADP-ribose) polymerase: a guardian of the genome that facilitates DNA repair by protecting against DNA recombination. Mol. Cell. Biochem. 193, 23-30.

Cherney, B. W., McBride, O. W., Chen, D. F., Alkhatib, H., Bhatia, K., Hensley, P., and Smulson, M. E. (1987). cDNA sequence, protein structure, and chromosomal location of the human gene for poly(ADP-ribose) polymerase. Proc. Natl. Acad. Sci. 84, 8370-4.

Chinnaiyan, A., Orth, K., O'Rourke, K., Duan, H., Poirier, G., and Dixit, V. (1996). Molecular ordering of the cell death pathway. Bcl-2 and Bcl-xL function upstream of the CED-3-like apoptotic proteases. J Biol Chem 271, 4573-6.

Cross, S., Sanchez, C., Morgan, C., Schimke, M., Ramel, S., Idzerda, R., Raskind, W., and Reid, B. (1995). A p53-dependent mouse spindle ckeckpoint. Science 267, 1353-1356.

De Benedictis, G., Carotenuto, L., Carrieri, G., De Luca, M., Falcone, E., Rose, G., Cavalcanti, S., Corsonello, F., Feraco, E., Baggio, G., Bertolini, S., Mari, D., Mattace, R., Yashin, A. I., Bonafe, M., and Franceschi, C. (1998). Gene/longevity association studies at four autosomal loci (REN, THO, PARP, SOD2). Eur J Hum Genet 6, 534-41.

de Murcia, J. M., Niedergang, C., Trucco, C., Ricoul, M., Dutrillaux, B., Mark, M., Oliver, F. J., Masson, M., Dierich, A., Le Meur, M., Walztinger, C., Chambon, P., and de Murcia, G. (1997). Requirement of poly(ADP-ribose) polymerase in recovery from DNA damage in mice and in cells. Proc Natl Acad Sci U S A 94, 7303-7.

Ding, R., Pommier, Y., Kang, V. H., and Smulson, M. (1992). Depletion of poly(ADP-ribose) polymerase by antisense RNA expression results in a delay in DNA strand break rejoining. J. Biol. Chem. 267, 12804-12.

Ding, R., and Smulson, M. (1994). Depletion of nuclear poly(ADP-ribose) polymerase by antisense RNA expression: influences on genomic stability, chromatin organization and carcinogen cytotoxicity. Cancer Research 54, 4627-4634.

du Manoir, S., Speicher, M., and Jovs, S. (1993). Detection of complete and partial chromosome gains and losses by comparative genomic in situ hybridization. Hum. Genet. 90, 590-610.

Eliasson, M. J., Sampei, K., Mandir, A. S., Hurn, P. D., Traystman, R. J., Bao, J., Pieper, A., Wang, Z. Q., Dawson, T. M., Snyder, S. H., and Dawson, V. L. (1997). Poly(ADP-ribose) polymerase gene disruption renders mice resistant to cerebral ischemia. Nat Med 3, 1089-95.

Endres, M., Wang, Z. Q., Namura, S., Waeber, C., and Moskowitz, M. A. (1997). Ischemic brain injury is mediated by the activation of poly(ADP- ribose)polymerase. J Cereb Blood Flow Metab 17, 1143-51.

Fisher, D. (1994). Apoptosis in cancer therapy: crossing the threshold. Cell 78, 539-42.

Fritsche, M., Haessler, C., and Brandner, G. (1993). Induction of nuclear accumulation of the tumor suppressor protein by p53 by DNA damaging agents. Oncogene 8, 307-318.

Fukasawa, K., Choi, T., Kuriyama, R., Rulong, S., and Vande Woude, G. (1996). Abnormal centrosome amplification in the absence of p53. Science 271, 1744-1747.

Grube, K., and Burkle, A. (1992). Poly(ADP-ribose) polymerase activity in mononuclear leukocytes of 13 mammalian species correlates with species-specific life span. Proc Natl Acad Sci U S A 89, 11759-63.

Herceg, Z., and Wang, Z. Q. (1999). Failure of Poly(ADP-ribose) polymerase cleavage by caspases leads to induction of necrosis and enhanced apoptosis [In Process Citation]. Mol Cell Biol 19, 5124-33.

Ikejima, M., Noguchi, S., Yamashita, R., Ogura, T., Sugimura, T., Gill, D. M., and Miwa, M. (1990). The zinc fingers of human poly(ADP-ribose) polymerase are differentially required for the recognition of DNA breaks and nicks and the consequent enzyme activation. Other structures recognize intact DNA. J Biol Chem 265, 21907-13.

Jacobson, M. K., and Jacobson, E. L. (1989). ADP-ribose transfer reactions: mechanisms and biological significance, 1st Edition, M. K. Jacobson and E. L. Jacobson, eds. (New York: Springer-Verlag).

Kallioniemi, A., Kallioniemi, O.-P., Sudar, D., Rutovitz, D., Gray, J., Waldman, F., and Pinkel, D. (1992). Comparative genomic hybridization for molecular cytogenetic analysis of solid tumors. Science 258, 818-821.

Kannan, P., Yu, Y., Wankhade, S., and Tainsky, M. A. (1999). PolyADP-ribose polymerase is a coactivator for AP-2-mediated transcriptional activation. Nucleic Acids Res 27, 866-74.

Kasid, U. N., Halligan, B., Liu, L. F., Dritschilo, A., and Smulson, M. (1989). Poly(ADP-ribose)-mediated post-translational modification of chromatin-associated human topoisomerase I. Inhibitory effects on catalytic activity. J. Biol. Chem. 264, 18687-92.

Kastan, M. B., Onyekwere, O., Sidransky, D., Vogelstein, B., and Craig, R. W. (1991). Participation of p53 protein in the cellular response to DNA damage. Cancer Research 51, 6304-11.

Kastan, M. B., Zhan, Q., El-Delry, W., S., Carrier, F., Jacks, T., Walsh, W. V., Plunkett, B., S., Vogelstein, B., and Fornace, A. J. J. (1992). A mammalian cell cycle checkpoint pathway utilizing p53 and GADD45 is defective in ataxia-telangiectasia. Cell 71, 587-97.

Kaufmann, S. H., Desnoyers, S., Ottaviano, Y., Davidson, N. E., and Poirier, G. G. (1993). Specific proteolytic cleavage of poly(ADP-ribose)polymerase: an early marker of chemotherapy-induced apoptosis. Cancer Research 53, 3976-85.

Kuo, M. L., Chau, Y. P., Wang, J. H., and Shiah, S. G. (1996). Inhibitors of poly(ADP-ribose) polymerase block nitric oxide-induced apoptosis but not differentiation in human leukemia HL-60 cells. Biochem Biophys Res Commun 219, 502-8.

Kupper, J., Muller, M., and Burkle, A. (1996). Trans-dominant inhibition of poly(ADP-ribosyl)ation potentiates carcinogen-induced gene amplification in SV40-transformed Chinese hamster cells. Cancer Res. 56, 2715-2717.

Le Rhun, Y., Kirkland, J. B., and Shah, G. M. (1998). Cellular responses to DNA damage in the absence of Poly(ADP-ribose) polymerase. Biochem Biophys Res Commun 245, 1-10.

Levine, A. (1997). p53,the cellular gatekeeper for growth and division. Cell 88, 323-331.

Li, X., and Coffino, P. (1996). Identification of a region of p53 that confers lability. J. Biol. Chem. 271, 4447-4451.

Lyn, D., Cherney, B., Lalande, M., Berenson, J., Lupold, S., Bhatia, K., and Smulson, M. (1993). A duplicated region is responsible for the poly(ADP-ribose) polyerase polymorphism, on chromosome 13 associated with a predisposition to cancer. Am. J. Hum. Genetics 52, 124-134.

Lyn, D., Deaven, L., Istock, N., and Smulson, M. (1993). The polymorphic poly(ADP-ribose) polymerase gene in humans interrupts an endogenous Pol-like element on 13q34. Genomics 18, 206-211.

Malanga, M., Pleschke, J., Kleczkowska, H., and Althaus, F. (1998). Poly(ADP-ribose) binds to specific domains of p53 and alters its DNA binding functions. J. Biol. Chem. 273, 11839-11843.

Mandir, A. S., Przedborski, S., Jackson-Lewis, V., Wang, Z. Q., Simbulan-Rosenthal, C. M., Smulson, M. E., Hoffman, B. E., Guastella, D. B., Dawson, V. L., and Dawson, T. M. (1999). Poly(ADP-ribose) polymerase activation mediates 1-methyl-4-phenyl-1, 2,3,6-tetrahydropyridine (MPTP)-induced parkinsonism. Proc Natl Acad Sci U S A 96, 5774-9.

Masutani, M., Suzuki, H., Kamada, N., Watanabe, M., Ueda, O., Nozaki, T., Jishage, K., Watanabe, T., Sugimoto, T., Nakagama, H., Ochiya, T., and Sugimura, T. (1999). Poly(ADP-ribose) polymerase gene disruption conferred mice resistant to streptozotocin-induced diabetes. Proc Natl Acad Sci U S A 96, 2301-4.

Meisterernst, M., Stelzer, G., and Roeder, R. G. (1997). Poly(ADP-ribose) polymerase enhances activator-dependent transcription in vitro. Proc Natl Acad Sci U S A 94, 2261-5.

Monti, D., Cossarizza, A., Salvioli, S., Franceschi, C., Rainaldi, G., Straface, E., Rivabene, R., and Malorni, W. (1994). Cell death protection by 3-aminobenzamide and other poly(ADP- ribose)polymerase inhibitors: different effects on human natural killer and lymphokine activated killer cell activities. Biochem Biophys Res Commun 199, 525-30.

Morgan, W., and Cleaver, J. (1982). 3-aminobenzamide synergistically increases sister-chromatid exchanges in cells exposed to methyl methanesulfonate but not to ultraviolet light. Mutat. res. 104, 361-366.

Muller, M., Wilder, S., Bannasch, D., Israeli, D., Lelbach, K., Li-Weber, M., Friedman, S., Galle, P., Stremmel, W., Oren, M., and Krammer, P. (1998). p53 activates the CD95 (APO-1/Fas) gene in response to DNA damage by anticancer agents. J. Exp. Med. 188, 2033-2045.

Nicholson, D. W., Ali, A., Thornberry, N. A., Vaillancourt, J. P., Ding, C. K., Gallant, M., Gareau, Y., Griffin, P. R., Labelle, M., Lazebnik, Y. A., Munday, N. A., Raju, S. M., Smulson, M. E., Yamin, T. T., Yu, V. L., and Miller, D. K. (1995). Identification and inhibition of the ICE/CED-3 protease necessary for mammalian apoptosis. Nature 376, 37-43.

Nosseri, C., Coppola, S., and Ghibelli, L. (1994). Possible involvement of poly(ADP-ribosyl) polymerase in triggering stress-induced apoptosis. Experimental Cell Research 212, 367-73.

O'Connor, P. M., Jackman, J., Jondle, D., Bhatia, K., Magrath, I., and Kohn, K. W. (1993). Role of the p53 tumor suppressor gene in cell cycle arrest and radiosensitivity of Burkitt's lymphoma cell lines. Cancer Research 53, 4776-80.

Oliver, F. J., de la Rubia, G., Rolli, V., Ruiz-Ruiz, M. C., de Murcia, G., and Murcia, J. M. (1998). Importance of poly(ADP-ribose) polymerase and its cleavage in apoptosis. Lesson from an uncleavable mutant. J Biol Chem 273, 33533-9.

Pieper, A. A., Brat, D. J., Krug, D. K., Watkins, C. C., Gupta, A., Blackshaw, S., Verma, A., Wang, Z. Q., and Snyder, S. H. (1999). Poly(ADP-ribose) polymerase-deficient mice are protected from streptozotocin-induced diabetes. Proc Natl Acad Sci U S A 96, 3059-64.

Poirier, G. G., de Murcia, G., Jongstra-Bilen, J., Niedergang, C., and Mandel, P. (1982). Poly(ADP-ribosyl)ation of polynucleosomes causes relaxation of chromatin structure. Proc. Natl. Acad. Sci. 79, 3423-3427.

Ramel, S., Sanchez, C., Schimke, M., Neshat, K., Cross, S., Raskind, W., and Reid, B. (1995). Inactivation of p53 and the development of tetraploidy in the elastase-SV40 Tantigen transgenic mouse pancreas. Pancreas 11, 213-222.

Rosenthal, D. S., Ding, R., Simbulan-Rosenthal, C. M. G., Vaillancourt, J. P., Nicholson, D. W., and Smulson, M. E. (1997a). Intact cell evidence fo the early snthesis, and subsequent late apopain-mediated suppression, of poly(ADP-ribose) during apoptosis. Exp. Cell Res. 232, 313-321.

Rosenthal, D. S., Ding, R., Simbulan-Rosenthal, C. M. G., Cherney, B., Vanek, P., and Smulson, M. E. (1997b). Detection of DNA breaks in apoptotic cells utilizing the DNA binding domain of poly(ADP-ribose) polymerase with fluorescence microscopy. Nucleic Acids Research 25, 1437-1441.

Salminen, A., Helenius, M., Lahtinen, T., Korhonen, P., Tapiola, T., Soininen, H., and Solovyan, V. (1997). Down-regulation of Ku autoantigen, DNA-dependent protein kinase, and poly(ADP-ribose) polymerase during cellular senescence. Biochem Biophys Res Commun 238, 712-6.

Satoh, M. S., and Lindahl, T. (1992). Role of poly(ADP-ribose) formation in DNA repair. Nature 356, 356-8.

Satoh, M. S., Poirier, G. G., and Lindahl, T. (1993). NAD+-dependent repair of damaged DNA by human cell extracts. J. Biol. Chem. 268, 5480-87.

Schreiber, V., Hunting, D., Trucco, C., Gowans, B., Grunwald, P., de Murcia, G., and de Murcia, J. (1995). A dominant negative mutant of human PARP affects cell recovery, apoptosis, and sister chromatid exchange following DNA damage. Proc. Natl. Acad. Sci. USA 92, 4753-4757.

Shi, Y., Naik, P., Dietrich, W., Gray, J., Hanahan, D., and Pinkel, D. (1997). DNA copy number changes associated with characteristic LOH in islet cell carcinomas of transgenic mice. Genes Chrom. & Cancer 2, 104-111.

Simbulan, C., Suzuki, M., Izuta, S., Sakurai, T., Savoysky, E., Kojima, K., Miyahara, K., Shizuta, Y., and Yoshida, S. (1993). Poly (ADP-ribose) polymerase stimulates DNA polymerase alpha. J. Biol. Chem. 268, 93-99.

Simbulan-Rosenthal, C. M., Rosenthal, D. S., Iyer, S., Boulares, A. H., and Smulson, M. E. (1998). Transient poly(ADP-ribosyl)ation of nuclear proteins and role for poly(ADP-ribose) polymerase in the early stages of apoptosis. J. Biol. Chem. 273, 13703-13712.

Simbulan-Rosenthal, C. M., Rosenthal, D. S., and Smulson, M. E. (1999). Poly(ADP-ribosyl)ation of p53 during apoptosis in human osteosarcoma cells. Cancer Res. 59, 2190-2194.

Simbulan-Rosenthal, C. M. G., Rosenthal, D. S., Hilz, H., Hickey, R., Malkas, L., Applegren, N., Wu, Y., Bers, G., and Smulson, M. (1996). The expression of poly(ADP-ribose) polymerase during differentiation-linked DNA replication reveals that this enzyme is a component of the multiprotein DNA replication complex. Biochemistry 35, 11622-11633.

Smith, S., Giriat, I., Schmitt, A., and de Lange, T. (1998). Tankyrase, a poly(ADP-ribose) polymerase at human telomeres [see comments]. Science 282, 1484-7.

Smulson, M., Istock, N., Ding, R., and Cherney, B. (1994). Deletion mutants of poly(ADP-ribose) polymerase support a model of cyclic association and dissociation of enzyme from DNA ends during DNA repair. Biochemistry 33, 6186-6191.

Szabo, C., Lim, L. H., Cuzzocrea, S., Getting, S. J., Zingarelli, B., Flower, R. J., Salzman, A. L., and Perretti, M. (1997). Inhibition of poly (ADP-ribose) synthetase attenuates neutrophil recruitment and exerts antiinflammatory effects. J Exp Med 186, 1041-9.

Tewari, M., Quan, L. T., O'Rourke, K., Desnoyers, S., Zeng, Z., Beidler, D. R., Poirier, G. G., Salvesen, G. S., and Dixit, V. M. (1995). Yama/CPP32b, a mammalian homolog of CED-3, is a crmA-inhibitable protease that cleaves the death substrate poly(ADP-ribose) polymerase. Cell 81, 801-809.

Trucco, C., Oliver, F, G, d. M., and de Murcia, J. (1998). DNA repair defect in PARP-deficient cell lines. Nuc. Acids Res. 26, 2644-2649.

van Steensel, B., and de Lange, T. (1997). Control of telomere length by the human telomeric protein TRF1. Nature 385, 740-3.

Vaziri, H., and Benchimol, S. (1996). From telomere loss to p53 induction and activation of a DNA-damage pathway at senescence: the telomere loss/DNA damage model of cell aging. Exp Gerontol 31, 295-301.

Vaziri, H., West, M., Allsop, R., Davison, T., Wu, Y., Arrowsmith, C., Poirier, G., and Benchimol, S. (1997). ATM-dependent telomere loss in aging human diploid fibroblasts and DNA damage lead to the post-translational activation of p53 protein involving poly(ADP-ribose) polymerase. EMBO J. 16, 6018-6033.

Waldman, A., and Waldman, B. (1991). Stimulation of intrachromosomal homologous recombination in mammalian cells by an inhibitor of poly(ADP-ribosyl)ation. Nucleic Acids Res. 19, 5943-5947.

Wang, Z., Stingl, L., Morrison, C., Jantsch, M., Los, M., Schulze-Osthoff, K., and Wagner, E. (1997). PARP is important for genomic stability but dispensable in apoptosis. Genes & Dev. 11, 2347-2358.

Wang, Z. Q., Auer, B., Stingl, L., Berghammer, H., Haidacher, D., Schweiger, M., and Wagner, E. F. (1995). Mice lacking ADPRT and poly(ADP-ribosyl)ation develop normally but are susceptible to skin disease. Genes & Development 9, 509-520.

Weaver, Z., McCormack, S., Liyanage, M., du Manoir, S., Coleman, A., Schrock, E., Dickson, R., and Ried, T. (1999). A recurring pattern of chromosomal aberrations in mammary gland of MMTV-cmyc transgenic mice. Genes Chrom. and Cancer 25, 251-260.

Wesierska-Gadek, J., Bugajska-Schretter, A., and Cerni, C. (1996). ADP-ribosylation of p53 tumor suppressor protein: mutant but not wild-type p53 is modified. J Cell Biochem 62, 90-101.

Wesierska-Gadek, J., Schmid, G., and Cerni, C. (1996). ADP-ribosylation of wild-type p53 in vitro: binding of p53 protein to specific p53 consensus sequence prevents its modification. Biochem. Biophys. Res. Commun. 224, 96-102.

Whitacre, C. M., Hashimoto, H., Tsai, M.-L., Chatterjee, S., Berger, S. J., and Berger, N. A. (1995). Involvement of NAD-poly(ADP-ribose) metabolism in p53 regulation and its consequences. Cancer Research 55, 3697-3701.

Yin, X., Grove, L., Datta, N., Long, M., and Prochownik, E. (1999). C-myc overexpression and p53 loss cooperate to promote genomic instability. Oncogene 18, 1177-1184.

Yonish-Rouach, E., Resnitzky, D., Lotem, J., Sachs, L., Kimchi, A., and Oren, M. (1991). Wild-type p53 induces apoptosis of myeloid leukaemic cells that is inhibited by interleukin-6. Nature 352, 345-7.

Yuan, J., Shaham, S., Ledoux, S., Ellis, H. M., and Horvitz, H. R. (1993). The C. elegans death gene ced-3 encodes a protein similar to mammalian interleukin-1-b-converting enzyme. Cell 75, 641-652.

Zingarelli, B., Szabo, C., and Salzman, A. L. (1999). Blockade of Poly(ADP-ribose) synthetase inhibits neutrophil recruitment, oxidant generation, and mucosal injury in murine colitis. Gastroenterology 116, 335-45.

© 2001 Elsevier Science B.V. All rights reserved.
The Role of DNA Damage and Repair in Cell Aging
B.A. Gilchrest and V.A. Bohr, volume editors.

DNA INSTABILITY, TELOMERE DYNAMICS, AND CELL TRANSFORMATION

Robert J. Shmookler Reis[1-4] *and Masood A. Shammas*[1,4]

Depts. of [1]Geriatrics, [2]Biochemistry & Molecular Biology, and
[3]Medicine, University of Arkansas for Medical Sciences, Little Rock, AR;
[4]J.L. McClellan Veterans Medical Center, Little Rock, AR

1. Cellular Mechanisms for DNA Rearrangement

In eukaryotes, DNA rearrangements on a gross scale include chromosomal duplications and losses, translocations and inversions, fragmentations and fusions, and large-block amplifications (typically of 0.3-10 megabase units). Smaller rearrangements include more limited duplication or further amplification of genes or intragenic regions, and deletions on a similar scale, which would not be detected cytologically but which can have profound effects on cell survival. All of the above are thought to be mediated predominantly by genetic recombination — either between essentially identical sequences on homologous chromosomes, or between regions of much more limited sequence identity such as interspersed repeats, but also (with lower frequency) between apparently unrelated sequences. We will not include among DNA rearrangements the many smaller perturbations of sequence — e.g. deletions, insertions or repetitions of short oligonucleotides, mini- and microsatellite instability, or triplet expansion — which are usually classed as mutations and which are understood to be mediated by polymerase errors and failures of mismatch repair (Kolodner 1996; Lengauer *et al.* 1998) rather than DNA strand exchange.

Genetic recombination refers to the creation of new gene combinations in a diploid cell or gamete, distinct from those of its progenitors, in particular those involving DNA strand exchange (breakage and reunion of strands). Of course, recombination occurs also in prokaryotes, but only at times and genomic regions which are present in two or more copies within the same cell, such as duplicate DNA behind the replication fork, or regions duplicated within the chromosome or episomally. Mechanistically, recombinations can be categorized into three classes: homologous recombination, site-specific recombination (including transposition), and illegitimate recombination.

Homologous recombination requires extensive sequence identity between the DNA duplexes to be recombined (Szostak *et al.* 1983; West 1994), and occurs at rates dependent on both the length of "homologous" sequences and the degree to which they are identical (Ayares *et al.* 1986; Liskay *et al.* 1987; Nassif and Engels 1993). *Site-specific recombination* occurs between DNA duplexes containing characteristic signal sequences that have little or no homology. Site-specific recombination events include rearrangements that generate immunoglobulin and T-cell receptor diversity (Oettinger *et al.* 1990), and insertion or excision of transposons and viral DNA. Aberrations of these processes would also qualify, such as chromosomal inversions and translocations in human cancers, primarily leukemias, in which immunoglobulin signal sequences may recombine with signal-related sequences in or near oncogenes (Leder *et al.* 1983). *Transposition,* usually regarded as a subclass of site-specific recombination, is the process by which blocks of DNA sequence move from one place to another in a genome. Transposable elements can cause insertional mutation of interrupted genes, and altered expression of nearby genes, and thus may contribute significantly to genomic evolution (Wichman *et al.* 1992) and somatic mutation (Nikitin and Shmookler Reis, 1997). *Illegitimate* or *nonhomologous recombination* does not require either a signal sequence or regions of extensive sequence identity. Although its frequency per kilobase of target DNA is much lower than that for homologous recombination, the total over the whole genome can be substantial (Mansour *et al.*, 1988, 1993). Thus, stable integration of transfected DNA occurs predominantly through illegitimate recombination (Waldman and Waldman 1990; Sweezy and Fishel 1994), even when homologous recombination is greatly enhanced by gene targeting with sequences identical to genomic DNA (Folger *et al.* 1985; Mansour *et al.*, 1988; Sedivy and Sharp 1989).

In differentiating eukaryotes, recombination occurs both in mitosis and meiosis, where it may involve overlapping but distinct genetic pathways. Meiotic recombination — in germ-line cells only — augments the reassortment of gene alleles accomplished by segregation of chromosomes at the first meiotic division (Baker *et al.* 1976). Because mitoses greatly exceed meioses in lineages leading to gamete formation, mitotic recombination also contributes greatly to germ-line reassortment and rearrangement of genes (Smith 1974, 1976). Moreover, meiotic recombination between homologous DNA sequences of multigene families plays a significant role in generating genetic diversity and facilitating the evolution of genes (Amstutz *et al.* 1985). Mitotic recombination plays a crucial role in repairing DNA damaged by a variety of agents (Friedberg *et al.* 1991), both in somatic tissues and in gametic lineages. Additionally, mitotic homologous recombination in somatic cells mediates immunoglobulin class switching (Davis *et al.* 1980; Early *et al.* 1980; Sakano *et al.* 1980), and may underlie the loss of heterozygosity which, in regions containing anti-oncogenes, contributes to carcinogenesis (Solomon *et al.* 1991).

2. Homologous Recombination is Critical to DNA Repair

Intrachromosomal homologous recombination, the generation of new assortments of genetic material by DNA strand exchange, is an essential genetic process which occurs in all living organisms. Mechanisms to facilitate homologous recombination must have arisen very early in evolution, very likely as mediators of DNA repair. The simplest models of homologous recombination involve an initial nicking of DNA duplexes, invasion of partially identical (often termed homologous) sequences in another duplex, displacement of the crossover point by branch migration, and resolution of the crossover intermediate by a second strand exchange (Holliday 1964; Meselson and Radding 1975). Initiation is now thought to occur primarily with a double-strand break in the "recipient" DNA duplex, followed by exonucleolytic recission of the broken ends and gap repair directed by a homologous donor duplex (Szostak *et al.* 1983). Double-strand break models explain otherwise puzzling features of homologous recombination, account for nonreciprocal information flow — including gene conversion, and place the recombination mechanism firmly in the realm of DNA repair.

Damage to DNA is inevitable, and its repair is thus a fundamental requirement for life. All organisms are exposed to DNA-damaging agents, including exogenous chemicals and radiation, and endogenously-generated oxygen radicals. The resulting lesions (Solomon *et al.* 1991) include missing bases (*e.g.,* nearly 10,000 spontaneous depurinations per mammalian cell per day), modified bases (due to alkylation and radiation ionization), and incorrect bases (due chiefly to deamination). DNA strands may be affected by insertions or deletions, largely spontaneous in origin, and by double- and single-strand breaks — either caused directly by ionizing radiation, or indirectly via DNA repair, elicited by cross-linking agents, UV-induced cyclobutyl dimers, oxidation, and large adducts. Unrepaired oxidative damage alone, in human cells, affects nearly 10^{-5} of nucleotides in nuclear DNA, and greater than 10^{-4} of bases in mitochondrial DNA, at any point in time (Richter *et al.* 1988; Loft and Poulsen 1996).

DNA lesions in bacteria are highly mutagenic if unrepaired prior to DNA replication, because they may be bypassed through error-prone (SOS) repair, an inducible *recA*-dependent pathway (Kornberg and Baker 1992) for which no counterpart in eukaryotes has yet been defined. In both prokaryotes and eukaryotes, post-replication repair can be effected with high fidelity by a pathway termed *recombinational repair,* wherein gap filling is achieved by recombination with homologous DNA sequences from the other branch of the replication fork. Either mode of post-replication repair is literally a "stopgap" measure, however, which fails to remove the mutagenic lesion but buys time — another cell cycle — for repair. Double-strand breaks pose a particular danger to genomic integrity, since exonucleolytic trimming of broken ends assures the loss of some genetic information from the duplex, which can only be repaired through recombination (Resnick *et al.*, 1989; Sweezy and Fishel 1994). The occurrence of genetic recombination in mitotic cells may thus be predominantly or entirely repair-related (Roca and Cox 1990). Surprisingly, recombinational repair of DNA damage appears to utilize both homologous and nonhomologous pathways (Sweezy and Fishel 1994). Moreover, the Rad52 protein, although intimately involved in homologous recombination through interaction with Rad51, is not required for intrachromosomal "gene conversion" or for repair of several types of DNA damage (Yamaguchi-Iwai *et al.* 1998). Post-replication repair in eukaryotic cells should be minimized by p53-mediated cell-cycle arrest, which allows repair to precede S phase, but may nevertheless occur to some extent in rapidly-cycling cells, and might be especially prevalent in p53-deficient cancer cells. The p53 protein is a key component of cellular defenses against genotoxic agents. Upon damage to DNA, p53 is activated and directs the cell either to apoptosis, or to transient cell-cycle arrest at the G_1/S boundary, by modulation of the transcription of multiple target genes (Kastan *et al.* 1991; Kuerbitz *et al.* 1992; Almasan *et al.* 1995). Apoptosis may be effected by p53-mediated transcriptional activation of "Bax", a member of the Bcl-2 family, which then induces apoptosis (Zhan *et al.* 1994; Larsen 1994; Almasan *et al.* 1995), but the mechanism for deciding this fate is not well understood (see El-Deiry *et al.* 1994; Kondo *et al.* 1998). It has been proposed that different classes of DNA damage could lead to the induction of specific subsets of modifying proteins, which activate p53 both by induction and post-translational modification (Nakamura 1998). For example, DNA damage induced by ionizing radiation results in the activation of ATM (mutated in Ataxia telangiectasia) kinase (Savitsky et al. 1995), probably through autophosphorylation (Nakamura 1998), which then activates p53 by phosphorylation at Ser[15] (Banin et al. 1998; Canman et al. 1998).

Activated p53 also mediates cell-cycle arrest through trans regulation of target gene expression. One target, p21, inhibits cyclin-dependent kinases (Xiong et al. 1993), thus preventing phosphorylation of Rb and subsequent release of E2F, which is required for transcriptional activation of S-phase-associated genes. Transcriptional activation of other downstream genes such as GADD45 (Kastan et al. 1992), and WAF1/CIP1 (El-Deiry et al. 1994) has also been implicated in p53-mediated cell cycle arrest. However, the observation that certain DNA-base-damaging agents, such as methyl methane sulfonate (MMS), can induce GADD45 in the absence of functional p53 (Fornace et al. 1989), implies that p53-independent pathways also exist for cell-cycle arrest after DNA damage (Kastan et al. 1992). The extent of p53 induction can vary greatly between different DNA damaging agents (Lu and Lane 1993), and/or cell types (Khanna and Lavin 1993), which also supports the hypothesis of multiple signal transduction pathways for cell-cycle arrest.

3. Implication of Recombinogenic Agents in Carcinogenesis

Most carcinogens are mutagenic when assayed in *Salmonella* HIS-reversion assays, but 30-50% are not (Ashby and Tennant 1988; Mason et al. 1990). Of those carcinogens for which point-mutagenicity is not commensurate with oncogenicity, some are active in other HIS⁻ strains requiring small frame-shift insertions or deletions for reversion. However, these assay systems also appear to be inefficient at detecting a subset of carcinogens (Ashby and Tennant 1988; Mason et al. 1990), which may act primarily by augmenting recombination. Several carcinogens with low mutagenicity in Ames tests have now been shown to be recombinogenic in bacteria (Luisi-Deluca et al. 1984) yeast (Schiestl et al. 1989) and transformed mammalian cells (Zhang and Jenssen 1994). Using the HPRT-reversion assay in untransformed human fibroblasts, we find substantial recombinogenic activity for 6/6 carcinogens tested, which increase reversion rates by 2.4- to 12-fold even without addition of a metabolic-activation extract, whereas five noncarcinogenic chemicals altered recombination by no more than 1.7-fold (Li et al. 1997).

Recombination can be elevated by a variety of agents — chemical carcinogens, radiation and oncogenic viruses (Radman et al. 1982; Sengstag 1994; Galli and Schiestl 1995; Cheng et al. 1997; Li et al. 1997) — which might elicit an increased abundance of substrates, and/or of enzymes mediating recombination. Substrate-induction of recombination pathways has yet to be demonstrated in higher eukaryotes, although it is well established in microbial systems (West 1994). The capacity for such induction is suggested by our observation of high *Rad51* transcript levels in immortal and pre-immortal human cells (Xia et al. 1997). Activation of recombinational pathways could serve to protect chromosomal structure, or may result in an increased number of chromatid breaks due to initial endonuclease activity. Although multiple mechanisms may be involved, the very high concordance between recombinogenesis and oncogenesis by genotoxic treatments (Galli and Schiestl 1995; Cheng *et al.* 1997; Li *et al.* 1997) would be difficult to explain without a *functional relationship* between the two processes.

4. Does Elevated Recombination Expedite Cell Immortalization and Cancer?

Cell immortality. Normal diploid cells decline progressively in cell cycling rate with increasing passage level, effectively limiting their *in vitro* life spans, whereas established cell lines are able to divide indefinitely and are thus termed "immortal". Transformation by a DNA tumor virus, such as SV40, papillomavirus, Epstein-Barr virus, or adenovirus (specific for the cell type affected), facilitates but is not generally sufficient for cell immortalization (Khoobyarian and Marczynska 1993). High levels of chromosomal aberration, characteristic of essentially all immortal cell lines (Hayflick 1977; Solomon *et al.* 1991; Pathak *et al.* 1994), are apparent soon after transformation (Stewart and Bacchetti 1991; Ray *et al.* 1990), although abnormal chromosomes are seen only rarely in mortal cells, despite many generations of growth selection *in vivo* and in culture (Shmookler Reis and Goldstein 1980; Srivastava *et al.* 1985; Tlsty 1990). Immortally-transformed human cell lines support increased amounts of plasmid homologous recombination (Finn *et al.* 1989; Xia *et al.* 1997) and also express elevated transcript levels for the *HsRAD51* recombinase compared to mortal cell strains (Xia *et al.* 1997). This striking association, between cell immortality and increased levels of both recombination and *HsRAD51* transcripts, implicates homologous recombination in the process leading to immortalization. Chromosomal homologous recombination is also elevated in extended-life-span, *mortal* clones of human fibroblasts upon transformation by, and dependent on expression of, SV40 large T antigen (Cheng *et al.* 1997; Shammas *et al.* 1997). Immortal subclones arise rarely ($<10^{-8}$ per cell generation) from such clones, presumably following subsequent mutations (Cheng *et al.* 1997). *Thus, recombinational activation precedes, and may predispose cells toward, escape from cellular senescence.*

Carcinogenesis. Tumorigenicity of cancer cell lines is highly correlated with their karyotypic instability, with the most tumorigenic cell lines displaying "new" chromosomes in each cell examined (Wiener *et al.* 1976; Gee and Harris 1979) — implying a role of genomic rearrangement in cancer etiology. Since genetic recombination has been implicated in chromosomal translocation (Cheng *et al.*, 1997; Shammas *et al.* 1997; Honma *et al.* 1997), loss of heterozygosity (Honma *et al.* 1997), and gene amplification (Windle *et al.* 1991), it may underlie a variety of chromosomal abnormalities observed with high incidence in many neoplasias. Loss of heterozygosity at loci encoding tumor suppressor genes is associated with many cancers (*e.g.,* Takita *et al.* 1997; Honma *et al.* 1997; Orntoft and Wolf 1998; Lee and Testa 1999), and is believed to occur by somatic gene conversion rendering an intact anti-oncogene identical to its defective homolog. Of course, the converse outcome is equally probable, but would have no phenotypic consequences.

Site-specific recombination may occasionally utilize incorrect target sequences, causing activation of oncogenes (Bishop, 1987) or deletion of functional alleles of tumor suppressor genes (Weinberg, 1995). These must be exceptional outcomes, even among the products of errant (not-so-site-specific) recombination, but are brought to our attention by the enhanced cell proliferation they elicit. This selection process vitiates any inferences regarding the frequencies of rare initiating or contributory events. The sequences at rearrangement junctions in a variety of lymphoblastoid cancers can implicate aberrant site-specific recombination (Bergsagel et al., 1996), but also homologous recombination (Super et al., 1997), and illegitimate recombination not involving site-specific signals (Super et al. 1997; Honma et al. 1997; Zucman-Rossi et al. 1998) in oncogenesis. The relative frequencies of these processes cannot be inferred, but must reflect the sequences in and around specific oncogenes and tumor-suppressor genes characteristically mutated in each cancer type, and perhaps the ensuing cell-growth advantage conferred.

Cancer is a multi-step process (Farber 1984; Weinberg 1988, 1995) wherein a normal cell acquires, by mutation, a variety of characteristics in progressing to hyperplasia, benign transformation, and malignancy. Most human tumors carry at least four independent mutations (Neiman and Hartwell 1991), and colorectal carcinoma is thought to require seven or more genetic events (reviewed in Tsancheva 1997). In the latter group of cancers, mutations in c-k-ras, c-myc, APC, MCC, DCC, and p53 genes are found very frequently. The most common subclass, hereditary nonpolyposis colorectal cancer, is associated with mutations in multiple DNA mismatch-correction and repair genes, including hMSH2, hMLH1, hPMS1, hPMS2, and hMSH6/GTBP (Lynch and Lynch 1998). Frequent mutations in p53 (Yamada et al. 1991), APC (Horii et al. 1992), and c-K-ras (Miki et al. 1991) have also been found in gastric cancer. Mutations to the p53 gene are probably the most common in human cancers (Hollstein et al. 1991; Suzuki et al. 1992), and may themselves be recombinogenic (Mekeel et al. 1997; Chang et al. 1997; Shammas et al. 1997).

Such successions of phenotypic changes could occur through a series of mutations and clonal expansions, but are normally impeded by the limited proliferative capacity of diploid somatic cells. In this sense, cellular senescence may serve as an anti-oncogenic defense, by limiting the "window of opportunity" for accrual of mutations jointly required for cancer. Both cellular escape from senescence, and subsequent selective progression of the tumor, would be rendered more likely and more rapid following the acquisition of a "hyper-mutator" phenotype (Finn et al. 1989; Neiman and Hartwell 1991; Xia et al., 1997; Loeb 1997, 1998; Jackson and Loeb 1998). Clinical and molecular studies lend support to this argument. In particular, genetic defects known to increase the rate of mutation are associated with high risk and early incidence of cancer (German 1980; Heddle 1991; Cheng et al. 1991; Digweed 1994). Examples include Xeroderma pigmentosum, comprising defects in seven excision repair genes (reviewed in Bootsma et al. 1995); hereditary non-polyposis colorectal cancer, defective in any of several mismatch repair genes (Nicolaides et al. 1994); Ataxia telangiectasia, in which mutations to a regulatory kinase lead to increases in both homologous and site-specific recombination (Savitsky et al. 1995); and Werner (Gray et al. 1997), Cockayne (van Gool et al. 1994) and Bloom syndromes (Ellis et al. 1995; Karow et al. 1997), each traced to defects in a putative helicase gene.

In the latter four syndromes, genetic mutability is detected by high rates of chromosomal aberration, and is associated with increased rates of homologous recombination in both Werner syndrome (Cheng et al. 1991) and Ataxia telangiectasia (Meyn 1993). Karyotypic alteration has also been implicated in many specific cancer types, such as gliomas (Kruse et al. 1998), retinoblastoma (Cavenee et al. 1983) and Burkitt's lymphoma (Leder *et al.* 1983). Genetic instability is seen on a much finer scale in those cancers featuring defects in nucleotide excision repair or mismatch repair, but these defects are not evident in the great majority of cancers which instead display chromosomal rearrangement (Lengauer *et al.* 1997, 1998). One interpretation would be that most cancers develop from hypermutable progenitors, which can arise by several alternative mechanisms. Apart from the chromosome-breakage syndromes, there are direct data indicating that cancer cells have increased frequency of karyotypic instability (Weiner *et al.* 1976; Gee and Harris 1979; Lengauer *et al.* 1997) and increased rates of loss for marker heterozygosity (Vogelstein *et al.* 1989; Phear *et al.* 1996). Observations that human cells in culture, soon after SV40-T antigen transformation but prior to immortalization, undergo abrupt increases in recombination (Cheng *et al.* 1997; Xia *et al.* 1997) and karyotypic abnormality (Ray *et al.* 1990; Stewart and Bacchetti 1991), suggest that chromosomal instability may also be a characteristic of precancerous cells which ultimately give rise to chromosomally-unstable tumors.

5. Telomeres, Cell Replication Potential, and DNA Damage

Telomeres, specialized nucleoprotein structures at the ends of chromosomes, contribute to genomic integrity by protecting genomic DNA from degradation and end-to-end joining of chromosomes (Day *et al.* 1993). The linear DNA duplex of each chromosome terminates at both ends in telomeric DNA — tandem arrays comprising hundreds to thousands of short oligonucleotide repeats (GGGTTA in the vertebrates), with a guanosine (G)-rich strand running 5' to 3' toward the terminus, extending beyond a cytosine (C)-rich complementary strand (Zakian 1989). The G-rich strand of telomere DNA is extended by telomerase, a ribonucleoprotein with reverse transcriptase activity. The RNA component of telomerase serves as the template for the addition of short G-rich repeats at the 3' end of telomeric DNA (Blackburn 1992).

Human germ-line cells possess telomerase activity and maintain telomere length (Kim *et al.* 1994), whereas somatic tissues can have low levels of telomerase or (more commonly) no detectable activity (Counter *et al.* 1995; Shay 1997), and their telomeres undergo a progressive shortening as a function of cumulative cell replication (Harley *et al.* 1990; Hastie *et al.* 1990). Telomerases are re-activated in most cancers (Shay and Bacchetti 1997) and immortalized cells (Counter *et al.* 1992; Avilion *et al.* 1996; Shay and Bacchetti 1997; Shay and Wright 1996), probably due to somatic mutations or epigenetic changes (*e.g.*, demethylations). However a subset of tumor and immortalized cells lack telomerase activity (Mayne *et al.* 1986; Wright *et al.* 1989; Wright and Shay 1992; Counter *et al.* 1995; Bryan *et al.* 1995), and typically but not invariably have very long telomeres (Bryan *et al.* 1995; Xia *et al.* 1996; Gollahon *et al.* 1998).

Transgenic expression of the human telomerase reverse transcriptase, hTERT, within normal human fibroblasts and retinal epithelial cells in culture, greatly extends their replicative life-spans (Bodnar *et al.* 1998), without inducing changes characteristic of malignant cells (Jiang *et al.* 1999). These observations establish the important point that deficiency of telomerase alone, and more precisely of the telomerase catalytic subunit, limits replicative potential in cultured human cells. The absence of "malignant" changes in hTERT-expressing cells is not surprising (from our perspective), since these cells have skipped the predisposing hyper-mutator stage of transformation, having instead directly "corrected" the telomerase deficiency which imposes a replicative limit on normal somatic cells. Among the phenotypes examined, karyotypic abnormality is an expected consequence of hyper-recombination, as discussed above. Other traits sought but lacking in hTERT-transformed cells — loss of contact inhibition, growth independent of serum factors and substrate anchorage, changes in status and response of anti-oncogenes and oncogenes, etc. — are phenotypes acquired variably and sporadically by neoplastic cells over extended periods of selection. The cells studied by Jiang *et al.* may not have traversed sufficient generations, in an appropriately selective environment, for the emergence of these "cancer-progression" traits. It is also possible that such traits *would* have arisen under these conditions, but only within a hyper-mutable cell. Distinction between these two alternatives has important consequences for the use of hTERT-transformed human cells for autologous or heterologous tissue replacement, since the safety of these procedures hinges on the question of whether such cells may eventually give rise to tumors.

Considerable effort has been devoted to the converse experiment — attempts to reverse the immortal state of transformed cell lines through inhibition of telomerase. Since most somatic cells lack telomerase activity anyway, it is plausible that the effects of telomerase inhibitors may be restricted to neoplasias, gametes and embryos, and therefore would be relatively innocuous to the patient. The caveat which should not be forgotten is that low-level telomerase activity detected in lymphocytes and bone marrow cells may have significance to normal immune function, since recurrently stimulated B or T cell lineages might otherwise exhaust their replicative capacity. Although a number of compounds have been reported to be effective telomerase inhibitors *in vitro* (Norton *et al.* 1996; Fletcher *et al.* 1996; Pitts and Corey 1998; Pai *et al.* 1998; Fedoroff *et al.* 1998; Wheelhouse *et al.* 1998), such studies by their nature do not address either cell toxicity of the treatment, or specificity of effects to telomerase alone. Moreover, for essentially all exogenous treatments, it has proven elusive to demonstrate *in vivo* reduction of telomere length, or of replicative potential, in treated cells. Several clones derived from two human cell lines, expressing a transgenic sequence complementary to full-length telomerase RNA, were reported to have reduced cell proliferation and shortened telomeres (Feng *et al.* 1995; Kondo *et al.* 1998). We have recently found that cellular immortality can also be reversed upon uptake into cultured human cells of peptide nucleic acids (PNAs) complementary to and avidly binding the templating region of telomerase RNA, whereas PNAs of altered sequence were much less effective (Shammas *et al.*, 1999).

Embryonic cells derived from knockout mice lacking the RNA component of telomerase, at generations one through six, are unimpaired in tumorigenic potential following transformation by oncogenes (SV40 T antigen and/or activated *ras*) and injection into immune-deficient *nude* mice (Blasco *et al.* 1997), whereas highly-proliferative normal tissues did exhaust their replicative potential by the sixth generation (Lee *et al.* 1997). This paradox may be reconciled if oncogene transformation generated *hyper-recombinant* tumor cells, which were then able to maintain telomeric arrays by inter-telomere recombination.

Telomere shortening may lead to chromosome loss or telomere fusion (Hastie and Allshire 1989), either of which can be a cell-lethal event. The regulation and maintenance of telomeric DNA length appears to involve telomere-specific binding proteins such as TRF1 and TRF2 (van Steensel and de Lange 1997; van Steensel et al. 1998; Broccoli et al. 1997; Griffith et al. 1998; reviewed in Smith and de Lange 1997). TRF2, for which the gene was cloned recently, may serve to protect chromosomes from end-to-end fusions (van Steensel et al. 1998), probably by enhancing formation or stability of D-loops at telomere termini (Griffith et al. 1999). Telomeric fusion, which is rarely seen in normal human cells, occurs frequently in cancer cells (Dhaliwal et al. 1994; Fitzgerald and Morris 1984; Hastie et al. 1990), and in several genetic disorders characterized by hypermutability [Xeroderma pigmentosum] and/or elevated recombination [Ataxia telangiectasia, Fanconi anaemia, and Bloom syndrome] (Digweed 1993; Kojis et al. 1989, 1991). The connection may be that telomeres below a critical length fail to bind TRF proteins, and would then resemble ragged DNA ends arising from double-strand breaks. "Repair" of two such telomeres via homologous recombination could then create telomere-telomere fusions.

Alternatively, failure to arrest cell cycling in response to telomere damage may permit telomeric fusions which would otherwise be avoided. Telomeric instability is a feature of genetic lesions in a gene family called the phosphoinositol kinase- (PIK-) related kinases (Keith and Schreiber 1995; Xia et al., 1996), which include the ATM gene mutated in Ataxia telangiectasia (Savitsky et al. 1995). Their gene products are involved in response to DNA damage through cell-cycle checkpoints, and have pronounced effects on meiotic and/or mitotic recombination levels (Keith and Schreiber 1995; Cimprich et al. 1996). Although the pathways are not well understood, the conjunction of single-gene effects on telomere array maintenance, DNA damage control, and the regulation of recombination, demonstrate a fundamental association between these processes in our cellular machinery. As noted above, DNA-damage response pathways arrest cell cycling in the presence of unrepaired lesions such as cross-links or double-strand breaks, which could otherwise lead to chromosome breakage or translocation, while other checkpoints may serve to prevent aberrant segregation of homologues when spindles or centrosomes are defective (Lengauer et al. 1998). Abrogation of either type of checkpoint could allow telomeric fusions and chromosome translocations, or aneuploidy, respectively. Nevertheless, the reported interactions among p53, SV40 T antigen, and HsRad51 (see Xia et al. 1997, and references therein) imply a more complex picture than increased recombination and telomeric fusion in checkpoint-defective cells arising solely as a byproduct of cell-cycle arrest failure.

6. Is Telomeric Recombination an Alternative Means to Justify the Ends?

Alternative mechanisms clearly must exist for the maintenance, or episodic expansion, of telomere length in telomerase-negative immortal cells. Although the process remains to be defined for mammalian cell lines, in yeast the regeneration of telomeric arrays can occur via a recombination mechanism apparently involving gene conversion (Wang and Zakian 1990). That a comparable pathway exists in mammals is suggested by detection of extrachromosomal telomeric DNA fragments in telomerase-negative immortal human cell lines, but not in telomerase-proficient lines or in normal human fibroblasts (Ogino *et al.* 1998). Such fragments provide circumstantial evidence of *intra*-telomere recombination, whereas *inter*-telomere recombination would also have to occur in order to extend telomeres and hence replicative potential. Note that this sort of exchange is not a zero-sum game: proliferative potential is likely to be limited by the shortest telomeres present in a cell, and homologous recombination with a longer telomere could add many generations to replicative potential, and hence the ability to predominate in a cell population.

Immortalization of human fibroblasts, whether telomerase-dependent or independent, is associated with mutation or loss of p53 genes (Gollahon *et al.* 1998). Since loss of p53 function is also implicated in both cell transformation and in the induction of homologous recombination (Xia *et al.* 1997; Mekeel *et al.* 1997), the higher levels of homologous recombination observed in immortal cells (Finn *et al.* 1989; Xia *et al.* 1997) are not entirely surprising. However, recombination between telomeres remains to be demonstrated as a mechanism for telomere extension in telomerase-negative immortal cell lines.

If telomere-telomere recombination can, like telomerase reactivation, extend the replicative life span of mammalian cells, it may provide an elusive link in the causal chain between elevated recombination and cancer. Although it seems plausible that hypermutability (including recombination) might precede cell transformation to immortality, a process requiring multiple mutations, this is an *a posteriori* argument. *It has not been obvious what the basis would be for selection favoring such hypermutable cells prior to oncogene activation or anti-oncogene inactivation.* Lacking such selection, or facing adverse selection, hypermutable cells might never accumulate in sufficient numbers to support complete transformation. As telomerase-negative diploid cells approach the end of their replicative life spans, however, they can avoid the consequences of telomere shortening either by reactivating telomerase or by inter-telomere recombination (or gene conversion). In the latter case, cell selection would directly favor those cells with the highest recombination rates. Once elevated, recombination could also provide new variants which may be selected for other properties, thereby fueling completion of carcinogenesis and tumor progression.

Acknowledgments

This work was supported by grants (to RJSR) from the U.S. Dept. of Veterans Affairs (Merit Review 001) and from the National Institutes of Health (AG-09413 and AG-13918).

References

Almasan, A, Linke, SP, Paulson, TG, Huang, L, and Wahl, GM: Genetic instability as a consequence of inappropriate entry into and progression through S-phase. Cancer Metas. Rev., 14: 59-73, 1995.

Amstutz, H., Munz, P, Heyer, WD, Leupold, U, Kohli, J: Concerted evolution of tRNA genes: intergenic conversion among three unlinked serine tRNA genes in S. pombe. Cell, 40: 879-886, 1985.

Ashby, J, and Tennant, R.W.: Chemical structure, Salmonella mutagenicity and extent of carcinogenicity as indicators of genotoxic carcinogenesis among 222 chemicals tested in rodents by the US NCI/NTP. Mutat.Res., 204: 17-115, 1988.

Avilion, AA, Piatyszek, MA, Gupta, J, Shay, JW, Bacchetti, S, and Greider, CW: Human telomerase RNA and telomerase activity in immortal cell lines and tumor tissues. Cancer Research, 56: 645-650, 1996.

Ayares D, Cherkuri L, Song K-Y, and Kucherlapati R. 1986. Sequence homology requirements for intermolecular recombination in mammalian cells. Proc. Natl. Acad. Sci. USA 83:5199-5203.

Baker, BS, Carpenter, TC, Esposito, MS, Esposito, RE, and Sandler, L: The genetic control of meiosis. Ann. Rev. Genet., 10: 53-134, 1976.

Banin, S, Moyal, L, Shieh, S, Taya, Y, Anderson, CW, Chessa, L, Smorodinsky, NI, Prives, C, Reiss, Y, Shiloh, Y, Ziv, Y: Enhanced phosphorylation of p53 by ATM in response to DNA damage. Science, 281: 1674-1677, 1998.

Bergsagel PL, Chesi M, Nardini E, Brents LA, Kirby SL, Kuehl WM Promiscuous translocations into immunoglobulin heavy chain switch regions in multiple myeloma. Proc Natl Acad Sci USA 93: 13931-13936, 1996.

Bishop, JM: The molecular genetics of cancer. Science, 253: 305-311, 1987.

Blackburn, EH: Telomerases. Ann. Rev. Biochem., 61: 113-129, 1992.

Blasco MA, Lee HW, Hande MP, Samper E, Lansdorp PM, DePinho RA, Greider CW: Telomere shortening and tumor formation by mouse cells lacking telomerase RNA. Cell 91: 25-34, 1997.

Bodnar AG, Ouellette M, Frolkis M, Holt SE, Chiu CP, Morin GB, Harley CB, Shay JW, Lichtsteiner S, and Wright WE: Extension of life-span by introduction of telomerase into normal human cells. Science, 279: 349-352, 1998.

Bootsma, D, Weeda, G, Vermeulen, W, van Vuuren, H, Troelstra, C, van der Spek, P, Hoeijmakers, J: Nucleotide excision repair syndromes: molecular basis and clinical symptoms. Phil. Trans. Roy. Soc. London B, 347: 75-81, 1995.

Broccoli, D, Smogorzewska, A, Chong, L, de Lange, T: Human telomeres contain two distinct Myb-related proteins, TRF1 and TRF2. Nature Genetics, 17: 231-235, 1997.

Bryan, TM, Englezou, A, Gupta, J, Bacchetti, S, Reddel, RR: Telomere elongation in immortal human cells without detectable telomerase activity. EMBO J., 14: 4240-4248, 1995.

Busser, MT, and Lutz, WK: Stimulation of DNA synthesis in rat and mouse liver by various tumor promoters. Carcinogenesis, 8: 1433-1437, 1987.

Canman, CE, Lim, DS, Cimprich, KA, Taya, Y, Tamai, K, Sakaguchi, K, Appella, E, Kastan, MB, Siliciano, JD: Activation of the ATM kinase by ionizing radiation and phosphorylation of p53. Science, 281: 1677-1679, 1998.

Cavenee, WK, Dryja, TP, Phillips, RA, Benedict, WF, Godbout, R, Gallie, BL, Murphree, AL, Strong, LC, and White, RL: Expression of recessive alleles by chromosomal mechanisms in retinoblastoma. Nature, 305: 779-784, 1983.

Cheng, RZ, Murano, S, Kurz, BW, and Shmookler Reis RJ: Homologous recombination is elevated in some Werner-like syndromes but not during normal in vitro or in vivo senescence of human cells. Mutat. Res., 237: 259-269, 1991.

Cheng, RZ, Shammas, MA, Li, J, and Shmookler Reis, RJ: Expression of SV40 large T antigen stimulates reversion of a chromosomal gene duplication in human cells. Exp. Cell Res., 234: 300-312, 1997.

Cimprich KA, Shin TB, Keith CT, and Schreiber SL: cDNA cloning and gene mapping of a candidate human cell cycle checkpoint protein. Proc. Natl. Acad. Sci. USA 93: 2850- 2855, 1996.

Counter, CM, Avilion, AA, LeFeuvre CE, Stewart, NG, Greider, CW, Harley, CB, and Bacchetti, S: Telomere shortening associated with chromosome instability is arrested in immortal cells which express telomerase activity. EMBO, J., 11: 1921-1929, 1992.

Counter CM, Gupta J, Harley CB, Leber B, and Bacchetti S: Telomerase activity in normal leukocytes and in hematologic malignancies. Blood, 85: 2315-2320, 1995.

Davis, MM, Calame, K, Early PW, Livant, DL, Joho, R, Weissman, IL, and Hood, L: An immunoglobulin heavy chain is formed by two recombinational events. Nature, 283: 733-739, 1980.

Day, JP, Marder, BA, Morgan, WF: Telomeres and their possible role in chromosome stabilization. Environ. Mol. Mutagen., 22: 245-249, 1993.

Dhaliwal, MK, Satya-Prakash, KL, Davis, PC, and Pathak, S: High frequency of telomeric association in a family with multiple congenital neoplasia. In Vivo, 8: 1023-1026, 1994.

Digweed, M: Human genetic instability syndromes: single gene defects with increased risk of cancer. Toxicol. Lett., 67: 259-281, 1993.

Early, P, Huang, H, Davis, M, Calame, K, and Hood, L: An immunoglobulin heavy chain variable region gene is generated from three segments of DNA: V_H, D, and J_H. Cell, 19: 981-992, 1980.

El-Deiry WS, Harper, JW, O'Connor, PM, Velculescu, VE, Canman, CE, Jackman, J, Pietenpol, JA, Burrell, M, Hill, DE, Wang, Y, Wiman, KG, Mercer, WE, Kastan, MB, Kohn, KW, Elledge, SJ, Kinzler, KW, and Vogelstein, B: WAF1/CIP1 is induced in p53-mediated G1 arrest and apoptosis. Cancer Res., 54: 1169-1174, 1994.

Ellis, NA, Groden, J, Ye, TZ, Straughen, J, Lennon, DJ, Ciocci, S, Proytcheva, M, German, J: The Bloom's syndrome gene product is homologous to RecQ helicases. Cell, 83: 655-666, 1995.

Farber, E: The multistep nature of cancer development. Cancer Res., 44: 4217-4223, 1984.

Fedoroff, OY, Salazar, M, Han, H, Chemeris, VV, Kerwin, SM, and Hurley, LH: NMR-based model of a telomerase-inhibiting compound bound to G-quadruplex DNA. Biochemistry, 37:12367-12374, 1998.

Feng, J, Funk, WD, Wang, S-S, Weinrich, SL, Avilion, AA, Chiu, CP, Adams, RR, Chang, E, Allsopp, RC, Yu, J, et al.: The RNA component of human telomerase. Science, 269: 1236-1241, 1995.

Feunteun, J: Breast Cancer and genetic instability: the molecules behind the scenes. Mol. Med. Today, 4: 263-267, 1998.

Finn, GK, Kurz, BW, Cheng, RZ, and Shmookler Reis, RJ: Homologous plasmid recombination is elevated in immortally transformed cells. Mol. Cell. Biol., 9: 4009-4017, 1989.

Fitzgerald PH, Morris CM: Telomeric association of chromosomes in B-cell lymphoid leukemia. Hum Genet. 67: 385-390, 1984.

Fletcher, TM, Salazar, M, and Chen, S-F: Human telomerase inhibition by 7-deaza-2'-deoxypurine nucleoside triphosphates. Biochemistry, 35: 15611-15617, 1996.

Folger, KR, Thomas, K, and Capecchi, MR: Nonreciprocal exchanges of information between DNA duplexes coinjected into mammalian cell nuclei. Mol. Cell. Biol., 5: 59-69, 1985.

Fornace, AJ, Jr., Nebert, DW, Hollander, MC, Luethy, JD, Papathanasiou, M, Fargnoli, J, and Holbrook, NJ: Mammalian genes coordinately regulated by growth arrest signals and DNA-damaging agents. Mol. Cell. Biol., 9: 4196-4203, 1989.

Friedberg, EC, Siede, W, Cooper, AJ: In: Broach J. R., Pringle, J. R., and Jones, E. W. (eds): The molecular and cellular biology of the yeast Saccharomyces: genome dynamics, protein synthesis and energetics, vol. 1, Cold Spring Harbor Laboratory, Cold Spring Harbor, New York, pp. 147-192, 1991.

Galli, A, and Schiestl, RH: On the mechanism of UV and gamma-ray-induced intrachromosomal recombination in yeast cells synchronized in different stages of the cell cycle. Mol. Gen. Genet., 248: 301-310, 1995.

Gee CJ, and Harris H: Tumorigenicity of cells transformed by Simian virus 40 and of hybrids between such cells and normal diploid cells. J. Cell Sci., 36: 223-240, 1979.

German J: Chromosome-breakage syndromes: different genes, different treatments, different cancers. Basic Life Sci. 15:429-439, 1980.

Gollahon, LS, Kraus, E, Wu, TA, Yim, SO, Strong, LC, Shay, JW, and Tainsky, MA: Telomerase activity during spontaneous immortalization of Li-Fraumeni syndrome skin fibroblasts. Oncogene 17: 709-717, 1998.

Gray, MD, Shen, JC, Kamath-Loeb, AS, Blank, A, Sopher, BL, Martin, GM, Oshima, J, and Loeb, LA: The Werner syndrome protein is a DNA helicase. Nature Genetics, 17: 100-103, 1997.

Griffith, J, Bianchi, A, de Lange, T: TRF1 promotes parallel pairing of telomeric tracts in vitro. J. Mol. Biol., 278: 79-88, 1998.

Griffith, JD, Comeau, L, Rosenfield, S, Stansel, RM, Bianchi, A, Moss, H, and de Lange, T: Mammalian telomeres end in a large duplex loop. Cell, 97: 503-514, 1999.

Harley, CB, Futcher, AB, Greider, CW: Telomeres shorten during ageing of human fibroblasts. Nature, 345: 458-460, 1990.

Hastie, ND, and Allshire, RC: Human telomeres: fusion and interstitial sites. Trends Genet., 5: 326-331, 1989.

Hastie, ND, Dempster, M, Dunlop, MG, Thompson, AM, Green, DK, and Allshire, RC: Telomere reduction in human colorectal carcinoma and with ageing. Nature, 346: 866-868, 1990.

Hayflick, L: The cellular basis for biological aging, in: Handbook of the Biology of Aging (Finch, CE, and Hayflick, L, eds.) Van Nostrand Reinhold Co., N.Y., pp. 159-186, 1977.

Heddle JA: Implications for genetic toxicology of the chromosomal breakage syndromes. Mutat. Res. 247:221-229, 1991.

Holliday, R: A mechanism for gene conversion in fungi. Genet. Res., 5: 282-304, 1964.

Hollstein, M, Sidransky, D, Vogelstein, B, and Harris, CC: p53 mutations in human cancers. Science, 253:49-53, 1991.

Honma, M, Zhang, LS, Hayashi, M, Takeshita, K, Nakagawa, Y, Tanaka, N, Sofuni, T. Illegitimate recombination leading to allelic loss and unbalanced translocation in p53-mutated human lymphoblastoid cells. Mol. Cell. Biology 17: 4774-4781, 1997.

Horii, A, Nakatsuru, S, Miyoshi, Y, Ichii, S, Nagase, H, Kato, Y, Yanagisawa, A, and Nakamura, Y: The APC gene, responsible for familial adenomatous polyposis, is mutated in human gastric cancer. Cancer Res., 52: 3231-3233, 1992.

Jackson, AL, Loeb, LA: The mutation rate and cancer. Genetics, 148: 1483-1490, 1998.

Karow, JK, Chakraverty, RK, and Hickson, ID: The Bloom's syndrome gene product is a 3'-5' DNA helicase. J. Biol. Chem., 272: 30611-30614, 1997.

Kastan, MB, Onyekwere, O, Sidransky, D, Vogelstein, B, Craig, R.W.: Participation of p53 protein in the cellular response to DNA damage. Cancer Res., 51: 6304-6311, 1991.

Kastan, MB, Zhan, Q, El-Deiry, WS, Carrier, F, Jacks, T, Walsh, WV, Plunkett, BS, Vogelstein, B, and Fornace, Jr., AJ: A mammalian cell cycle checkpoint pathway utilizing p53 and GADD45 is defective in ataxia-telangiectasia. Cell, 71: 587-597, 1992.

Keith CT, and Schreiber SL: PIK-related kinases: DNA repair, recombination, and cell cycle checkpoints. Science, 270: 50-51, 1995.

Khanna, KK, and Lavin, MF: Ionizing radiation and UV induction of p53 protein by different pathways in ataxia-telangiectasia cells. Oncogene, 8: 3307-3312, 1993.

79.   Khoobyarian, N, and Marczynska, B: Cell immortalization: the role of viral genes and carcinogens. Virus Res., 30: 113-128, 1993.

Kim, NW, Piatyszek, MA, Prowse, KR, Harley, CB, West, MD, Ho, PL, Coviello, GM, Wright, WE, Weinrich, SL, Shay, JW: Specific association of human telomerase activity with immortal cells and cancer. Science, 266: 2011-2015, 1994.

Kojis, TL, Schreck, RR, Gatti, RA, and Sparkes, RS: Tissue specificity of chromosomal rearrangements in Ataxia telangiectasia. Hum. Genet., 83: 347-352, 1989.

Kojis, TL, Gatti, RA, and Sparkes, RS: The cytogenetics of Ataxia telangiectasia. Cancer Genet. Cytogenet., 56: 143-156, 1991.

Kolodner, R: Biochemistry and genetics of eukaryotic mismatch repair. Genes Dev. 10: 1433-1442, 1996.

Kondo, S, Tanaka, Y, Kondo, Y, Hitomi, M, Barnett, GH, Ishizaka, Y, Liu, J, Haqqi, T, Nishiyama, A, Villeponteau, B, Cowell, JK, and Barna, BP: Antisense telomerase treatment: induction of two distinct pathways, apoptosis and differentiation. FASEB J, 12: 801-811, 1998.

Kornberg, A, and Baker, T: DNA Replication, 2nd edition, 1992; W.H. Freeman & Co.

Kruse, CA, Varella-Garcia, M, Kleinschmidt-Demasters, BK, Owens, GC, Spector, EB, Fakhrai H, Savelieva, E, Liang, BC: Receptor expression, cytogenetic, and molecular analysis of six continuous human glioma cell lines. In Vitro Cell. Devel. Biol. Animal. 34: 455-62, 1998.

Kuerbitz, SJ, Plunkett, BS, Walsh, WV, and Kastan, MB: Wild type p53 is a cell cycle check point determinant following irradiation. Proc. Natl. Acad. Sc. USA, 89: 7491-7495, 1992.

Larsen, CJ: The BCL2 gene, prototype of a gene family that controls programmed cell death (apoptosis). Annales de Genetique, 37: 121-134, 1994.

Leder, P, Battey, J, Lenoir, G, Moulding, C, Murphy, W, Potter, H, Stewart, T, Taub, R: Translocations among antibody genes in human cancer. Science, 222: 765-771, 1983.

Lee HW, Blasco MA, Gottlieb GJ, Horner JW 2nd, Greider CW, and DePinho RA: Essential role of mouse telomerase in highly proliferative organs. Nature, 392: 569-574, 1998.

Lee WC, and Testa JR: Somatic genetic alterations in human malignant mesothelioma. Int. J. Oncol., 14: 181-188, 1999.

Lengauer, C, Kinzler, KW, and Vogelstein, B: Genetic instabilities in colorectal cancers. Nature, 386: 623-627, 1997.

Lengauer, C, Kinzler, KW, and Vogelstein, B: Genetic instabilities in human cancers. Nature, 396: 643-649, 1998.

Li, J, Ayyadevara, R, and Shmookler Reis, RJ: Carcinogens stimulate intrachromosomal homologous recombination at an endogenous locus in human diploid fibroblasts. Mutation Res., 385: 173-193, 1997.

Liskay, RM, Letsou, A, and Stachelek, JL. Homology requirement for efficient gene conversion between duplicated chromosomal sequences in mammalian cells. Genetics, 115:161-167, 1987.

Loeb, LA: Transient expression of a mutator phenotype in cancer cells. Science, 277: 1449-1450, 1997.

Loeb, LA: Cancer cells exhibit a mutator phenotype. Adv Cancer Res., 72: 25-56, 1998.

Loft, S, and Poulsen, HE: Cancer risk and oxidative DNA damage in man. J Molec Med, 74: 297-312, 1996.

Lu, X, Lane, DP: Differential induction of transcriptionally active p53 following UV or ionizing radiation: defects in chromosome instability syndromes. Cell, 75:765-778, 1993.

Luisi-Deluca, C, Porter, RD, and Taylor, WD: Stimulation of recombination between homologous sequences on plasmid DNA and chromosomal DNA in Escherichia coli by N-acetoxy-2-acetyl-aminofluorene. Proc. Natl. Acad. Sci. USA, 81: 2831-2835, 1984.

Lynch, HT, and Lynch, JF: Genetics of colonic cancer. Digestion, 59: 481-492, 1998.

Mansour SL, Thomas KR, and Capecchi MR: Disruption of the proto-oncogene int-2 in mouse embryo-derived stem cells: a general strategy for targeting mutations to non-selectable genes. Nature, 336:348-352, 1988.

Mansour SL, Goddard JM, and Capecchi MR: Mice homozygous for a targeted disruption of the proto-oncogene int-2 have developmental defects in the tail and inner ear. Development, 117:13-28, 1993.

Mason, JM, Langenbach, R, Shelby, MD, Zeiger, E, and Tennant, R.W.: Ability of short-term tests to predict carcinogenesis in rodents. Annu. Rev. Pharmacol. Toxicol., 30: 149-168, 1990.

Mayne, LV, Priestley, A, James, MR, Burke, JF: Efficient immortalization and morphological transformation of human fibroblasts by transfection with SV40 DNA linked to a dominant marker. Exper. Cell Res., 162: 530-538, 1986.

Mekeel, KL, Tang, W, Kachnic, LA, Luo, CM, DeFrank, JS, and Powell, SN: Inactivation of p53 results in high rates of homologous recombination. Oncogene, 14: 1847-1857, 1997.

Meselson, MS, and Radding, CM: A general model for genetic recombination. Proc. Natl. Acad. Sciences USA, 72: 358-361, 1975.

Meyn, MS: High spontaneous intrachromosomal recombination rates in ataxia-telangiectasia. Science, 260: 1327-1330, 1993.

Miki, H, Ohmori, M, Perantoni, AO, and Enomoto, T: K-ras activation in gastric epithelial tumors in Japanese. Cancer Lett., 58: 107-113, 1991.

Morales CP, Holt SE, Ouellette M, Kaur KJ, Yan Y, Wilson KS, White MA, Wright WE, Shay JW: Absence of cancer-associated changes in human fibroblasts immortalized with telomerase. Nat Genet 21: 115-118, 1999.

Nakamura, Y: ATM: the p53 booster. Nature Medicine, 4: 1231-1232, 1998.

Nassif, N, and Engels, W: DNA homology requirements for mitotic gap repair in Drosophila. Proc. Natl. Acad. Sci. USA, 90: 1262-1266, 1993.

Neiman, PE, and Hartwell, LH: Malignant instability. New Biologist, 3: 347-351, 1991.

Nicolaides, NC, Papadopoulos, N, Liu, B, Wei, YF, Carter, KC, Ruben, SM, Rosen, CA, Haseltine, WA, Fleischmann, RD, Fraser, CM, et al. Mutations of two PMS homologues in hereditary nonpolyposis colon cancer. Nature, 371: 75-80, 1994.

Nikitin, AG, and Shmookler Reis, RJ: Role of transposable elements in age-related genomic instability. Genet. Res., Camb., 69: 183-195, 1997.

Norton, JC, Piatyszek, MA, Wright, WE, Shay, JW, and Corey, DR: Inhibition of human telomerase activity by peptide nucleic acids. Nature Biotech., 14: 615-619, 1996.

Oettinger, MA, Schatz, DA, Gorka, C, and Baltimore, D: RAG-1 and RAG-2, adjacent genes that synergistically activate V(D)J recombination. Science, 248: 1517-1523, 1990.

Ogino, H, Nakabayashi, K, Suzuki, M, Takahashi, E, Fujii, M, Suzuki, T, and Ayusawa, D: Release of telomeric DNA from chromosomes in immortal human cells lacking telomerase activity. Bioch. Biophy. Res. Comm., 248: 223-227, 1998.

Orntoft, TF, and Wolf, H: Molecular alterations in bladder cancer. Urol. Res. 26: 223-233, 1998.

Pai, RB, Pai, SB, Kukhanova, M, Dutschman, GE, Guo, X, and Cheng, YC: Telomerase from human leukemia cells: properties and its interaction with deoxynucleoside analogues. Cancer Res., 58: 1909-1913, 1998.

Pathak, S, Dave, BJ, and Gagos, S: Chromosome alterations in cancer development and apoptosis. In Vivo, 8: 843-850, 1994.

Phear, G, Bhattacharyya, NP, and Meuth, M. Loss of heterozygosity and base substitution at the APRT locus in mismatch-repair-proficient and -deficient colorectal carcinoma cells. Mol. Cell. Biol., 16: 6516-6523, 1996.

Pitts, AE, and Corey, DR: Inhibition of human telomerase by 2'-O-methyl-RNA. Proc. Natl. Acad. Sci. U.S.A., 95: 11549-11554, 1998.

Radman, M, Jeggo, P, and Wagner, R: Chromosomal rearrangement and carcinogenesis. Mut. Res., 98: 249-264, 1982.

Ray, FA, Peabody, DS, Cooper, JL, Cram, LS, and Kraemer, PM: SV40 T antigen alone drives karyotypic instability that precedes neoplastic transformation of human diploid fibroblasts. J. Cell. Biochem., 42: 13-31, 1990.

Ray, S, Anderson, ME, and Tegtmeyer, P: Differential interaction of temperature-sensitive SV40 T antigens with tumor suppressors pRb and p53. J. Virol., 70: 7224-7227, 1996.

Resnick, MA, Skaanild, M, and Nilsson, TT: Lack of DNA homology in a pair of divergent chromosomes greatly sensitizes them to loss by DNA damage. Proc. Natl. Acad. Sci. USA, 86: 2276-2280, 1989.

Richter, C, Park, JW, and Ames, BN: Normal oxidative damage to mitochondrial and nuclear DNA is extensive. Proc. Natl. Acad. Sci. USA, 85:6465-6467, 1988.

Roca, AI, and Cox, MM: The RecA protein: structure and function. Crit. Rev. Bioch. Mol. Biol., 25: 415-456, 1990.

Sakano, H, Maki, R, Kurosawa, Y, Roeder, W, and Tonegawa, S: Two types of somatic recombination are necessary for the generation of complete immunoglobulin heavy chain genes. Nature, 286: 676-683, 1980.

Savitsky, K, Bar-Shira, A, Gilad, S, Rotman, G, Ziv, Y, Vanagaite, L, Tagle, DA, Smith, S, Uziel, T, Sfez, S, Ashkenazi, M, Pecker, I, Frydman, M, Harnick, R, Patanjali, SR, Simmons, A, Clines, GA, Sartiel, A, Gatti, RA, Chessa, L, Sanal, O, Lavin, MF, Jaspers, NGJ, Taylor, MR, Arlett, CF, Miki, T, Weissman, SM, Lovett, M, Collins, FS, Shiloh, Y: A single ataxia telangiectasia gene with a product similar to PI-3 kinase. Science, 268: 1749-1753, 1995.

Schiestl, RH, Gietz, RD, Mehta, RD, and Hastings, PJ: Carcinogens induce intrachromosomal recombination in yeast. Carcinogenesis, 10: 1445-1455, 1989.

Scully, R, Chen, J, Plug, A, Xiao, Y, Weaver, D, Feunteun, J, Ashley, T, and Livingston, DM: Association of BRCA1 with Rad51 in mitotic and meiotic cells. Cell, 88: 265-275, 1997.

Sedivy, JM, and Sharp, PA: Positive genetic selection for gene disruption in mammalian cells by homologous recombination. Proc. Natl. Acad. Sci. USA, 86: 227-231, 1989.

Sengstag, C: The role of mitotic recombination in carcinogenesis. Crit. Rev. Toxicol., 24: 323-353, 1994.

Shammas, MA, Xia, SJ, and Shmookler Reis, RJ: Induction of duplication reversion in human fibroblasts, by wild-type and mutated SV40 T antigen, covaries with the ability to induce host DNA synthesis. Genetics, 146: 1417-1428, 1997.

Shammas, MA, Simmons C, Corey DR, and Shmookler Reis, RJ: Telomerase inhibition by peptide nucleic acids reverses "immortality" of transformed human cells. Oncogene Oncogene 46: 6191-6200, 1999.

Shay, JW: Telomerase in human development and cancer. J. Cell. Physiol., 173: 266-270, 1997.

Shay, JW, and Bacchetti, S: A survey of telomerase activity in human cancer. Euro. J. Cancer, 33: 787-791, 1997.

Shay, JW, and Wright, WE: The reactivation of telomerase activity in cancer progression. Trends in Genetics, 12: 129-131, 1996.

Shmookler Reis RJ, Goldstein S: Loss of reiterated DNA sequences during serial passage of human diploid fibroblasts. Cell, 21: 739-749, 1980.

Smith, GP: Unequal crossover and the evolution of multigene families. Cold Spring Harbor Symposia on Quantitative Biology. 38: 507-513, 1974.

Smith GP: Evolution of repeated DNA sequences by unequal crossover. Science. 191: 528-535, 1976.

Smith, S, de Lange, T: TRF1, a mammalian telomeric protein. Trends in Genetics. 13: 21-26, 1997.

Solomon, E, Barrow, J, and Goddard, AD: Chromosome aberrations and cancer. Science, 254: 1153-1160, 1991.

Srivastava, A, Norris, JS, Shmookler Reis, RJ, and Goldstein, S: c-Ha-ras-1 proto-oncogene amplification and overexpression during the limited replicative life span of normal human fibroblasts. J. Biol. Chem., 260: 6404-6409, 1985.

Stewart, N, and Bacchetti, S: Expression of SV40 large T antigen, but not small t antigen, is required for the induction of chromosomal aberrations in transformed human cells. Virology 180: 49-57, 1991.

Stürzbecher, H-W, Donzelmnn, B, Henning, W, et al.: P53 is linked directly to homologous recombination processes via RAD51/RecA protein interaction. EMBO J., 15: 1992-2002, 1996.

Super HG, Strissel PL, Sobulo OM, Burian D, Reshmi SC, Roe B, Zeleznik-Le NJ, Diaz MO, Rowley JD: Identification of complex genomic breakpoint junctions in the t(9;11) MLL-AF9 fusion gene in acute leukemia. Genes Chromo. Cancer 20: 185-195, 1997.

Suzuki, H, Takahashi, T, Kuroishi, T, Suyama, M, Ariyoshi, Y, Takahashi, T, and Ueda, R: p53 mutations in non-small cell lung cancer in Japan: association between mutations and smoking. Cancer Res., 52: 734-736, 1992.

Sweezy, MA, and Fishel, R: Multiple pathways leading to genomic instability and tumorigenesis. Ann. N. Y. Acad. Sci., 726: 165-177, 1994.

Szostak, JW, Orr-Weaver, TL, Rothstein, RJ, and Stahl, FW: The double-strand-break repair model for recombination. Cell, 33: 25-35, 1983.

Takita J, Hayashi Y, and Yokota J: Loss of heterozygosity in neuroblastomas--an overview. Eur J Cancer, 33:1971-1973, 1997.

Tlsty, TD: Normal diploid human and rodent cells lack a detectable frequency of gene amplification. Proc. Natl. Acad. Sci., 87: 3123-3136, 1990.

Tsancheva, M: The molecular biology and genetics of colorectal carcinoma. Khirurgiia, 50: 40-44, 1997.

van Gool, AJ, Verhage, R, Swagemakers, SM, van de Putte, P, Brouwer, J, Troelstra, C, Bootsma, D, and Hoeijmakers, JH: RAD26, the functional S. cerevisiae homolog of the Cockayne syndrome B gene ERCC6. EMBO J., 13: 5361-5369, 1994.

van Steensel, B, and de Lange, T: Control of telomere length by the human telomeric protein TRF1. Nature, 385: 740-743, 1997.

van Steensel, B, Smororzewska, A, and de Lange, T: TRF2 protects human telomeres from end-to-end fusions. Cell, 92: 401-413, 1998.

Vogelstein, B, Fearon, ER, Kern, SE, Hamilton, SR, Preisinger, AC, Nakamura, Y, White, R: Allelotype of colorectal carcinomas. Science, 244: 207-211, 1989.

Waldman, BC, and Waldman, AS: Illegitimate and homologous recombination in mammalian cells: differential sensitivity to an inhibitor of poly(ADP-ribosylation). Nuc. Acids Res., 18: 5981-5988, 1990.

Wang, SS, and Zakian, VA: Telomere-telomere recombination provides an express pathway for telomere acquisition. Nature, 345: 456-458, 1990.

Weinberg RA.: The genetic origins of human cancer. Cancer. 61:1963-1968, 1988.

Weinberg RA.: The molecular basis of oncogenes and tumor suppressor genes. Ann N Y Acad Sci., 758:331-338, 1995.

West, SC: The processing of recombination intermediates: mechanistic insights from studies of bacterial proteins. Cell, 76: 9-15, 1994.

Wheelhouse, RT, Sun, D, Han, H, Han, FX, and Hurley, LH: Cationic porphyrins as telomerase inhibitors: the interaction of tetra-(N-methyl-4-pyridyl)porphine with quadruplex DNA. J. Amer. Chem. Soc., 120: 3261-3262, 1998.

Wichman, HA, Van Den Bussche, RA, Hamilton, MJ, Baker, RJ: Transposable elements and the evolution of genome organization in mammals. Genetica, 86: 287-293, 1992.

Wiener F, Klein G, Harris H: Tumorigenicity of L cell derivatives and hybrid cells derived from them. Cancer Lett. Mar., 1: 207-210, 1976.

Windle, B, Draper, BW, Yin, YX, O'Gorman, S, Wahl, GM: A central role for chromosome breakage in gene amplification, deletion formation, and amplicon integration. Genes & Development, 5: 160-174, 1991.

Wright, WE, Pereira Smith, OM, and Shay, JW: Reversible cellular senescence: implications for immortalization of normal human diploid fibroblasts. Mol. Cell Biol., 9: 3088-3092, 1989.

Wright, WE, and Shay, JW: The two-stage mechanism controlling cellular senescence and immortalization. Exper. Girandole., 27: 383-389, 1992.

Xia, SJ, Shammas, MA, and Shmookler Reis, RJ: Reduced telomere length in ataxia-telangiectasia fibroblasts. Mutat. Res., 364: 1-11, 1996.

Xia, S, Shammas, MA, Shmookler Reis, RJ: Elevated recombination in immortal human cells is mediated by HsRAD51 recombinase. Mol. Cell. Biol., 17: 7151-7158, 1997.

Xiong, Y, Hannan, GJ, Zhang, H, Canso, D, Kobayashi, R, and Beach, D: p21 is a universal inhibitor of cyclin kinases. Nature, 366: 701-704, 1993.

Yamada, Y, Yoshida, T, Hayashi, K, Sekiya, T, Yokota, J, Hirohashi, S, Nakatani, K, Nakano, H, Sugimura, T, Terada M: p53 gene mutations in gastric cancer metastases and in gastric cancer cell lines derived from metastases. Cancer Res., 51: 5800-5805, 1991.

Yamaguchi-Iwai, Y, Sonoda, E, Buerstedde, JM, Bezzubova, O, Morrison, C, Takata, M, Shinohara, A, Takeda S: Homologous recombination, but not DNA repair, is reduced in vertebrate cells deficient in RAD52. Mol. Cell. Biol., 18: 6430-6435, 1998.

Yusof, YA, and Edwards, AM: Stimulation of DNA synthesis in primary rat hepatocyte cultures by liver tumor promoters: interactions with other growth factors. Carcinogenesis, 11: 761-770, 1990.

Zakian, VA: Structure and function of telomeres. Annu. Rev. Genet., 23: 579-604, 1989.

Zhan, Q, Fan, S, Bae, I, Guillouf, C, and Liebermann, DA, O'Connor, PM, Fornace, AJ, Jr.: Induction of bax by genotoxic stress in human cells correlates with normal p53 status and apoptosis. Oncogene, 9: 3743-3751, 1994.

Zhang, LH, and Jenssen, D: Studies on intrachromosomal recombination in SP5/V79 Chinese hamster cells upon exposure to different agents related to carcinogenesis. Carcinogenesis, 15: 2303-2310, 1994.

Zucman-Rossi J, Legoix P, Victor JM, Lopez B, Thomas G: Chromosome translocation based on illegitimate recombination in human tumors. Proc. Natl. Acad. Sci. USA, 95: 11786-11791, 1998.

© 2001 Elsevier Science B.V. All rights reserved.
The Role of DNA Damage and Repair in Cell Aging
B.A. Gilchrest and V.A. Bohr, volume editors.

COMMITMENT SIGNALLING FOR APOPTOSIS, OR DNA REPAIR AND ITS RELEVANCE TO AGING AND AGE-DEPENDENT DISEASES

Eugenia Wang[1,2], Richard Marcotte[1], Harry T. Papaconstantinou[3] and John Papaconstantinou[4]

[1]The Bloomfield Center for Research in Aging, Lady Davis Institute, Jewish General Hospital;
[2]Department of Medicine, McGill University, Montreal, Canada;
[3]Deparment of Surgery, University of Cincinnati, OH;
[4]Department of Human Biological Chemistry and Genetics of
the University of Texas Medical Branch, Galveston, TX

1. Introduction

One of the most recent developments in the biomedical sciences, relating directly to understanding the cellular mechanisms controlling aging and age-dependent disease, is the emergence of the complex regulation of programmed cell death or apoptosis, a cellular self-eliminating program. The physiological need for apoptosis is of major biological significance. Developmentally, it is the process that eliminates extra cells during organogenesis; in the post-natal and mature organism it plays a key role in tissue homeostasis to maintain optimal organ function throughout the life cycle. More importantly, the maintenance of a threshold critical number of functional cells is exquisitely controlled throughout life; dysregulation of this upkeep, either favoring too many or too few cells, plays a key role in the functional senescence of the organ. Therefore, for any given organ, the interplay of cell replication, apoptosis, and repair may be the key process that determines the optimal number of cells needed for normal organ maintenance and function. The biological interactions that determine the basic mechanism(s) of susceptibility or resistance to apoptotic signals, and the relationship of the function of the DNA repair mechanism to apoptosis, are key factors in understanding the stages at which the fate of each cell to live or die is determined. We and others have proposed that the initiation of apoptosis closely resembles early cell cycle events, and a critical phase for this determination may occur at the G_1/S border [24, 71, 94, 113]. Resistance to apoptosis, therefore, may be regulated by three important physiological states: 1. failure to initiate apoptosis, by remaining in the growth-arrested state; 2. failure to respond to apoptotic signals that trigger its further progression at the G_1/S border; and 3. failure to commit to the final stage of apoptotic death, which involves activation of specific processes which are components of the CASPASE cascade. These three steps function in a highly coordinated fashion to orchestrate the signal transduction pathway leading to the G_1/S border, by first exiting from growth arrest by losing growth arrest-specific gene expressions and acquiring immediate early gene expressions; second, at the G_1/S border, acquiring the annulment of survival factor function; and third, a cascade of CASPASE proteolysis. All or part of the above biochemical changes converge to precipitate death by apoptosis. Absence of

153

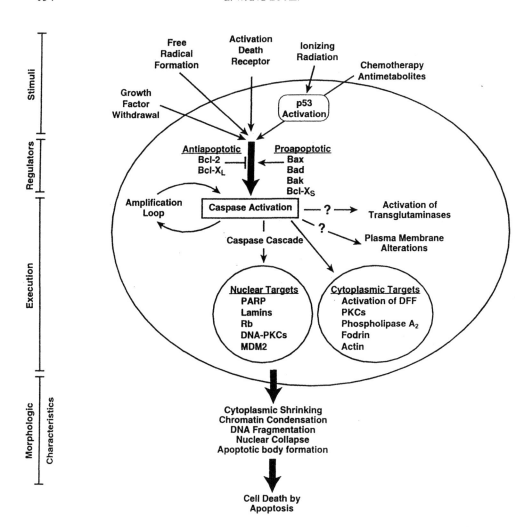

Figure 1. A summary of the anti-apoptotic and pro-apoptotic pathways. The diagram outlines factors that stimulate the pro-apoptotic pathway, the components of regulation and execution of the pathway; the nuclear and cytoplasmic targets of the CASPASE cascade, and the resultant morphological characteristics of the apoptotic cell.

these changes may result in resistance to apoptosis, and indefinite survival of such cells. In this Chapter we propose to describe the molecular characteristics of these three physiological states, and how they relate to apoptosis and aging.

2. Mechanism for Apoptosis: Global Characteristics

In this chapter, apoptosis is defined as the genetically regulated process of cellular self-demise. Characterization of the molecular and genetic processes of programmed cell death (PCD) has been accomplished by intense efforts to identify the genes and their products whose functions mediate the prime suicidal action in unwanted or damaged cells, resulting in morphological and biochemical features of apoptotic cells (Figure 1; [35, 36, 88]). The morphological and biochemical characteristics of apoptotic cells include the appearance of chromosomal condensation, membrane blebbing and phospholipid rearrangements, endonucleolytic DNA cleavage, and possibly a need for RNA and protein synthesis [160, 163, 165, 166]. The final products are apoptotic bodies, round refractile cell corpses which are eventually phagocytosed by neighboring cells, without subsequent inflammation or "secondary necrosis" [36].

Physiologically, tissue homeostasis is maintained through the regulation of apoptosis by environmental signals including hormones, growth factors, mediators of inflammation (biological and chemical toxins), UV irradiation, and nutritonal status [140, 175]. Cell death by apoptosis differs from cell death by necrosis in that the plasma and organelle membrane of apoptotic cells remain intact. Necrotic cells exhibit breakdown of the plasma membrane with cytoplasmic swelling and cell lysis, resulting in an intense inflammatory response with infiltration of lymphocytes. Therefore, the regulation of tissue homeostasis by apoptosis is a physiological process that is sensitive to many extrinsic (environmental) and intrinsic (hormones, growth factor) factors.

3. The Anti-Apoptotic and Pro-Apoptotic Gene Families: "Death and Anti-Death Genes"

Many genes such as TRPM-2, *ced3, 4,* and *9,* SGP-2, *etc.,* have been implicated in the activation of anti-apoptotic (survival) or pro-apoptotic (cell death) pathways [8, 10, 17, 37, 54, 101, 106, 142, 176], and have thus been classified into the categories of "death" or "anti-death" genes [21, 152]. Whether cells remain alive by taking the survival pathway, or die by taking the cell death pathway, depends upon the balance between the pool levels of the proteins whose synthesis and activities are regulated by environmental and physiological factors. Thus, the balance of pool levels of members of the family of anti-apoptotic proteins, known as *Bcl-2, BclX$_L$, etc.* with pro-apoptotic proteins, *Bax, Bad, etc.,* plays a key role in selecting between the survival and cell death pathways. These pathways are determined by protein-protein interactions between these family members, as well as the structural changes caused by modification that facilitate these protein-protein interactions. The protein structural characteristics of members of the family that are anti-apoptotic or pro-apoptic are shown in reference 1.

The identification of the anti-apoptotic and pro-apoptotic genes and their protein products has been greatly facilitated by studies with the nematode worm *C. elegans*. A comparison of the pathway to cell death in *C. elegans* and mammals is shown in Figure 2. For example, ICE, a homologue of the nematode *ced3* gene, as well as its sister genes YAMA/CPP32β, *etc.* [37, 101, 142, 176], were shown to function as death genes, whose products become killers following activation by proteolytic cleavage from pro-ICE to the ICE form [10, 106, 119]. *Bcl-2* in mammalian cells is homologous to the *C. elegans ced9*, and functions as an anti-apoptotic or cell survival factor by forming a heterodimer with a homologous pro-apoptotic gene, *Bax*. Thus, the pool levels of anti- and pro-apoptotic proteins and their interactions to form dimers are factors that regulate survival or apoptosis. These interactions and their balancing are seen with other anti- and pro-apoptotic proteins. Thus, coupled regulation occurs when the *ced3* killing activity is suppressed by heterodimerization with *ced9*, or the similar activity of ICE is suppressed by *crmA in vitro* [44, 76, 118,]. This counterbalancing by protein-protein interaction of the anti- and pro-apoptotic proteins clearly suggests specific funtions of these proteins with respect to enabling the cell to either remain in a survival pathway or proceed to the next phase of the apoptotic pathway [117]. Further details of mechanisms and interactions that lead to survival or apoptosis are discussed in a model presented below.

4. Apoptosis Initiation, Departure from Growth Arrest State, and Replicative Senescence

One of the factors contributing to senescent fibroblasts' resistance to apoptosis is that they are permanently locked into the growth-arrested state [18, 32, 48, 137]. Studies focused upon understanding the molecular and genetic bases of fibroblast senescence have shown significant changes in activity of proteins that regulate cell proliferation [2, 112, 122, 128, 129]. For example, the findings that c-*fos* is repressed [130] and Rb remains unphosphorylated [139] in senescent fibroblasts suggest the involvement of oncogene loss, and gain of Rb tumor-suppressing function, as important functional changes that lead to the acquisition of the growth-arrested phenotype. Moreover, the finding that the cell cycle inhibitor, Sdi1 (a 21-kDa protein also known as Pic1, Cip1, or WAF1) [34, 51, 89, 91, 107, 168] is activated in part by p53 suggests that fibroblast senescence is the manifest destiny of "anti-oncogenesis" [83, 91]. This multifunctional protein affects cell proliferation through its ability to inhibit *cdks* enzyme activity (a function of its N-terminal end) and bind to PCNA [19, 89, 161]. However, although the lack of c-*fos* expression and Rb phosphorylation, as well as the permanent presence of statin [156, 158], suggest changes in cell cycle regulation in senescent cells, these biochemical charactersistics may also confer an "anti-death" status upon senescent fibroblasts, since the initiation of apoptosis shares the same need for the expression of these and other hallmark G_1 molecular changes [24, 71, 96, 113, 115].

Pathways to Cell Death in C. elegans and Mammals

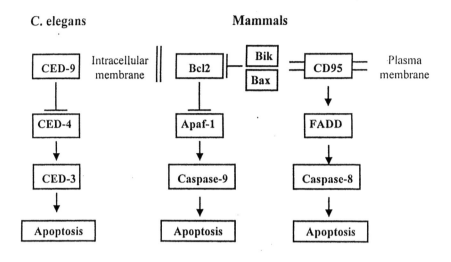

Figure 2. An overview of the pathways to cell death in C. elegans and mammals. The CED-9/Bcl-2 family integrates positive and negative signals and arbitrates whether apoptosis should occur. Activation of CED-4/Apaf-1 commits to apoptosis, and CED-3/CASPASE-9 mediates the cell death process. In mammalian cells, signals from diverse cytotoxic stimuli such as cytokine deprivation, DNA damage or staurosporine are regulated via Bcl-2. The signal induced by engagement of the "death receptor" CD95 proceeds through the adaptor FADD, which directly activates CASPASE-8 and largely bypasses the Bcl-2 family (taken from Adams, J. M., and Cory, S. Science, 281, 1323, 1998).

5. Activation of G_1 Traverse and Apoptosis Checkpoint at G_1/S

Many proto-oncogenes, such as c-*Myc*, c-*fos*, and c-*abl*, have been implicated in apoptosis [20, 38, 39, 136]. Recently it was realized that shortly after activating the apoptosis pathway, these genes' expressions are actually required, suggesting that the initiating phase for this type of death resembles closely the characteristics of G_1 cell cycle traverse [24, 71, 96, 113]. Thus, the disappearance of statin [160], a marker for the nonproliferating state, and the sequential expression of c-*Myc*, c-*fos*, cdc2, c-*jun*, PCNA, and Rb phosphorylation, which are characteristics of progression through G_1 to the G_1/S border, are often followed by abortive DNA synthesis in the early phases of apoptosis [4, 53].

It has also been proposed that c-*Rel*, a member of the NFκB/*Rel* family of transcription factors, may play a role in inducing apoptosis, based on observations that high levels of expression of c-Rel induce apoptosis. This pro-apoptotic function of c-*Rel* is associated with cell-cycle arrest at G_1/S, and correlates with inhibition of E2F DNA-binding activity, accumulation of hypophosphorylated Rb, inhibition of Cdk2 kinase, and accumulation of transcripts encoding Cdk inhibitor p21Wafl. Interestingly, *RelA*, another member of the NFκB/Rel family, exhibits anti-apoptotic function. It remains to be understood how RelA can protect from apoptosis, while its potential dimerization partner c-*Rel* induces apoptosis. One explanation could be that the kB motifs in pro-apoptotic and anti-apoptotic genes selectively bind complexes with c-*Rel* or *RelA* respectively. It is also conceivable that overexpression of c-*Rel* prevents formation of *RelA* dimer combinations that are required for inducing anti-apoptotic genes. Thus, an important factor to consider is the regulation of *Rel* pool levels or dimerization capabilities. Pool levels are important factors in "pushing" pro- and anti-apoptotic pathways: This may be an important factor in the participation of NFκB/*Rel* protein in apoptosis.

A checkpoint for apoptosis seems to occur at the G_1/S boundary, where cells are committed (triaged) either to successfully enter S-phase and the survival pathway, or to drop out and die via the apoptosis pathway. Interestingly, if indeed cells must progress through G_1 to the G_1/S phase to proceed with apoptosis, the arrest of senescent cells in the early G_1 phase or G_0 may be a basic factor in their resistance to apoptosis.

The *Bcl*-2 family has a strong impact on the cell cycle. *Bcl*-2 family proteins can modulate cell cycle progression [1, 14, 15, 84, 86, 94, 109, 121, 148, 151, 171]. Under suboptimal growth conditions *Bcl*-2 promotes entry into quiescence, and retards re-entry into the cycle [56, 134]. This effect is separable from *Bcl*-2's survival function, in that cell cycle inhibition but not pro-survival function is ablated by deletion in the non-conserved loop or mutation of tyrosine-28 [58, 147]. It is proposed that the inhibitory effect may involve a protein such as the phosphatase calcineurin, which binds to that region of *Bcl*-2; the ability of *Bcl*-2 to sequester calcineurin reduces the nuclear translocation of NFAT, a transcription factor that co-migrates to the nucleus with calcineurin. Since NFAT is required for progression into S-phase, attenuation of its nuclear translocation will affect S-Phase. Thus, T-cells expressing *Bcl*-2 make less IL-2, which is required for progression into S-phase, apparently because of the *Bcl*-2 mediated reduced nuclear translocation of NFAT [86].

6. p53, RB Phosphorylation, and Multiple Apoptosis Pathways

Of all the genes implicated in apoptosis, p53 is the best known, since it is suggested as a master switch for apoptosis [9, 23, 132, 141, 154, 162, 173]. However, we have recently found that fibroblasts derived from p53 knock-out mice are capable of undergoing apoptotic death upon total serum withdrawal [57]. This finding suggests that there may be at least two different apoptotic pathways, distinguished by their dependence upon p53 [57, 97]. It may be that apoptotic death stimulated by DNA-damaging agents follows the p53-dependent pathway, while withdrawal of growth factors or removal or addition of cytokines follows the p53-independent path. In this case our system, using serum deprivation with high cell density, *i.e.* confluent, cultures as the apoptosis stimulus, falls into the p53-independent pathway, categorized by the re-entry to G_1, progression to G_1/S, but triage to a path leading to death due to the absence of functional survival factors. Here Rb's function may be more important to the success of apoptosis, since its phosphorylation may allow cells to leave growth arrest and re-enter G_1, necessary to initiate apoptosis [24, 71, 96, 113].

7. Commitment to Apoptosis and Proteolysis

Our kinetic studies show that subsequent to the G_1/S checkpoint, serum-deprived cells are committed to apoptosis [53, 113]. Thus, at some checkpoint between entry into G_1 and prior to arrival at G_1/S, serum-deprived cells cannot be rescued from death by returning them to serum-containing media; this commitment point is marked by the proteolytic cleavage of a precursor protein, terminin (Tp90), to form the lower molecular weight product, terminin Tp30 [99, 100]. The cleavage of Tp90 to form Tp30 is a powerful biochemical marker for the irreversible commitment to apoptotic death [53, 99, 100, 115, 159]. The cleavage of Tp90 to Tp30 closely resembles the noted apoptosis-dependent proteolysis of pro-ICE to ICE [10, 37, 101, 106, 119, 142, 176], and thus suggests proteolysis as an integral component of the pathway leading to ultimate cell death. The role of proteolytic cleavage has been the recent subject of intense study on the apoptotic pathway. These studies have defined specific proteolytic CASPASE cascades that exhibit selective activation in response to specific apoptotic initiating signals.

8. Upstream Regulators for Apoptosis Commitment, and Negative Environment for DNA Replication

As described above, immediately preceding the commitment to apoptosis, marked by the proteolytic production of Tp30, an entire repertoire of immediate early genes of the cell cycle is expressed [24, 71, 96, 113]; there is even an attempt at low level, but significant, DNA synthesis activity [4, 53]. Ultimately this activity, though abortive in nature, suggests that cells are following the path leading to DNA replication. Perhaps the signals for apoptosis somehow trigger the machinery preparing for S-phase DNA replication. However, in the milieu of a negative environment such as the absence of functional "anti-death" factors like *Bcl-2*, DNA replication is aborted, which in turn sends an alternate signal activating the death route.

9. *Bcl-2* and Oxidative Stress in Apoptosis

What makes the apoptotic attempt at DNA replication abortive? or what enables the apoptotic pathway to impose an abortive effect on DNA replication? The clue may lie in part in another complex biochemical response, oxidative homeostasis. Oxygen free radicals induce features similar to the hallmarks of apoptosis, such as DNA strand breaks, membrane blebbing, *etc.* [6, 79]; oxidative stress may be the deleterious change underlying the entire pathway leading to cell demise. We have shown that manganese superoxide dismutase (MnSOD) activity is up-regulated early during apoptosis as a response to this stress. *Bcl-2* is a strong candidate to combat oxidative stress; the specific *Bcl-2* function involved, whether reducing lipid peroxidation [56] or maintaining glutathione (GSH) levels by binding to GSH peroxidase [65], may counteract the lethal action of reactive oxygen species during apoptosis [65, 134]. The finding that phosphorylated *Bcl-2* cannot inhibit lipid peroxidation in *in vitro* assays [49] certainly supports the notion that the non-phosphorylated form of *Bcl-2* plays a role in combating oxidative stress. In apoptotic cells, as suggested above, activation of the phosphorylation of *Bcl-2* may leave the cells without protection from increased levels of oxygen free radicals, so that they cannot help but abort the attempt at DNA replication and die.

It has been proposed that ceramide levels could act as a general "rheostat" controlling cell survival by regulating PI(3)K anti-apoptotic effector mechanisms (Figure 3; [3, 90, 179]). Thus, this ceramide-dependent "rheostat" may affect the PI(3)K signalling pathway that is linked to the regulation of *Bcl-2* family activity, as well as the pathway that regulates intermediates of cell proliferation (DNA synthesis – DNA damage and repair). Many stimuli that induce apoptosis generate the lipid second messenger ceramide. Cellular ceramide levels increase within 30-60 minutes in response to such DNA-damaging agents as oxidative stress, UV and ionizing radiation, chemotherapeutic agents as well as cytokines, through the activation of acid and neutral sphingomyelinases [3]. Thus, the regulation of ceramide generation plays a central and necessary role in apoptosis induced by stress. Genetic evidence for this is derived from Nieman-Pick (NP) patients, whose cells lack sphingomyelinase [16] and are more refractory to dying by apoptosis when exposed to genotoxic stress [126]. Furthermore, lymphoblasts from NP patients exhibit apoptotic resistance to ceramide-generating stresses, but not to ceramide-induced cell death [126]. This is attributed to their very low acid-sphingomyelinase activity, which would explain the observation that γ-irradiation, UV-C and hyperosmotic stress do not induce a rapid decrease in PI(3)K activity, and that these cells are resistant to stress-induced apoptosis. These studies show that the PI(3)K down-regulation by stress is dependent upon the generation of ceramide *via* acid-sphingomyelinase activity [126]. In addition, cells transformed by oncogenic Ras exhibit decreased sphingomyelinase activity [80], and resistance to radiation-induced apoptosis in Burkitt's lymphoma is associated with defective ceramide signalling [98].

Certain growth factors such as insulin-like growth factor-I (IGF-I) and insulin, as well as matrix attachment, have an anti-apoptotic effect on cells of mesenchymal origin (Figure 4; [52]). It has now been shown that a PI(3)K modulated pathway may be responsible for this anti-apoptotic effect [42, 50]. This pathway is as follows: (Figure 4A; [42]). (a) PI(3)K consists of a regulatory subunit (p85) that binds to an activated, phosphorylated growth factor receptor/cytokine receptor; (b) this results in the activation of PI(3)K catalytic subunit (p110) [123]; (c) a product of PI(3)K, PIP$_2$,

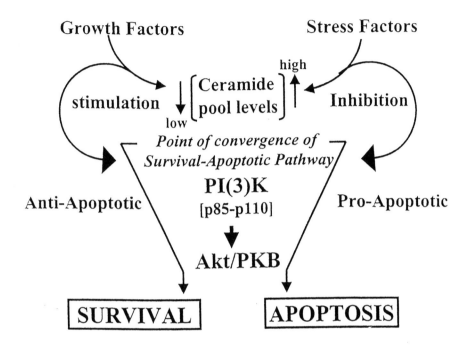

Figure 3. The convergence of the survival or apoptosis pathways that are regulated by changes in ceramide levels. The pathway shows that ceramide levels are regulated by growth factors that lead to the survival pathway, or stress factors that lead to the apoptotic pathway. The components of this site of convergence are PI(3)K and Akt/PKB.

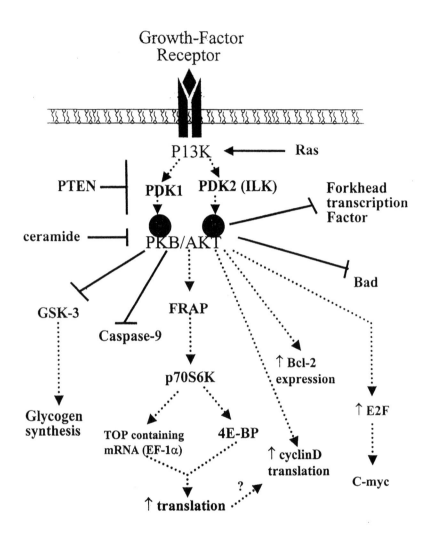

Figure 4. IGF-1 and PI(3)K-directed signal transduction pathway. This diagram presents a simplified model of multiple cascades of signalling, directed from the PI(3)K kinase, through the 'hub' of PKB/Akt, into the pathways leading to translational regulation, glycogen synthesis, and transcriptional regulation of E2F and c-Myc.

facilitates recruitment of the protein kinase, *Akt*, to the plasma membrane [43]; (d) *Akt* is activated by protein kinases [3]. In general these data suggest that PI(3)K-*Akt* serve as a point of convergence of the anti-apoptotic and pro-apoptotic signals, and that it is at this point that the pathway to survival (DNA synthesis and repair) or apoptosis (proteolysis and DNA degradation) is determined. The fact that overexpression of PI(3)K or *Akt* results in a decreased rate of apoptosis in response to serum/growth factor deprivation, UV-B irradiation or loss of matrix attachment suggests that this may indeed be the site of convergence and distribution of pro- and anti-apoptotic signals [66, 67, 69, 77]. These studies support the hypothesis that cells undergoing apoptosis or surviving possess a mechanism that can inhibit or activate PI(3)K [179]; that stress-mediated increase in ceramide levels plays a role in regulation of apoptosis *via* its effect on PI(3)K activity; and the decision of whether cells take the pathway of survival or of apoptosis occurs through the regulation of PI(3)K activity.

Studies have indicated that *Bad*, a member of the *Bcl-2* family, may mediate the anti-apoptotic effects of the PI(3)K pathway (Figure 5; [26, 27, 30]). In this mechanism, phosphorylation of *Bad* by *Akt* causes it to be sequestered by 14-3-3 proteins, thereby preventing its binding to *Bcl-2* [177]. Thus, since *Bad* is a downstream effector of *Akt*, inhibition of *Akt* activity by ceramide should also result in inhibition of *Bad* phosphorylation. Ceramide also inhibits *Bad* phosphorylation, directly linking ceramide-mediated apoptosis to the cell death effector pathway [179]. Taken together, these results suggest that ceramide promotes the induction of apoptosis by inhibiting the constitutively active anti-apoptotic pathway regulated by PI(3)K and *Akt*. This is done by activating the release of *Bad* from 14-3-3 proteins and its interaction with *Bcl-2* (Figure 6). A similar mechanism is depicted for *Bcl-X$_L$-Bad* interactions in the pro and anti-apoptotic responses to growth factors.

The fine-tuned, "rheostatic" role of ceramide in the regulation of anti-apoptotic effects of PI(3)K by growth or stress factors provides new insights into how aging may affect these processes, and how lipid second messengers control cell fate in response to both stress and oncogenic transformation. It is important, therefore, to understand whether ceramide production is altered by aging. The identification of the p110 subunit of PI(3)K as the Age-1 gene of *C. elegans*, as well as other genes linked to the PI(3)K/*Akt* pathway, suggests a basic importance of the PI(3)K/*Akt* pathway as a regulator of life-span (70, 85, 103, 110). If constitutive ceramide levels are reduced in aging, then PI(3)K activity would be altered accordingly, and would be a basic factor in age-associated changes in apoptosis and tissue homeostasis. Thus, reduced ceramide levels, or a failure to respond to stress factors in aging, would attenuate apoptosis and thus the replacement of damaged cells. Alternatively, if ceramide levels are high, then the apoptosis pathway is accelerated and may result in a "tissue wasting syndrome." Thus, the ability of aged tissues to respond to stress factors may reflect directly on ceramide generation. Whether this occurs, and whether this plays a role in age-associated changes in tissue apoptosis, remains to be determined. In degenerative diseases in which ceramide levels are elevated, survival mechanisms are constitutively down-regulated, leading to increased cell death. Thus, in normal aged tissues, which

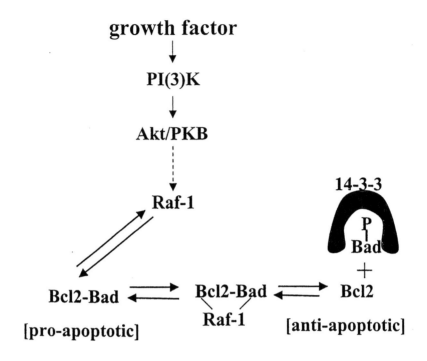

Figure 5. The growth factor-mediated pro-apoptotic. and anti-apoptotic pathway that involves the interaction (heterodimerization) of Bcl-2-Bad.

exhibit chronic levels of stress, constitutively increased ceramide levels may play a role in accelerated tissue destruction and remodeling—homeostasis.

Ceramide mediates a variety of other lipid-mediated signalling mechanisms, such as phospholipase D and protein kinase C [22, 63, 81, 127, 138, 153] and can also regulate receptor- mediated signalling in addition to its role in regulating the PI(3)K pathway. This clearly suggests a global effect of ceramide upon such processes as cell proliferation and metabolism, as well as response to stress.

Because ceramide modulates the PI(3)K response at extremely early times following stress, it is feasible that the effect of ceramide on p85 could modulate p85's affinity for p110, and thus make it available for binding to other potential targets that could be pro-apoptotic. In the unchecked tumor state, this affinity may be such that it prevents the p110 interaction, thus favoring apoptotosis. This could, therefore, be a determining factor in the resistance to chemotherapy/radiotherapy and apoptosis.

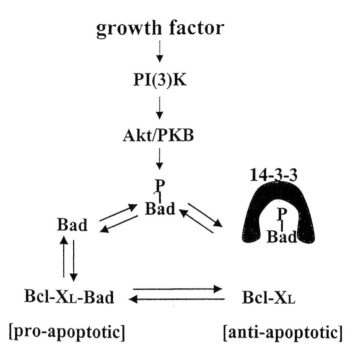

Figure 6. The growth factor-mediated pro-apoptotic and anti-apoptotic pathway that involves the interaction (heterodimerization) of Bcl-X$_L$ and Bad.

10. p21, p53, DNA Replication and Repair, and Apoptosis

Another possible factor in abortive DNA replication may involve the up-regulation of p21 during apoptosis. Since the initial discovery of its inhibitory action on DNA synthesis, p21 has been thought to exert a growth arrest function [19, 34, 89, 161, 168]. Thus, we expect p21 to inhibit apoptosis, since growth arrest would halt the cells from entering the G$_1$ phase, a necessary initiating event for apoptosis. Based on this rationale, p21 should be down-regulated during apoptosis, and the introduction of antisense p21 should facilitate apoptosis. Surprisingly, our results show the contrary: p21 is upregulated during apoptosis, and antisense p21 rescues cells from death [29, 33].

The apoptosis-associated up-regulation of p21 has characteristics similar to those associated with its increase in response to mitogen stimulation in late G$_1$ [83, 105], thought to be controlled by TGF-β, rather than the usual regulator, p53 [29]. Why should the p21 level increase here? The answer may lie in the occurrence of parallel apoptotic signals that converge to a common determining check-point factor. In mitogen-stimulated cells, an increase in p21 allows the accumulation of cyclin-dependent kinases (*cdks*) and proliferating cell nuclear antigen (PCNA) during mid- and late G_1, before entry into S-phase [83], thus preventing premature progression of the G_1 phase. This notion is valid only if the binding of p21 to *cdks* and PCNA is reversible, which would allow eventual entry to S-phase. There may also be alternative properties of p21 binding between G_1 and the growth-arrested state; the inhibitory effect on DNA synthesis may apply only to the latter. Another point of interest may be the temporal kinetics of *cdks* and PCNA expression; upon serum withdrawal, both *cdc2* and PCNA expression last for 18 hours, *i.e.*, until shortly before death. This prolonged period of PCNA expression may be a key factor in p21 binding selectivity. Taken together, the binding properties, temporal kinetics of partner relationships, and p21 increase simultaneously with *Bcl-2* phosphorylation, may demarcate two types of p21 regulation: in dying cells, this p21 increase may contribute to the final abortive DNA replication.

It is well established that a major series of biological responses to DNA damage in mammalian cells involves a cascade of protein kinases related to the 2nd messenger signalling molecule PI(3)K [169]. Two major protein kinases are ATM, mutated in ataxia telangiectasia, and DNA-dependent protein kinase (DNA-PK), whose activity is associated with V(D)J recombination of immunoglobulin genes. ATM and DNA-PK are the mammalian homologs of the yeast RAD-3/MEC1 kinases that likewise mediate DNA damage repsonses. Both ATM and DNA-PK trigger cascades of cellular responses that include activation of cell cycle checkpoints and growth arrest, repair and apoptosis [169].

Under normal conditions, p53 tumor suppressor is maintained at low levels through its interaction with Mdm-2. Both p53 and Mdm-2 are targets of ATM and DNA-PK [64, 93,133, 169], so that DNA-damage-induced phosphorylation of either p53 or Mdm-2 prevents the two proteins from interacting, thus stabilizing and activating p53.

Activation of p53 is mediated through DNA damage, and regulates two cellular responses, arrest (in cell cycle stages G_1 and G_2) and apoptosis. Thus, any changes in Mdm-2 levels can lead to p53 inactivation. Substantial evidence suggests that a major part of p53-mediated growth arrest proceeds through the transcriptional activation of p21, allowing time for DNA repair. In contrast, the mechanism by which p53 promotes apoptosis is not clearly understood, although there are specific apoptosis-related targets that differ from those implementing growth arrest [179]. For example, transcription targets of p53 implicated in apoptosis are *Bax* [102, 172], the insulin-like-growth factor receptor (IGF-IR) [116], and the binding protein IGF-BP3. Although p53 is a major regulator of apoptosis, its activation does not universally lead to cell death. The factors that determine whether damaged cells should apoptose, or whether DNA damage should be repaired and the cells survive, are unknown. It has been suggested that p53 response levels determine this differential function: Low to moderate p53 activation results in cell cycle arrest with DNA repair and cell survival, while a high level of p53 response results in cell death by apoptosis. It is likely that the level of repsonse is directly related to degree of DNA damage. The greater the degree of DNA damage, the less advantageous for the cell to repair.

Oncogenes such as c-*Myc* have been implicated in the control of proliferation, and have also been shown to promote apoptosis [13, 38, 125]. Interestingly, the mitogenic and pro-apoptotic properties of c-*Myc* are genetically inseparable, in that both require the N-terminal transcriptional activation domain, DNA-binding and dimerization domains, and interaction with the *Myc* partner protein, *Max* [38]. This presents the interesting concept that the control of proliferation and apoptosis may utilize the same "machinery," and that the difference between the two pathways may involve regulation of levels of the components of this machinery. Thus, when *Myc*-induced apoptosis is suppressed by the IGF-I triggered survival pathway, this anti-apoptotic route proceeds through the *Ras*, PI(3)K, *Akt* pathway that ultimately impacts on *Bad,* which is a key modulator of *Bcl-2* [61, 72, 73]. E1A, like *Myc,* is also a potent inducer of apoptosis. Both growth promoting and apoptotic functions map to the N-terminal region [120], which is involved in binding the retinoblastoma protein (Rb), and the transcriptional corepressor p300 (CBP) [41]. Interaction of both Rb and p300 is required for E1A-induced cell proliferation and apoptosis. Thus, as with *Myc*, the balance of anti-apoptotic *versus* pro-apoptotic pathways may involve post-translational processes of protein modification and protein-protein interactions which are regulated by signal activity and pool levels of the interactants.

We presented the argument above that the processes of cell proliferation or renewal and cell death are tightly linked. This was suggested by early reports of apoptotic properties of c-*Myc* and the adenovirus oncoprotein E1A. Both proteins are potent inducers of proliferation and apoptosis. Many known potent promoters of cell proliferation also possess pro-apoptotic activity [179]. An explanation of this link between cell proliferation and cell death is that the tendency to undergo apoptosis is a normal consequence of engaging the cell's proliferative machinery, so that cell proliferation and apoptotic pathways are coupled. The regulatory determinants of these divergent pathways remains to be discovered; however, we hypothesize that one of those factors may be ceramide.

Another regulator of cell cycle progression, the tumor suppressor Rb, functions by binding to and thereby inactivating E2F proteins (transcription factors). The E2F proteins regulate expression of genes required for G_1 phase cell cycle progression [62]. However, changes in E2F expression, such as ectopic expression or mutation, accelerate S-phase entry and apoptosis. Thus, the regulatory role of Rb in restraining E2F action is a major factor in the selection of either survival or the apoptotic pathway. Apparently the deregulated entry into the cell cycle, by mutation or overexpression of E2F and inactivation of Rb, favors apoptosis. Interestingly, the nervous system, liver, lens and skeletal muscle of Rb (-/-) null mutants which die *in utero* all show excessive proliferation and apoptosis [60].

It has been proposed that oncoprotein-induced apoptosis may reflect the fact that the machinery mediating growth and apoptosis are coupled processes. This has been presented as *The Dual Signal Model* [59, 179]. In this model it is proposed that the activation of cell proliferation primes the cellular apoptotic program, and unless contermanded by appropriate survival signals, automatically initiates removal of the affected cell. Survival signals, provided by neighboring cells, ensure that somatic cells remain mutually interdependent for survival. By this mechanism the possible proliferative autonomy of any individual cell is limited, and it is the balance between pro-apoptotic growth processes and anti-apoptotic survival signals that determines whether a cell proliferates or dies.

Another interesting hypothesis has proposed that growth-deregulating mutations do not themselves induce apoptosis, but merely *sensitize cells* to other apoptotic triggers. This may explain how in clonal, synchronized cell populations, apoptosis is asynchronous. It is speculated that oncoprotein sensitization of cells is caused by a variety of minor insults. Thus, a low level of chronic DNA damage, which is repaired in normal cells, might trigger apoptosis in an oncoprotein-sensitized cell. In addition, in place of the conventional idea that *Myc* causes p53 to trigger apoptosis, *Myc* would sensitize cells to p53-induced death signals; in a cell harboring activated p53 in response to low levels of DNA-damage, c-*Myc* would cause apoptosis by raising sensitivity to the p53 trigger. In such cells, p53 would appear to be necessary for *Myc*-induced apoptosis. By a similar mechanism, E1A would cause apoptosis by raising the sensitivity to the p53 trigger.

The hypothesis requires that there be an effector through which oncoproteins generate sensitivity to apoptosis. The recent isolation of an E1A-induced CASPASE activity suggests that the components of the intracellular "apoptosome" are likely candidates for effectors. The apoptosome is a complex composed of Ced4 protein, Apaf-1, holo-cytochrome C (released from mitochondria), ATP and pro-CASPASE 9. Activation of these apoptosome components could generate the requisite oncogene-dependent sensitized cell.

11. Sequential Apoptotic Steps, Growth Arrest, and Neoplastic Transformation

A successful serum deprivation-induced apoptotic pathway may consist of three sequential steps: 1. *initiation,* marked by departing from growth arrest; 2. *the triage checkpoint*, the determination or sorting of pathways, marked by Bcl-2 phosphorylation and p21 up-regulation; and 3. *commitment*, the apoptotic process, marked by proteolysis of Tp90 terminin to form Tp30 terminin. Quiescent, contact-inhibited young human fibroblasts or their senescent counterparts are resistant to apoptosis while in growth arrest, as long as neither Tp30 nor phosphorylated Bcl-2 is produced. In cycling cells, resistance to apoptosis may be established by escaping the *two-prong pincer mechanism* composed of: (1) p21 up-regulation, to bind both *cdk*s and PCNA, and inhibition of DNA replication, and (2) attenuating *Bcl-2*'s function by phosphorylation. This escape not only supports successful S-phase entry and DNA replication, but also avoids triage to the death path, marked by subsequent Tp30 production.

Regulation of ceramide metabolism has been implicated as a contributory factor to tumor progression and resistance to cancer chemotherapy. Under normal conditions of low ceramide levels, activity of the PI(3)K-*Akt* pathway leads to cell proliferation, *i. e.*, to the survival pathway. During tumorigenesis, ceramide levels become limiting, and

Precursor and Proteolytic Maturation of Caspases

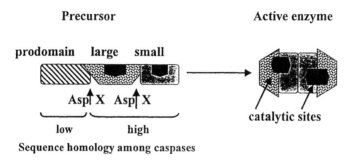

Precursor   Active enzyme

Figure 7. A diagram of the proposed precursor and proteolytic maturation of CASPASEs. CASPASEs are synthesized as precursors that undergo proteolytic maturation. (Modification of figure from Thornberry, N. A. and Lazebnik, Y. Science 281, 1312, 1998)

PI(3)K activity is unchecked, leading to increased survival and resistance to chemo- and radiotherapy. Tumors possessing activated signalling mechanisms that utilize the PI(3)K catalytic subunit or its downstream effectors, such as *Akt*, could also bypass the regulatory mechanism of ceramide, generated during apoptotic stress. Thus, modulators of ceramide metabolism represent an important class of anti-cancer and disease-modifying pharmaceuticals that have yet to be exploited. This *two-prong pincer mechanism* may operate to triage cycling cells at the G_1/S border to the path where key proteolytic steps can occur, resulting in final death; for senescent fibroblasts and their growth-arrested young counterparts, an added step of leaving the growth-arrested state and acquiring the marching order of G_1 events may also be in operation.

12. The Caspase Pathway

ICE (interleukin-converting enzyme) and CPP32 are the first two identified cysteine proteases, or CASPASEs, which specifically cleave proteins after aspartate residues during proteolytic apoptosis. The map in Fig. 7 shows the CASPASE precursor and proteolytic maturation [144] and Table 1 shows the classification of caspase family members [5, 25]. ICE is homologous to *ced3* of *C. elegans*, and is a five-member family, with ICE α, β, and γ being the major functional members, ICE δ being a minor one, and ICE ε being inhibitory to ICE α and β by forming a heterodimer to inactivate the latter's enzymatic action. In general, all apoptosis-associated cysteine proteases are expressed in the pro-enzyme form, such as p45 of ICEα; when apoptosis is activated, the p45 is cleaved to p30, which is then cleaved to p20 and p10. Once this proteolysis occurs, cells are committed to apoptotic death [25].

CASPASEs 1, 3, and 8 are probably most relevant to the cell type studied here, *i.e.* peripheral lymphocytes. CASPASE 1 (ICE), first identified in lymphocytes, primarily functions in pro-cytokine activation by making a single cleavage in each pro-cytokine, and thus activating them. CASPASEs 3 (CPP32) and 8 promote the downstream apoptosis pathways. We suggest a tentative cascade pathway is shown in Fig. 8, which is restricted to molecules involved in the *Fas*/TNF specific apoptosis pathway. Most likely, the arrangement of all three CASPASEs is such that CASPASE 8 is activated by ligand receptor binding (*Fas*L/*Fas*R) by an adapter protein, *Fas* activated death domain (FADD). This initial event leads to the downstream activation of CASPASE 8, which is then followed by CASPASE 3 (CPP32). By no means is this the complete picture; we do not know how many enzymatic steps lie between each of the steps described, nor how signalling events control each step. However, the overall pattern of biochemical cascade follows the general pattern of the reactions depicted in Fig. 8.

Three apoptosis activating factors, Apaf-1, 2, and 3, are equally important to CASPASE activation. Apaf-1 is homologous to *ced4* of *C. elegans*, and Apaf-3 is equivalent to CASPASE 9, while Apaf-2 is cytochrome C. When released from mitochondria, Apaf-1 and 3 collaborate with Apaf-2 (cytochrome C) to induce proteolytic processing and activation of CASPASE 3 (CPP32). The release of cytochrome C activates Apaf-1, which then activates Apaf-3 (CASPASE 9), which in turn activates CASPASE 3 (CPP32) [82, 87]. Recent studies have demonstrated that cytochrome C release from mitochondria contributes to apoptosis. In the cytosol, cytochrome C associates with a complex of Apaf-1 and CASPASE-9, and thereby induces the activation of CASPASE 3 [82, 178]. The induction of apoptosis is associated with CASPASE-3-mediated cleavage of poly(ADP-ribose) polymerase, protein kinase C-(PKC), and other proteins [27, 47, 114]. Importantly, overexpression of the anti-apoptotic *Bcl*-2 and *Bcl*-X$_L$ proteins blocks cytochrome C release and the activation of CASPASE-3 [68, 74, 170]. In addition, *Bcl*-2 exhibits a downstream CASPASE-inhibiting activity [124]. *Bcl*-2 is thought to act as the channel protein directly monitoring the gate traffic for ions or cytochrome C from mitochondria. However, this mitochondrial role may not always be the case; the primary event of FAS/TNF-induced apoptosis can activate CASPASEs without the elaborate Apaf involvement. Anyhow, the mitochondria-associated oxidative stress, *via* either cytochrome C release or malfunctioning electron chain transport, certainly serves to amplify apoptosis events.

CASPASE-8 has been shown to activate CASPASEs through two pathways: mitochondria-dependent and mitochondria-independent [78]. In the absence of mitochondria (*Xenopus* cell-free system requiring organelles), activation of a CASPASE cascade by CASPASE-8 produces only a partial apoptotic phenotype in nuclei added to the extract. In contrast, the mitochondria-dependent pathway, which involves the release of cytochrome C from mitochondria into the cytosol, triggers full nuclear apoptosis. Moreover, engagement of the mitochondria-dependent pathway is more efficient than the mitochondria-independent pathway, as it can be activated even by small amounts of CASPASE-8. Thus, mitochondria provide an efficient means ofamplifying the apoptotic signal transduced by CASPASE-8. For the complete apoptotic phenotype, including disruption of the nuclear envelope, the release of cytochrome C from mitochondria is required.

The inability of CASPASE-8 to produce nuclear membrane fragmentation in the absence of mitochondria argues that CASPASE-8 by itself can induce the activation of only some of the downstream effectors of apoptosis. CASPASE-6, while having no apoptosis-inducing activity of its own, acts synergistically with CASPASE-8 to produce complete nuclear fragmentation in *Xenopus* extracts lacking mitochondria.

These studies link a proximal signalling event, the activation of CASPASE-8 at the plasma membrane, with a distal apoptotic pathway involving cytochrome C release from mitochondria, and subsequent activation of executioner CASPASEs. It is proposed that the cell uses this indirect pathway (as opposed to a direct activation of some downstream CASPASEs) for the following three reasons: (a) cytochrome C can simplify the effect of small amounts of active CASPASE-8; (b) the participation of mitochondria would allow the cell greater flexibility in controlling the apoptotic process, through modulating the events leading to cytochrome C release; (c) the effects on mitochondria (including but not necessarily limited to cytochrome C release) may lead to cell death even in the absence of functional executioner CASPASEs, as has been shown for *Bax*-induced cell killing [167].

The release of cytochrome C appears to be induced by only certain forms of cellular stress, such as DNA-damaging agents and staurosporine [68, 87]. Distinct mechanisms for inhibition of cytochrome C release are also restricted to certain apoptotic stimuli. In this context, cellular resistance to vincristine and doxorubicin is associated with induction of Bcl-X_L (and not Bcl-2 expression) [28]. This suggests that Bcl-X_L contributes to inhibition of cytochrome C release in acquired multidrug resistance. In fact, acquired resistance to cytotoxic drugs is associated with inhibition of cytochrome C release, and this occurs via a Bcl-X_L-dependent mechanism that is induced during selective response to cytotoxic agents, thereby conferring resistance to apoptosis. Thus, the abrogation of mitochondrial cytochrome C release and CASPASE-3 and 9 activation are a part of the mechanism of acquired multidrug resistance [75].

Yeast, *Drosophila* and *C. elegans* have been shown recently to be three of the most powerful models for the study of aging. As described above, genes such as *daf*s, identified in *C. elegans* as associated with extension or shortening of life-span, are homologous to key members of the PI(3)K pathway. The SGS1 gene, the yeast homologue to the Werner's syndrome gene, is associated with the yeast replicative senescence program [135, 174]. CASPASEs in mammalian systems certainly share a rich repertoire of homologues in *C. elegans*, with *ced3* as the equivalent of CASPASE 1 (ICE) as the most famous example. Hence, this linkage between genes associated with mammalian systems and *C. elegans*, has the immediate advantage of experimentally testing our hypothesis of the functional impact on aging processes and the etiology of

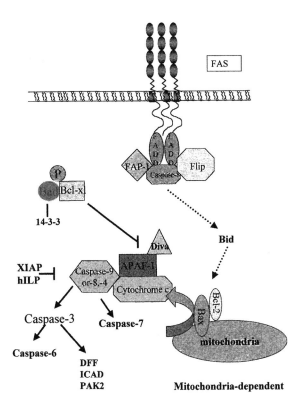

Figure 8. Signal transduction pathway of mitochondria-dependent CASPASE activation. The initiation of the pathway activation may start with either the FAS ligand binding, downstream targeting the Bid and mitochondria-associated Bcl-2 components, or the activation of a Bad/14-3-3 complex directed from the PI(3)K kinase 'hub', as indicated in Figure 1. The latter pathway is then the major route for the autocatalytic activation of CASPASEs, targetting mitochondrial involvement.

age-associated diseases, based on the rationale that some genes are evolutionarily conserved.

Recently, a model has been presented which describes a *Bcl-2*-CASPASE-apoptosome modality of cell death [54].

13. Are Senescent Fibroblasts Resistant to Apoptosis?

Since the first report that senescent human fibroblasts are resistant to growth-factor deprivation-induced apoptosis, additional studies have confirmed or refuted this finding; does replicative senescence indeed leave human fibroblasts not only in a permanent growth-arrested state, but also incapable of committing gene-directed suicide? This dispute resulted from two main factors: 1. the recent rapid increase in knowledge of mechanisms regulating apoptosis; and 2. the realization of the interconnecting networking of the signal transduction pathways, as described in the next section (Figure 9). It is well established that apoptosis can be induced by a variety of factors, ranging from serum deprivation (removal of growth factors) to treatment with metabolic toxic agents such as peroxide, AraC, or staurosporin . Therefore, even though replicatively senescent fibroblasts may not undergo apoptosis induced by growth factor deprivation, they can not nevertheless escape death induced by metabolic toxins. As depicted in Figs. 8 and 10, there are several pathways that can activate CASPASE activity. The signals induced in each of these pathways can simultaneously activate the targetted translational regulation, which provides *de novo* protein synthesis required for death, such as the actions of CASPASEs and c-*Myc*, as well as glycogen synthesis (Figure 8). However, another signalling cascade that leads to apoptosis may also play into this scenario, *i.e.* the FAS- and TNF-directed chain reactions which can be separated into mitochondria-independent or -dependent routes (Figure 8 & 10). Interestingly, with each of the steps involving activation of CASPASE activity, there is always an accompanying activation of *Bcl-2* and its sister genes, to provide the anti-apoptotic states. Ultimately, death occurs only as the outcome of the molecular tug-of-war between the gene forces pro- *versus* anti-apoptotic death. Overlaying this web of gene expressions determining life or death there is another layer of control, DNA repair mechanisms. The end-point decision whether to rescue cells from death or to commit them to die is dependent on the needs of their host tissues. In the brain, rescuing irreplacable neurons is the ever-present priority of the pro-survival program. In postmenopausal women, the aggressive pro-apoptosis program in no longer needed breast ductal epithelial cells may be the foremost task, to prevent their prolonged survival from posing the future threat of developing into cancerous tissue. However, the decision between pro- and anti-apoptosis programs may not be a simple black-and-white issue. Although it may seem vital to save every possible neuron from pro-apoptotic death, damaged and functionally compromised neuronal perseverance may be as detrimental as cell loss to the entire brain. Therefore, to repair or not repair is an intriguing molecular policy, as important as the strategy to die or not to die.

14. Signal Transduction Pathways and Apoptosis

Apoptotic death is a critical feature in embryonic development and tissue homeostasis [146]. An individual cell must integrate signals delivered by death-inducing and survival-promoting receptors to determine its fate in the context of the organism as a

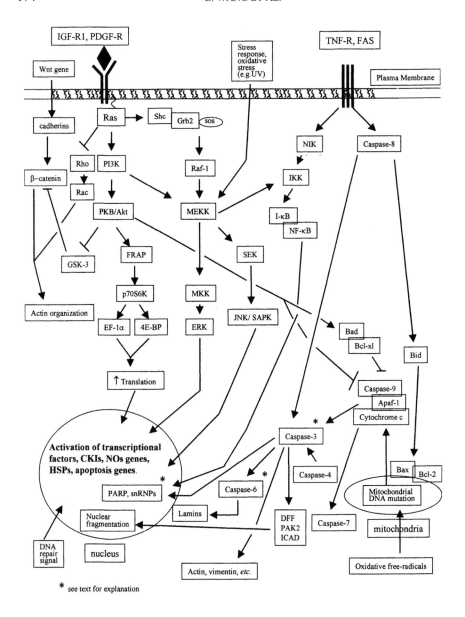

Figure 9. A bird's-eye view of four independent signal transduction pathways, leading to the activation of either apoptosis or DNA repair. The complex signals are depicted here as interweaving networks starting from the plasma membrane, as the signal-generating sites, to the nucleus as the final signal-targetting site. There are many 'hubs' serving as points of action for cross-talk among different pathways, such as MAP kinase (MEKK), CASPASE-3, and IKK of the NFκB pathway. There are also dominant strategic master points of reference, such as the PI(3)K kinase point and the mitochondria-associated actions.

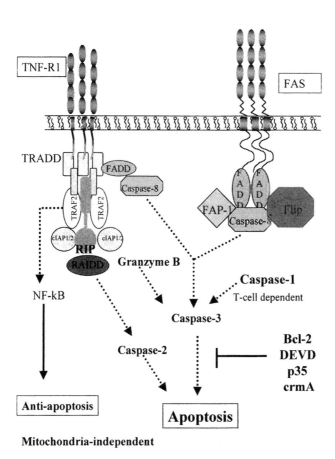

Figure 10. Signal transduction pathway of mitochondria-independent caspase activation. The initiation of the pathway activation starts with either TNF-receptor ligand binding and/or FAS ligand binding and converge to caspase 3 & casapse 2 for pro-apoptotic state, or via NF-kB for anti-apoptotic state. The entire signalling process does not involve the activation of mitochondria-associated biochemical action.

whole. Recent studies have clearly shown the importance of interactions of survival genes such as *Bcl-2* and *BclX*$_L$, and pro-apoptotic genes, such as *Bax* and *Bad*, in the determination of whether a cell takes a survival pathway or an apoptotic pathway [111].

Studies have indicated the importance of signalling pathway intermediates in the post-translational modification of *Bcl-2*-related proteins, which may play an important role in promoting cell survival or cell death. These observations suggest that signal transduction triggered by extracelluar survival or apoptotic signals can modify the function of the *Bcl-2* family proteins. These are as follows: (a) The ability of the chemotherapeutic agent taxol to induce phosphorylation of *Bcl-2* suggests this protein modification is a part of the mechanism for the induction of cell death [49]. This is consistent with the observation that deletion of the negative regulatory loop that contains the major serine/threonine phosphorylation sites in *Bcl-2* enables it to promote cell survival under conditions where it is normally inactive. (b) *Bcl-2* is cleaved by proteases activated during apoptosis, which may explain the death of lymphocytes infected with HIV. (c) A third mechanism involves post-translational modification, *i.e.,* serine phosphorylation of the pro-apoptotic *Bcl-2* family member, *Bad*, which results in its release form being sequestered by 14-3-3 proteins, and its ability to interact with *Bcl-2* (Figs. 5 and 6).

Interactions of *Bcl-2* family members by dimerization can affect survival promoting function [111]. These dimerizations are determined by phosphorylation or dephosphorylation reactions, mediated by signalling pathway intermediates. The pro-apoptotic member, *Bad,* can dimerize with *Bcl-2* and *Bcl-X*$_L$. *Bax*, another pro-apoptotic protein, also forms heterodimers with *Bcl-2* and *Bcl-X*$_L$. The ratio of *Bcl-2* to *Bax* determines the fate of transfected cells, *i. e.,* an excess of *Bcl-2* results in survival, but an excess of *Bax* results in death. Therefore, the balance between pro- and anti-apoptotic regulators of the *Bcl-2* family are yet another "rheostat" model of apoptosis regulation.

15. The *Bcl-2* Family Model

A model for the regulation of cell survival by *Bcl-2* family members has been suggested in which: (a) *Bax* homodimers comprise an active trigger for cell death. In the presence of excess *Bcl-2* or *Bcl-X*$_L$, heterodimerization with *Bax* prevents formation of toxic *Bax* homodimers, resulting in survival. (b) The presence of *Bad* would disrupt *Bcl-2/Bax* and *Bcl-X*$_L$ heterodimers, liberating *Bax* to homodimerize and promote cell death. (c) Alternatively, dominant heterodimerization with *Bax* or *Bad* may block *Bcl-2* and/or *Bcl-X*$_L$ from preventing apoptosis. All of the dimerization reactions are dependent upon post-translational modifications, *i.e.,* phosphorylation and dephosphorylation mediated by signalling proteins – kinases and phosphatases.

The relationship of *Bcl-2* function and receptor-mediated signal transduction, although not well worked out, involves: (a) *Bcl-2* phosphorylation at serine residues, in response to a variety of stimuli [49]. It is proposed that phosphorylation inactivates *Bcl-2*; (b) Phosphorylation of *Bcl-2* depends on the presence of a flexible loop domain which, when deleted, results in augmentation of *Bcl-2*'s protective function. This suggests that the loop domain may have a negative regulatory function, dependent upon serine phosphorylation. (c) Thus, the phosphorylation of *Bcl-2* lowers the apoptotic threshold, and promotes cell death, despite the presence of *Bcl-2*. It is the specific serine phosphorylation that attenuates the protective effect of *Bcl-2*. (d) *Bcl-2* and *Bcl-X*$_L$ have hydrophobic C-terminal domains that facilitate their distribution to the mitochondrial membrane and endoplasmic reticulum. (e) *Bcl-2* coprecipitates with, but may not be a substrate for, *Raf-1*. Interaction of *Bcl-2* with p21^{Ras} and with p23$^{R\text{-}Ras}$ has also been reported [31,55,95,108,149,150].

Similar post-translational modifications, such as phosphorylation, are important in the function of the pro-apoptotic protein, *Bad* [45]. *Bad* is a pro-apoptotic member of the *Bcl-2* family that is also phosphorylated on serine residues in response to IL-3, and is dephosphorylated upon withdrawal of growth factors. Non-phosphorylated *Bad* associates with *Bcl-X*$_L$, thereby neutralizing its anti-apoptotic activity. Phosphorylated *Bad* is unable to bind to *Bcl-X*$_L$, but does associate with 14-3-3 proteins. The Bad-14-3-3 complex resides, therefore, in the cytosol, thus preventing interaction of *Bad* with *Bcl-X*$_L$ at the mitochondrial membrane [104]. Mutation of the *Bad* serine phosphorylation site results in failure to bind 14-3-3, and accelerates apoptosis upon withdrawal of IL-3. Therefore, growth factors promote survival through induction of phosphorylation of *Bad*, resulting in its association with 14-3-3, which sequesters and inhibits its pro-apoptotic activity, *i.e.*, phosphorylation of *Bad* prevents *Bad-Bcl-X*$_L$ interaction, and frees *Bcl-X*$_L$ to carry out its anti-apoptotic function (Figure 5&6; 177).

Raf-1, a serine/threonine kinase, interacts with the mitochondrial membrane-bound *Bcl-2-Bad* complex to phosphorylate the serine residue in *Bad*, which is released from the complex. Thus, one mechanism by which growth factors promote survival is through the ability to induce phosphorylation of *Bad*, resulting in its association with *Bcl-2*. This sequesters *Bad* away from *Bcl-X*$_L$, freeing *Bcl-X*$_L$ to exert its anti-apoptotic effect. Finally, withdrawal of growth factors promotes apoptosis by activating the dephosphorylation of *Bad*, *via* action of a phosphatase which allows *Bad* to dimerize with *Bcl-X*$_L$ (177).

With respect to the identity of the kinases and phosphatases involved in these modifications, it has been shown that overexpression of *Bcl-2* can induce localization of *Raf-1*, a serine/threonine kinase, to mitochondria. Thus, one role of *Bcl-2* may be to target *Raf-1* to the outer mitochondrial membrane. In fact, mitochondrial localization of *Raf-1* is necessary for the phosphorylation of Bad, even though it (*Raf-1*) does not directly phosphorylate *Bad* (177). In summary, the signalling pathways to apoptosis are as follows: (a) *Bcl-2*-related proteins may serve as signalling receptors for intracellular organelles; (b) Serine kinases activated by pro-apoptotic stimuli lead to *Bcl-2* phosphorylation, to inactivate its ability to serve as a survival receptor; (c) These kinases could be counterbalanced by the activity of specific phosphatases; (d) Phosphorylation of *Bad* may provide a key survival-promoting decision point. By maintaining growth factor-mediated *Bad* phosphorylation, *Bad* remains associated with 14-3-3 in the cytosol, and is unavailable for interaction with *Bcl-2* or *Bcl-X*$_L$. This allows the latter proteins to execute anti-apoptotic function; (e) A potential role for *Raf-*

1 is to interact with *Bcl-2/Bad* at the mitochondrial membrane, thereby recruiting a signalling complex that phosphorylates *Bad* and facilitates its release. *Bad* then binds to 14-3-3 and is inactivated.

A humbling lesson for us biologists is the realization that the space between plasma and nuclear membranes is not just a cytoplasmic soup, with a few solid components such as cytoskeletal elements, mitochondria, and endoplasmic reticulum. The explosion of the last two decades' advances has helped us realize that the expression of 3 billion base pairs of the human genome is orchestrated in a most exquisitely controlled order of operation within this space, to link the signals from the extracellular environment to the nucleus, so that the cell may live and properly execute its function. Thus, the ever-increasing reports of new genes and proteins allow us to begin to formulate some conceptual understanding of the framework by which each of the signal pathways may be defined. The activation of CASPASE activity is no exception to this rule, and the connection of the CASPASE network with the broader network of signal transduction pathways (Fig. 9) provides a glimpse of how a particular event such as apoptosis is linked to the bigger picture of signal transduction pathways. Fig. 9 certainly does not represent the totality of all the genes involved in the cascade signalling for the pro- or anti-apoptotic states, nor does it represent all the pathways leading to successful apoptotic death.

At a glance, Fig. 9 presents a challenge to any biologist who attempts to understand the signal transduction pathways leading to apoptosis. Worst of all, one may be further discouraged by the fact that there may be hundreds of genes yet to be discovered in this complex network, as well as the two-dimensional presentation of a multi-dimensional actual operational mode within a cell. Alas, how can one begin to understand the mind of a cell in executing its wish to die, or to live, or to repair? The task of understanding this complex cellular orchestration needs to follow the principle of "divide-and-conquer". If one examines the complexity of Fig. 9 in all its cascade pathways, the final decision for death may be the combination of all the cascades working in unison. Deletion of any crucial components of these pathways may result in the failure or attenuation of signal processing for eventual death. By the same token, addition of new components to these pathways may accelerate the rate of death. What remains unknown, however, is the redundancy or frequency of each element in each pathway for other physiological events. Obviously, the MAPKinase pathway is not exclusively related to the need for apoptosis pathways, and the p70 S6 Kinase pathway is not playing the tune only of the process leading to self-directed death events. We do not know just how redundantly the same function can be fulfilled by different genes, or how frequently the same gene functions in different pathways. Obviously, with the completion of the Human Genome Project, one may begin to grasp the identities of the genes that may be involved, and thus pave the way to begin understanding how these genes function in the events governing apoptosis, and how aging and age-dependent diseases may be affected by dysfunctions of apoptosis.

Nevertheless, Fig. 9 does provide us a glimpse of the possible existence of so-called "Master Genes". Simply put, these genes are those strategically positioned at the crossroads of several pathways, for example, PI(3)K, *Akt*, MEKK, NF-kB, *etc.* A number of genetic studies in model systems already demonstrate the importance of PI(3)K in life-span determination. For example, cells derived from Nieman-Pick patients, which lack sphingomyelinase, are refractory to dying by apoptosis when exposed to genotoxic stress [126]. Interestingly, the null mutants of the p85 subunit of PI(3)K are resistant to oxidative apoptosis in a PI(3)K-independent, p53-dependent fashion. These data also suggest that the p85 subunit plays a key role in apoptosis; its phosphorylation may be important for its interaction and stability with p110. We should not be surprised if future experiments show that other genes fill the glamorous molecular status of "Master Gene". Biologists love to debate the existence of "master genes", with the subconscious wish that once their identities are known and their functions are understood, we can solve all the problems associated with diseases, and even the unwanted aging phenotypes. Therefore, we would like to know the number and the identity of these genes; are there five, or a dozen, a hundred, or a thousand? However, whether a gene's function happens for the moment to be "master" or "peon", it must be thought of in the context of the entire network of signal transduction. A gene's function may be "master" in one process, but inconsequential in another. What we do not know is whether, even as important as PI(3)K is, its function may be superceded by another yet-undiscovered "master gene" in another dominant pathway. The ultimate safeguard of a fault-proof signal transduction pathway may be the design of molecular flexibility in the execution of commands for cells to live, die, or repair. Evolution of the aging phenotype or age-dependent diseases may then demonstrate failures of such flexibility, and the loss of operational fluidity with considerable redundancy and parallelism.

It has been well documented that certain members of the NFκB/*Rel* family play a key role in the regulation of apoptosis [7]. A major cause of death in embryonic mice lacking the *Rel*A subunit is a massive apoptosis of liver cells [7, 12]. These observations suggest a role for *Rel*A-containing NFκB dimers in protecting cells from pro-apoptotic stimuli produced in the embryonic liver. Other studies support this observation, showing that the absence or inhibition of NFκB/*Rel* in cultured cells potentiates apoptosis in response to TNFα or T-cell activation. That the *Rel*A subunit plays such a role is indicated in macrophages and fibroblasts of *Rel* (-/-) mice; TNF readily kills these cells through TNF type 1 receptor, which can be prevented by transfecting *Rel*A [7]. A similar phenotype was observed using cell lines stably transfected with dominant negative IκBα proteins, produced by point mutations of phosphorylation sites or deletion of the regulatory N-terminus. The phenotype of these cells with respect to apoptotic resistance was very similar to that of *Rel*A(-/-). When a dominant negative IκBα mutant was expressed in transgenic mice, a loss of CD8+ cells was observed in the thymus [7]. These observations suggest that pro-apoptotic extracellular signals can also induce NFκB, which in turn induces expression of genes that are anti-apoptotic. Viruses that can induce NFκB could therefore protect against apoptotic elimination of infected cells.

Evidence of a pro-apoptotic function of *Rel*A is suggested by the observation that serum starvation of 293 cells causes cell death, accompanied by activation of *Rel*A-containing NFκB. *Bcl*-2 or a dominant negative *RelA* mutant can prevent cell death and kB-dependent reporter gene activity. A pro-apoptotic role of NFκB/*Rel* is also suggested by the study of radiation-sensitive fibroblasts from ataxia telangiectasia (AT) patients [7]. Radiation-induced apoptosis in AT cells is reduced by a dominant negative IκBα protein. Perhaps *RelA* can control apoptosis in quite opposite manners, depending on both cell type and the nature of the apoptotic stimulus.

There are studies suggesting that IkB can be a target for apoptosis-induced proteases. For example, induction of apoptosis by inactivation of a temperature-sensitive v-*Rel* mutant in chicken B cells causes an N-terminal truncation of chicken IκBα (pp40). This processing of IκBα can be prevented by the viral protease inhibitor *CrmA* in intact cells, and can be catalyzed *in vitro* by the ICE-like protease CPP32. A complete degradation of IκBα (chicken) is observed during apoptosis induced by the ts-v-*Rel* mutant. This is consistent with a persistent IκBα depletion (and NFκB activation) in 293 cells that are induced to undergo apoptosis by serum starvation.

16. Functional Genomic Studies, Apoptosis, DNA Repair, and Longevity

Ample examples from human medicine have shown that gains or losses of function associated with genetic mutations can cause fatal diseases and thus shorten life-span. Among them, the most famous are the loss of DNA repair ability in the recently identified Werner's syndrome gene [174], and the gain of function associated with the expansion of tri-nucleotide repeats found in Huntington's disease. The known examples of life-span extension in *C. elegans* mutants demonstrate that the mutations causing the reduction of function of *daf2*, *daf16,* and *daf23* genes extend the worms' life-span [70, 85, 103, 110]. Do aging processes or life-span determination share the same pattern of a loss of "bad" gene functions, or the opposite, the gain of "good" gene functions? For example, is the development of resistance to oxidative stress and longevity an example of the "gain of good gene fucntion" [79]? The answer may be both; as we suggested earlier, multiple molecular pathways comprise a combination of gains and losses of function. The functional operation of these genes may then be an orchestration into a symphonic masterpiece, extension of longevity.

Gain or loss of function for longevity may involve as many as a hundred genes in the decision at the triage point, including those involved in pro- or anti-apoptosis, DNA repair, pro- and anti-oxidative processes, in addition to transcription and translation factors functioning in the production of newly synthesized messages, and proteins needed for these processes. For example, in the context of cultured human fibroblasts, subsequent to change in any environmental stimuli, there are three options for cell response, triaging either to: (a) apoptotic cell suicide; (b) entering S-phase, to engage in DNA replication; or (c) remaining in a stable, non-apoptotic and non-DNA replication phase. Obviously, neither apoptosis nor unscheduled DNA replication is desirable in general, in the context of tissue homeostasis, since either avenue would cause loss or gain of cell mass in tissues. Thus, investigating the identity of genes influencing aging or age-dependent disease processes, and how their functions are coupled with those of many other genes in signalling cells to die, survive, or repair, is probably a key to obtaining genetic signatures for aging and age-dependent diseases [92, 143]. Clearly, environmental factors such as radiation or oxygen free radicals may trigger small changes, on the order of half-fold or less difference, in gene expression; over time, accumulation of these minor changes can cause damage and functional compromise [6, 11]. Therefore, genetic signatures for hereditary and nonhereditary factors are equally important in our quest to understand the regulation of aging and age-dependent diseases.

Obviously, developing gene profiling capability, with the technological capability to produce gene-specific DNA, RNA, or protein chips, will allow us to use rapid gene-screening processes to determine the expression profiles of multiple genes associated with each defined cellular state of a given tissue, utilizing their DNA and RNA as the screening materials. This approach will allow us to determine the inter- and intra-specific tissue variance in multiple gene expressions, including those involved in DNA replication and repair processes, transcriptional and translational regulation, regulation for signal transduction, oxidative defense, and apoptosis. Application of multiple gene-specific biochips will allow us to determine genetic profiles regulating the aging process, and those aberrations leading to the likelihood of developing age-dependent diseases.

In all, the approach of obtaining genetic signatures for aging and age-dependent diseases may identify groups of genes whose expressions are *qualitatively* up-regulated during critical phases of our life-span, such as menopause. However, a more important fact is that life-long experience may generate a genetic signature composed of *quantitative* changes in groups of genes undergoing small incremental expressions over the long duration. Identification of the genetic signatures for both qualitative and quantitative changes will create a molecular road-map to help us understand the functional genomic status of cells, contributing to our well being throughout our lifetime.

Acknowledgement

The authors would like to thank Mr. Alan Bloch for critical reading and editing of this manuscript. The work is supported by grants from National Institute on Aging to E. W. (AG09278 & AG07444) and to J.P. (P02AG10514) and from the defense Advance Research Project Agency (DARPA) of Department of Defense of the USA to E. W.

References

1. Adams, J. M. and Cory, S. 1998. The Bcl-2 protein family: arbiters of cell survival. Science 281: 1322-1326.
2. Alcorta, D. A., Xiong, Y., Phelps, D., Hannons, G., Beach, D., and Barrett, J. C. 1996. Involvement of the cyclin-dependent kinase inhibitor p16 (INK4a) in replicative senescence of normal human fibroblasts. Proc. Natl. Acad. Sci. USA 93: 13742-13747.
3. Alessi, D., James, S., Downes, C., Holmes, A., Gaffney, P., Reese, C., and Cohen, P. 1997. Characterization of a 3-phosphoinositide-dependent protein kinase which phosphorylates and activates protein kinase B alpha. Curr. Biol. 7: 261-269.

4. Almasan, A., Yin, Y., Kelly, R. E., Lee, E. Y. H., Bradley, A., Li, W. W., Bertino, J. R., and Wahl, G. M. 1995. Deficiency of retinoblastoma protein leads to inappropriate S-phase entry, activation of E2F-responsive genes, and apoptosis. Proc. Natl. Acad. Sci. USA 92: 5436-5440.

5. Alnemri, E. S., Livingston, D. J., Nicholson, D. W., Salvesen, G., Thornberry, N. A., Wong, W. W., and Yuan, J. 1996. Human ICE/CED-3 protease nomenclature. Cell 87: 171.

6. Ames, B. N., Shigenaga, M. K., and Hagen, T. M. 1993. Oxidants, antioxidants, and the degenerative diseases of aging. Proc. Natl. Acad. Sci. USA 90: 7915-7922.

7. Baeuerle, P. A. and Baltimore, D. 1996. NFκB: ten years after. Cell 87:13-20.

8. Bandyk, M. G., Sawczuk, I. S., Olsson, A., Katz, A. E., and Buttyan, R. 1990. Characterization of the products of a gene expressed during androgen-programmed cell death and their potential use as a marker of urogenital injury. J. Urology 143: 407-413.

9. Barinaga, M. 1993. Death gives birth to the nervous system; but how? Science 259: 762-763.

10. Barinaga, M. 1994. Cell suicide: by ICE, not Fire. Science 263: 754-756.

11. Barry, M. A., Behnke, C. A., and Eastman, A. 1990. Activation of programmed cell death (apoptosis) by cisplatin, other anticancer drugs, toxins and hyper-thermia. Biochem. Pharmacol. 40: 2353-2363.

12. Beg, A. A., Sha, W.C., Bronson, R.T., Ghosh, S. and Baltimore, D. 1995. Embryonic lethality and liver degeneration in mice lacking the RelA componentof NF-kappa B. Nature 376:167-170.

13. Bissonnette, R. P., Echeverri, R., Mahboubi, A., and Green, D. R. 1992. Apoptotic cell death induced by c-Myc is inhibited by Bcl-2. Nature 359: 552-554.

14. Brady, H. J., Gil-Gomez, G., Kirberg, J. and Berns, A. J. 1996. Bax alpha perturbs T cell development and affects cell cycle entry of T cells. EMBO J. 15: 6991-7001.

15. Brady, H. J., Salomons, G. S., Bodeldijk, R. C. and Berns, A. J. 1996. T cells from Bax alpha transgenic mice show accelerated apoptosis in response to stimuli but do not show restored DNA damage-induced cell death in the absence of p53 gene product. EMBO J. 15: 1221-1230.

16. Brady, R. O., Kanfer, J. N., Mock, M. B., and Fredrickson, D. S. 1966. The metabolism of sphingomyelin. II. Evidence of an enzymatic deficiency in Niemann-Pick disease. Proc. Natl. Acad. Sci. USA 55: 366-369.

17. Buttyan, R., Olsson, C. A., Pintar, J., Chang, C., Bandyk, M., Ng, P.Y., and Sawczuk, I. S. 1989. Induction of the TRPM-2 gene in cells undergoing programmed cell death. Molecular Cell Biol. 9(8): 3473-3481.

18. Campisi, J. 1996. Replicative senescence: an old lives' tale. Cell 84: 497-500.

19. Chen, J., Jackson, P. K., Kirschner, M. W., and Dutta, A. 1995. Separate domains of p21 involved in the inhibition of cdk kinase and PCNA. Nature 374: 386-388.

20. Chen, Y.-Y. and Rosenberg, N. 1992. Lymphoid cells transformed by Abelson virus require the v-abl protein-tyrosine kinase only during early G$_1$. Proc. Natl. Acad. Sci. USA 89: 6683-6687.

21. Chittenden, T., Harrington, E. A., O'Connor, R., Flemington, C., Lutz, R. J., Evan, G. I., and Guild, B. C. 1995. Induction of apoptosis by the Bcl-2 homologue bak. Nature 374: 733-736.

22. Chmura, S. J., Nodzenski, E., Crane, M. A., Virudachalam, S., Hallahan, D. E., Weichselbaum, R., R. and Quintans, J. 1996. Cross-talk between ceramide and PKC activity in the control of apoptosis in WEHI-231. Adv. Exp. Med. Biol. 406: 39-55.

23. Collins, M. K. L. and Rivas, A. L. 1993. The control of apoptosis in mammalian cells. TIBS 18: 307-309.

24. Colombel, M., Olsson, C. A., Ng, P. Y., and Buttyan, R. 1992. Hormone regulated apoptosis results from reentry of differentiated prostate cells into a defective cell cycle. Cancer Res. 50: 4313-4319.

25. Cryns, V. and Yuan, J. 1998. Proteases to die for. Genes & Dev. 12: 1551-1570.

26. Datta, R. R., Dudek, H., Tao, X., Masters, S., Fu, H., Gotoh, Y., and Greenberg, M. E. 1997. Akt phosphorylation of Bad couples survival signals to the cell-intrinsic death machinery. Cell 91: 231-241.

27. Datta, R., Kojima, H., Yoshida, K., and Kufe, D. 1997. CASPASE-3-mediated cleavage of protein kinase C theta in induction of apoptosis. J. Biol. Chem. 272: 20317-20320.

28. Datta, R., Manome, Y., Taneja, N., Boise, L. H., Weichselbaum, R. R., Thompson, C. B., Slapak, C. A. and Kufe, D. 1995. Overexpression of Bcl-XL by cytotoxic drug exposure confers resistance to ionizing radiation-induced internucleosomal DNA fragmentation. Cell Growth Differ 6: 363-370.

29. Datto, M. B., Li, Y., Panus, J. F., Howe, D. J., Xiong, Y., and Wang, X. F. 1995. Transforming growth factor B induces the cyclin-dependent kinase inhibitor p21 through a p53-independent mechanism. Proc. Natl. Acad. Sci. USA 92: 5545-5549.

30. Del Peso, L. ,Gonzalez-Garcia, M., Page, C., Herrera, R., and Nunez, G., 1997. Interleukin-3-induced phosphorylation of Bad through the protein kinase Akt. Science 278: 687-689.

31. Deng, G. and Podack, E. R. 1993. Suppression of apoptosis in a cytotoxic T-cell line by interleukin 2-mediated gene transcription and deregulated expression of the protooncogene Bcl-2. Proc. Natl. Acad. Sci. USA 90: 2189-2193.

32. Dice, J. F. 1993. Cellular and molecular mechanisms of aging. Physiological Review 74: 149-159.

33. Duttaroy, A., Qian, J.-F., Smith, J. S., and Wang, E. 1997. Up-regulated P21^{cip1} expression is part of the regulation quantitatively controlling serum deprivation-induced apoptosis. J. Cell. Biochem. 64: 434-446.

34. El-Deiry, W. S., Tokino, T., Velculescu, V. E., Levy, D. B., Parsons, R., Trent, J. M., Lin, D., Mercer, W. E., Kinzler, K. W., and Vogelstein, B. 1993. WAF1, a potential mediator of p53 tumor suppression. Cell 75: 817-825.

35. Ellis, R. E. and Horvitz, H. R. 1986. Genetic control of programmed cell death in the nematode C. elegans. Cell 44: 817-829.

36. Ellis, R. E., Jacobson, D. M., and Horvitz, H. R. 1991. Genes required for the engulfment of cell corpses during programmed cell death in Caenorhabditis elegans. Genetics 129: 79-94.

37. Enarl, M., Hug, N., and Nagata, S. 1995. Involvement of an ICE-like protease in Fas-mediated apoptosis. Nature 375: 78-81.

38. Evan, G. I., Wyllie, A. H., Gilbert, C. S., Littlewood, T. D., Land, H., Brooks, M., Waters, C. M., Penn, L. Z., and Hancock, D. C. 1992. Induction of apoptosis in fibroblasts by c-Myc protein. Cell 69: 119-128.

39. Evan, G., Harrington, E., Fanidi, A., Land, H., Amati, B., and Bennett, M. 1994. Integrated control of cell proliferation and cell death by the c-Myc oncogene. Phil. Trans. Royal Soc. London – Series B: Biological Sciences 345: 269-275.

40. Fanidi, A., Harrington, E. A., and Evan, G. I. 1992. Cooperative interaction between c-Myc and Bcl-2 proto-oncogenes. Nature 359: 554-556.

41. Flint, J. and Shenk, T. 1997. Viral transactivating proteins. Ann. Rev. Genet. 31: 177-212.

42. Franke, T., Kaplan, D., and Cantley, L. 1997. PI3K: Downstream AKTion blocks apoptosis. Cell 88: 435-437.

43. Franke, T., Kaplan, D., Cantley, L., and Toker, A. 1997. Direct regulation of the Akt proto-oncogene product by phosphatidylinositol-3,4-bis-phosphate. Science 275: 665-689.

44. Gagliardini, V., Fernandez, P. A., Lee, R. K. K., Drexler, H. C. A., Rotello, R. J., Fishman, M. C., and Yuan, J. -Y. 1994. Prevention of vertebrate neuronal death by the crmA gene. Science 263: 826-828.

45. Gajewski, T. F. and Thompson, C. B. 1996. Apoptosis meets signal transduction: Elimination of a Bad influence. Cell 87: 589-592.

46. Garcia, I., Martinou, I., Tsujimoto, Y., and Martinou, J. C. 1992. Prevention of programmed cell death of sympathetic neurons by the Bcl-2 proto-oncogene. Science 258: 302-304.

47. Ghayur, T., Hugunin, M., Talanian, R.V., Ratnofsky, S., Quinlan, C., Emoto, Y., Pandey, P., Datta, R., Kharbanda, S., Allen, H., Kamen, R., Wong, W. and Kufe, D. 1996. Proteolytic activation of protein kinase C delta by an ICE/CED 3-like protease induces characteristics of apoptosis. J. Exp. Med. 184: 2399-2404.

48. Goldstein, S. 1990. Replicative senescence: the human fibroblast comes of age. Science 249: 1129-1133.

49. Haldar, S., Jena, N., and Croce, C. M. 1995. Inactivation of Bcl-2 by phosphorylation. Proc. Natl. Acad. Sci. USA 92: 4507-4511.

50. Hara, K., Yonezawa, K., Sakaue, H., Kotani, K., Kojima, A., Waterfield, M., and Kasuga, M. 1995. Normal activation of p70 S6 kinase by insulin in cells overexpressing dominant negative 85kD subunit of phosphoinositide 3-kinase. Biochem. Biophys. Res. Commun. 208: 735-741.

51. Harper, J. W., Adami, G. R., Wei, N., Keyomarsi, K., and Elledge, S. J. 1993. The p21 Cdk-Interacting protein Cip1 is a potent inhibitor of G_1 cyclin-dependent kinases. Cell 75: 805-816.

52. Harrington, E., Bennett, M., Fanidi, A., and Evan, G. 1994. C-Myc-induced apoptosis in fibroblasts is inhibited by specific cytokines. EMBO J. 13: 3286-3295.

53. Hébert, L., Pandey, S., and Wang, E. 1994. Commitment to cell death is signalled by the appearance of a terminin protein of 30k. Exp. Cell Res. 210: 10-18.

54. Hengartner, M. O., Ellis, R. E., and Horvitz, H. R. 1992. Caenorhabditis elegans gene ced-9 protects cells from programmed cell death. Nature 356: 494-499.

55. Hermeking, H. and Eick, D. 1994. Mediation of c-Myc-induced apoptosis by p53. Science 265: 2091-2093.

56. Hockenbery, D. M., Oltvai, Z. N., Yin, X. M., Milliman, C. L., and Korsmeyer, S. J. 1993. Bcl-2 functions in an antioxidant pathway to prevent apoptosis. Cell 75: 241-251.

57. Houweling, A., He, X., and Wang, E. 1995. An apoptosis pathway regulated by the potential for proliferation rather than dependence on p53 gene expression. 35th Annual Meeting of the American Society for Cell Biology, Washington, D.C.

58. Huang, D. C., O'Reilly, L. A., Strasser, A. and Cory, S. 1997. The anti-apoptosis function of Bcl-2 can be genetically separated from its inhibitory effect on cell cycle entry. EMBO J. 16: 4628-4638.

59. Hueber, A.-O., Zornig, M., Lyon, D., Suda, T., Nagata, S., and Evan, G. I. 1997. Requirement for the CD95 receptor-ligand pathway in c-Myc-induced apoptosis. Science 278: 1305-1309.

60. Jacks, T., Fazeli, A., Schmitt, E. M., Bronson, R. T., Goodell, M. A., and Weinberg, R. A. 1992. Effects of an Rb mutation in the mouse. Nature 359: 295-300.

61. Janicke, R. U., Lee, F. H. H., and Porter, A. G. 1994. Nuclear c-Myc plays an important role in the cytoxicity of tumor necrosis factor alpha in tumor cells. Mol. Cell. Biol. 14: 5661-5670.

62. Johnson, D. G., Pierce, A. M., Philhower, J. L., and Schneider-Broussard. R. 1998. Differential activities of E2F family members: unique functions in regulating transcription. Mol. Carcin. 22:190-198.

63. Jones, M. and Murray, A. 1995. Evidence that ceramide selectively inhibits protein kinase C-alpha translocation and modulates bradykinin activation of phospholipase D. J. Biol. Chem. 270: 5007-5013.

64. Kamijo, T., Weber, J. D., Zambetti, G., Zindy, F., Roussel, M. F., and Sherr, C. J. 1998. Functional and physical interactions of the ARF tumor suppressor with p53 and Mdm2. Proc. Natl. Acad. Sci. USA 95: 8292-8297.

65. Kane, D. J., Sarafian, T. A., Anton, R., Hahn, H., Gralla, E. B., Valentine, J. S., Ord, T., and Bredesen, D. E. 1993. Bcl-2 inhibition of neural death: decreased generation of reactive oxygen species. Science 262: 1274-1277.

66. Kauffmann-Zeh, A., Rodriguez-Vicciana, P., Urlich, E., Gilbert, C., Coffer, P., Downward, J., and Evan. G. 1997. Suppression of c-Myc-induced apoptosis by Ras signalling through PI(3)K and PKB. Nature 385: 544-548.

67. Kennedy, S., Wagner, A., Conzen, S., and Jordan, J. 1997. The PI3-kinase/Akt signalling pathway delivers an anti-apoptotic signal. Genes & Dev. 11: 701-713.

68. Kharbanda, S., Pandey, P., Schofield, L., Israels, S., Roncinske, R., Yoshida, K., Bharti, A., Yuan, Z.-M., Saxena, S., Weichselbaum, R., Nalin, C., and Kufe, D. 1997. Role for Bcl-X_L as an inhibitor of cytosolic cytochrome C accumulation in DNA damage-induced apoptosis. Proc. Natl. Acad. Sci. USA 94: 6939-6942.

69. Khwaja, A., Rodriguez-Viciana, P., Wennstrom, S., Warne, P. H., and Downward, J. 1997. Matrix adhesion and Ras transformation both activate a phosphinositide 3-OH kinase and protein kinase B/Akt cellular survival pathway. EMBO J. 16: 2783-2793.

70. Kimura, K. D., Tissenbaum, H. A., Liu, Y., and Ruvkun, G. 1997. daf-2, an insulin receptor-like gene that regulates longevity and diapause in Caenorhabditis elegans. Science 277: 942-946.

71. King, K. L. and Cidlowski, J. A. 1995. Cell cycle and apoptosis: common pathways to life and death. J. Cell. Biochem. 58: 175-180.

72. Klefstrom, J., Arighi, E., Littlewood, T., Jaattela, M., Saksela, E., Evan, G. I., and Alitalo, K. 1997. Induction of TNF-sensitive cellular phenotype by c-Myc involves p53 and impaired NF-kB activation. EMBO J 16: 7382-7392.

73. Klefstrom, J., Vastrik, I., Saksela, E., Valle, J., Eilers, M., and Alitalo, K. 1994. c-Myc induces cellular susceptibility to the cytotoxic action of TNF-alpha. EMBO J. 13: 5442-5450.

74. Kluck, R. M., Bossy-Wetzel, E., Green, D. R., and Newmeyer, D. D. 1997. The release of cytochrome C from mitochondria: a primary site for Bcl-2 regulation of apoptosis. Science 275: 1132-1136.

75. Kojima, H., Endo, K., Moriyama, H., Tanaka,Y., Alnemri, E. S., Slapak, C. A., Teicher, B., Kufe, D. and Datta, R. 1998. Abrogation of mitochondrial cytochrome C release and CASPASE-3 activation in acquired multidrug resistance. J. Biol. Chem. 273: 16647-16650.

76. Komiyama, T., Ray, C. A., Pickup, D. J., Howard, A. D., Thornberry, N. A., Peterson, E. P., and Salvesen, G. 1994. Inhibition of interleukin-1B converting enzyme by the cowpox virus serpin CrmA. J. Biol. Chem. 269: 19331-19337.

77. Kulik, G., Klippel, A., and Weber, M. 1997. Antiapoptotic signalling by the insulin-like growth factor I receptor, phosphatidylinositol 3-Kinase, and Akt. Mol. Cell. Biol. 17: 1595-1606.

78. Kuwana, T., Smith, J. J., Muzio, M., Dixit, V., Newmeyer, D. D. and Kornbluth, S. 1998. Apoptosis induction by CASPASE-8 is amplified through the mitochondrial release of cytochrome C. J. Biol. Chem. 273: 16589-16594.

79. Larsen, P. L. 1993. Aging and resistance to oxidative damage in Caenorhabditis elegans. Proc. Natl. Acad. Sci. USA 90: 8905-8909.

80. Laurenz, J., Gunn, J., Jolly, C., and Chapkin, R. 1996. Alteration of glycerolipid and sphingolipid-derived second messenger kinetics in ras transformed 3T3 cells. Biochim. Biophys. Acta 1299: 146-154.

81. Lee, J., Hannun, Y., and Obeid, L. 1996. Ceramide inactivates cellular protein kinase C alpha. J. Biol. Chem. 271: 13169-13174.

82. Li, P., Nijhawan, D., Budihardjo, I., Srinivasula, S. M., Ahmad, M., Alnemri, E. S., and Wang, X. 1997. Cytochrome C and dATP-dependent formation of Apaf-1/CASPASE-9 complex initiates an apoptotic protease cascade. Cell 91: 479-489.

83. Li, Y., Jenkins, C. W., Nichols, M. A., and Xiong, Y. 1994. Cell cycle expression and p53 regulation of the cyclin-dependent kinase inhibitor p21. Oncogene 9: 2261-2268.

84. Lin, E. Y., Orlofsky, A., Berger, M. S., and Prystowsky, M. B. 1993. Characterization of A1, a novel hemopoietic-specific early-response gene with sequence similarity to Bcl-2. J. Immunol. 151: 1979-1988.

85. Lin, K., Dorman, J. B., Rodan, A., and Kenyon, C. 1997. Daf-16: an HNF-3/forkhead family member that can function to double the life-span of Caenorhabditis elegans. Science 278: 1319-1322.

86. Linette, G. P., Li, Y., Roth, K., and Korsmeyer, S. J. 1996. Cross talk between cell death and cell cycle progression: BCL-2 regulates NFAT-mediated activation. Proc. Natl. Acad. Sci. USA 93: 9545-9552.

87. Liu, X., Kim, C. N., Yang, J., Jemmerson, R., and Wang, X. 1996. Induction of apoptotic program in cell-free extracts: requirement for dATP and cytochrome C. Cell 86: 147-157.

88. Lockshin, R. A. and Zakeri, Z. 1991. Programmed cell death and apoptosis. In: Apoptosis: The Molecular Basis of Cell Death, L. D. Tomei and F. O. Cope, eds. Cold Spring Harbor, New York: Cold Spring Harbor Laboratory, pp 47-60.

89. Luo, Y., Hurwitz, J., and Massague, J. 1995. Cell-cycle inhibition by independent CDK and PCNA binding domains in p21Cip1. Nature 375: 159-161.

90. Marte, B. M. and Downward, J. 1997. PKB/Akt: Connecting phosphoinositide 3-kinase to cell survival and beyond. Trends Biochem. Sci. 22: 355-358.

91. Marx, J. 1993. How p53 suppresses cell growth. Science 262: 1644-1645.

92. May, P. C., Lampert-Elchelis, M., Johnson, S. A., Poirier, J., Masters, J. N., and Finch, C. E. 1990. Dynamics of gene expression for a hippocampal glycoprotein elevated in Alzheimer's disease and in response to experimental lesions in rat. Neuron 5: 831.

93. Mayo, L. D., Turchi, J. J. and Berberich, S. J. 1997. Mdm-2 phosphorylatin by DNA-dependent protein kinase prevents interaction with p553. Cancer Res. 57: 5013-5016.

94. Mazel, S., Burtrum, D., and Petrie, H. T. 1996. Regulation of cell division cycle progression by Bcl-2 expression: a potential mechanism for inhibition of programmed cell death. J. Exp. Med. 183: 2219-2226.

95. McDonnell, T. J., Deane, N., Platt, F. M., Nuñez, G., Jaeger, U., McKearn, J. P., and Korsmeyer, S. J. 1989. Bcl-2-immunoglobulin transgenic mice demonstrate extended B cell survival and follicular lymphoproliferation. Cell 57: 79-88.

96. Meikrantz, W. and Schlegel, R. 1995. Apoptosis and the cell cycle. J. Cell. Biochem. 58: 160-174.

97. Merlo, G. R., Basolo, F., Fiore, L., Duboc, L., and Hynes, N. E. 1995. p53-dependent and p53-independent activation of apoptosis in mammary epithelial cells reveals a survival function of EGF and insulin. J. Cell Biol. 128: 1185-1196.

98. Michael J. M., Lavin, M. F., and Watters, D. J. 1997. Resistance to radiation-induced apoptosis in Burkitt's lymphoma cells is associated with defective ceramide signalling. Cancer Res. 57: 3600-3605.

99. Mitmaker, B., Baytner, S., and Wang, E. 1993. Temporal relationship of statin and terminin expression in ventral lobe of rat prostate following castration. European J. of Histochem. 37: 295-301.

100. Mitmaker, B., Teng, C. J., Miller, M. M., and Wang, E. 1995. Characterization of the issue regression process in the uterus of older mice as apoptotic by the presence of Tp30, an isoform of terminin. European J. of Histochem. 39: 91-100.

101. Miura, M., Zhu, H., Rotello, R., Hartwieg, E. A., and Yuan, J. 1993. Induction of apoptosis in fibroblasts by IL-1β-converting enzyme, a mammalian homologue of the C. elegans cell death gene ced-3. Cell 75: 653-660.

102. Miyashita, T. and Reed, J. 1995. Tumor suppressor p53 is a direct transcriptional activator of the human Bax gene. Cell 80: 293-299.

103. Morris, J. Z., Tissenbaum, H. A., and Ruvkun, G. 1996. A phosphatidylinositol-3-OH kinase family member regulating longevity and diapause in Caenorhabditis elegans. Nature 382: 536-539.

104. Muslin, A. J., Tanner, J. W., Allen, P. M., and Shaw, A.S. 1996. Interaction of 14-3-3 with signalling proteins is mediated by the recognition of phosphoserine. Cell 84: 889-897.

105. Nakanishi, M., Adami, G. R., Robetorye, R. S., Noda, A., Venable. S. F., Dimitrov, D., Pereira-Smith, O. M., and Smith, J. R. 1995. Exit from G_0 and entry into the cell cycle of cells expressing p21 Sdi1 antisense RNA. Proc. Natl. Acad. Sci. USA 92: 4352-4356.

106. Nicholson, D. W., All. A., Thornberry, N. A., Vaillancourt, J. P., Ding, C. K., Gailant, M., Gareau, Y., Griffin, P. R., Labelle, M., Lazebnik, Y. A., Munday, N. A., Raju, M. S., Smulson, M. E., Yamin, T. T., Yu, V. L., and Miller, D. K. 1995. Identification and inhibition of the ICE/CED-3 protease necessary for mammalian apoptosis. Nature 376: 37-43.

107. Noda, A., Ning, Y., Venable, S. F., Pereira-Smith, O. M., and Smith, J. R. 1994. Cloning of senescent cell-derived inhibitors of DNA synthesis using an expression screen. Exp. Cell Res. 211: 90-98.

108. Nuñez, G., Hockenbery, D., McDonnell, T. M., Sorensen, C. M., and Korsmeyer, S. J. 1990. Deregulated Bcl-2 gene expression selectively prolongs survival of growth factor-deprived hematopoietic cell lines. J. Immunol. 144: 3602-3610.

109. O'Reilly, L. A., Harris, A. W., and Strasser, A. 1996. Bcl-2 transgene expression promotes survival and reduces proliferation of CD3-CD4-CD8-T cell progenitors. Int. Immun. 9: 1291-1301.

110. Ogg, S., Paradis, S., Gottlieb, S., Patterson, G. I., Lee, L., Tissenbaum, H. A., and Ruvkun, G. 1997. The fork head transcription factor DAF-16 transduces insulin-like metabolic and longevity signals in C. elegans. Nature 389: 994-999.

111. Oltvai, Z. N., Millman, C. L., and Korsmeyer, S. L. 1993. Bcl-2 heterodimerizes in vivo with a conserved homolog, Bax, that accelerates programmed cell death. Cell 74: 609-619.

112. Palmero, I., McConnell, B., Parry, D., Brooke, S., Hara, E., Bates, S., Jat, P., and Peters, G. 1997. Accumulation of p16ink4a in mouse fibroblasts as a function of replicative senescence and not of retinoblastoma gene status. Oncogene 15: 495-503.

113. Pandey, S. and Wang, E. 1995. Cells en route to apoptosis are characterized by the up-regulation of c-fos, c-Myc, c-jun, cdc-2, and Rb phosphorylation, resembling events of early cell-cycle traverse. J. Cell. Biochem. 58: 135-150.

114. Porter, A. G., Ng, P., and Janicke, R. U. 1997. Death substrates come alive (Review) Bioessays 19: 501-507.

115. Prabhakar, S., Antel, J. P., McLaurin, J., Schipper, H. M., and Wang, E. 1995. Phenotypic and cell cycle properties of human oligodendrocytes in vitro. Brain Research 672: 159-169.

116. Prisco, M., Hongo, A., Rizzo, M. G., Sacchi, A., and Baserga, R. 1997. The insulin-like growth factor I receptor as a physiologically relevant target of p53 in apoptosis caused by interleukin-3 withdrawal. Mol. Cell. Biol. 17: 1084-1092.

117. Raff, M. 1992. Social controls on cell survival and cell death. Nature 356: 397-400.

118. Ray, C. A., Black, R. A., Kronheim, S. R., Greenstreet, T. A., Sleath, P. R., Salvesen, G. S., and Pickup, D. J. 1992. Viral inhibtion of inflammation: cowpox virus encodes an inhibitor of the interleukin-1B-converting enzyme. Cell 69: 597-604.

119. Raybuck, S. A. and Livingston, D. J. 1994. Structure and mechanism of interleukin-1B converting enzyme. Nature 370: 270-275.

120. Raychaudhuri, P., Bagchi, S., Devoto, S. H., Kraus, V.,B., Morgan, E., and Nevins, J. R. 1991. Domains of the adenovirus E1A protein required for oncogenic activity are also required for dissociation of E2F transcription factor complexes. Genes & Dev. 5: 1200-1211.

121. Reed, J. C. 1994. Bcl-2 and the regulation of programmed cell death. J. Cell Biol. 124: 1-6.
122. Robles, S. J. and Adami, G. R. 1998. Agents that cause DNA double strand breaks lead to p16ink4a enrichment and the premature senescence of normal fibroblasts. Oncogene 16: 1113-1123.
123. Rodriguez-Viciana, P., Marte, B., Warne, P., and Downward, J. 1996. Phosphatidylinositol 3'-kinase: One of the effectors of Ras. Philos. Trans. R. Soc. Lond. B. Biol. Sci. 351: 225-231.
124. Rosse, T., Olivier, R., Monney, L., Rager, M., Conus, S., Fellay, I., Jansen, B., and Borner, C. 1998. Bcl-2 prolongs cell survival after Bax-induced release of cytochrome C. Nature 391: 496-499.
125. Sakamuro, D., Eviner, V., Elliot, K. J., Showe, L., White, E., and Prendergast, G. C. 1995. C-Myc induces apoptosis in epithelial cells by both p53-dependent and p53-independent mechanisms. Oncogene 11: 2411-2418.
126. Santana, P., Pena, L., Haimovitz-Friedman, A., Martin, S., Green, D., McLoughlin, M., Cordon-Cardo, C., Schuchman, E., Fuks, Z., and Kolesnick, R. 1996. Acid sphingomyelinase-deficient human lymphoblasts and mice are defective in radiation-induced apoptosis. Cell 86: 189-199.
127. Sawai, H., Okazaki, T., Takeda, Y., Tashima, M., Sawads, H., Okuma, M., Kishi, S., Umehara, H., and Domae, N. 1997. Ceramide-induced translocation of protein kinase C-delta and -epsilon to the cytosol. Implications in apoptosis. J. Biol. Chem . 272: 2452-2458.
128. Serrano, M., Hannon, G. J., and Beach, D. 1993. A new regulatory motif in cell-cycle control causing specific inhibition of cyclin D/CDK4. Nature 366: 704-707.
129. Serrano, M., Lin, A. W., McCurrach, M. E., Beach, D., and Lowe, S. W. 1997. Oncogenic ras provokes premature cell senescence associated with accumulation of p53 and p16INK4a. Cell 88: 593-602.
130. Seshadri, T. and Campisi, J. 1990. Repression of c-fos transcription and altered genetic program in senescent human fibroblasts. Science 247: 205-209.
131. Sester, U., Sawada, M., and Wang, E. 1990. Purification and biochemical characterization of statin, a nonproliferation specific protein from rat liver. J. Biol. Chem. 265: 19966-19972.
132. Shaw, P., Bovly, R., Tardy, S., Sahu, R., Sorda, B., and Costa, J. 1992. Induction of apoptosis by wild type p53 in a human colon tumor derived cell line. Proc. Natl. Acad. Sci. USA 89: 4495-4499.
133. Shieh, S.Y., Ikeda, M., Taya, Y. and Prives, C. 1997. DNA damage-induced phosphorylation of p53 alleviates inhibition by MDM2. Cell 91: 325-334.
134. Shimizu, S., Eguchi, Y., Kosaka, H., Kamiike, W., Matsuda, H., and Tsujimoto, Y. 1995. Prevention of hypoxia-induced cell death by Bcl-2 and Bcl-xL. Nature 374: 811-813.
135. Sinclair, D, A., Mills, K., and Guarente, L. 1997. Accelerated aging and nucleolar fragmentation in yeast sgs1 mutants. Science 277: 1313-1316.
136. Smeyne, R. J., Vendrell, M., Hayward, M., Baker, S. J., Miao, G. G., Schilling, K., Robertson, L. M., Curran, T., and Morgan, J. I. 1993. Continuous c-fos expression precedes programmed cell death in vivo. Nature 363: 166-169.
137. Smith, J. R. and Pereira-Smith, O. M. 1996. Replicative senescence: implications for in vivo aging and tumor suppression. Science 273: 63-67.
138. Spiegel, S. and Merrill, A. J. 1996. Sphingolipid metabolism and cell growth regulation. FASEB J. 10: 1388-1397.
139. Stein, G. H., Beeson, M., and Gordon, L. 1990. Failure to phosphorylate the retinoblastoma gene product in senescent human fibroblasts. Science 249: 666-669.
140. Sulston, J. E. and Horvitz, H. R. 1977. Post-embryonic cell lineages of the nematode Caenorhabditis elegans. Dev. Biol. 56: 110-156.
141. Symonds, H., Krall, L., Remington, L., Saenz-Robeles, M., Lowe, S., Jacks, T., and Van Dyke, T. 1994. p53-dependent apoptosis suppresses tumor growth and progression in vivo. Cell 73: 703-711.
142. Tewart, M., Quan, L. T., O'Rouke, K., Desnoyers, S., Zeng, Z., Baidler, D. R., Poirier, G. G., Salvesen, G. S., and Dixit, V. M. 1995. Yama/CPP32B, a mammalian homolog of ced-3, is a crmA-inhibitable protease that cleaves the death substrate poly(ADP) polymerase. Cell 81: 801-809.
143. Thompson, C. B. 1995. Apoptosis in the pathogenesis and treatment of diseases. Science 267: 1456-1462.
144. Thornberry, N. A. and Lazebnik, Y. 1998. CASPASEs: Enemies within. Science 281: 1312-1316.

145. Tsujimoto, Y., Gorham, J., Cossman, J., Jaffe, E., and Croce, C. M. 1985. The t(14;18) chromosome translocations involved in B-cell neoplasms result from mistakes in VDJ joining. Science 229: 1390-1393.

146. Ucker, D. S. 1991. Death by suicide - one way to go in mammalian cellular development. The New Biologist 3(2): 103-109.

147. Uhlmann, E. J., D'Sa-Eipper, C., Subramanian, T., Wagner, A. J., Hay, N., and Chinnadurai, G. 1996. Deletion of a nonconserved region of Bcl-2 confers a novel gain of function: suppression of apoptosis with concomitant cell proliferation. Cancer Res. 56: 2506-2509.

148. Vairo, G., Innes, K. M., and Adams, J.M. 1996. Bcl-2 has a cell cycle inhibitory function separable from its enhancement of cell survival. Oncogene 13: 1511-1519.

149. Vaux, D. L. 1993. Toward an understanding of the molecular mechanisms of physiological cell death. Proc. Natl. Acad. Sci. USA 90: 786-789.

150. Vaux, D. L., Aguila, H. L., and Weissman, I. L. 1992. Bcl-2 prevents death of factor-deprived cells but fails to prevent apoptosis in targets of cell-mediated killing. Int. J. Immunol. 4: 821-4.

151. Vaux, D. L., Cory, S., and Adams, J. M. 1988. Bcl-2 gene promotes haemopoietic cell survival and cooperates with c-Myc to immortalize pre-B cells. Nature 335: 440-442.

152. Vaux, D. L., Weissman, I. L., and Kim, S. K. 1992. Prevention of programmed cell death in Caeno-rhabditis elegans by human Bcl-2. Science 258: 1955-1957.

153. Venable, M., Blobel, G., and Obeid, L. 1994. Idenitification of a defect in the phospholipase D/diacylglycerol pathway in cellular senescence. J. Biol. Chem. 269: 26040-26044.

154. Vogelstein, B. and Kinzier, K. W. 1992. p53 function and dysfunction. Cell 70: 523-526.

155. Wagner, A. J., Kokontis, J. M., and Hay, N. 1994. Myc-mediated apoptosis requires wild-type p53 in a manner independent of cell cycle arrest and the ability of p53 to induce p21wafl/cip1. Genes & Dev. 8: 2817-2830.

156. Wang, E. 1989. Statin, a nonproliferation-specific protein, is associated with the nuclear envelope and is heterogeneously distributed in cells leaving quiescent state. J. Cell. Phys. 140: 418-426.

157. Wang, E. 1995. Failure to undergo programmed cell death in senescent human fibroblasts is related to inability to down-regulate Bcl-2 presence. Cancer Res. 55: 2284-2292.

158. Wang, E. and Lin, S. L. 1986. Disappearence of statin, a protein marker for non-proliferating and senescent cells, following serum-stimulated cell cycle entry. Exp. Cell. Res. 167: 135-143.

159. Wang, E. and Liu, D. 1995. Characterization of senescence- and apoptosis-dependent forms of terminin as posttranslational modification of a single polypeptide. J. Cell. Biochem. 60: 107-120.

160. Wang, E. and Pandey, S. 1995. Disappearance of statin, a nonproliferation specific nuclear protein, after induction of apoptosis in density-arrested mouse fibroblasts. J. Cell. Physiol. 163: 155-163.

161. Warbrick, E., Lane, D. P., Glover, D. M., and Cox, L. S. 1995. A small peptide inhibitor of DNA replication defines the site of interaction between the cyclin- dependent kinase inhibitor p21WAF1 and proliferating cell nuclear antigen. Cur. Biol. 5: 275-282.

162. Wu, X. and Levine, A. J. 1994. p53 and E2F-1 cooperate to mediate apoptosis. Proc. Natl. Acad. Sci. USA 91: 3602-3606.

163. Wyllie, A. H. 1980. Glucocorticoid-induced thymocyte apoptosis is associated with endogenous endonuclease activation. Nature 284: 555-556.

164. Wyllie, A. H. 1992. Apoptosis and the regulation of cell numbers in normal and neoplastic tissues: An overview. Cancer Metastasis Rev. 11:95-103.

165. Wyllie, A. H., Kerr, J. F. R., and Currie, A. R. 1980. Cell death: the significance of apoptosis. Int. Rev. Cytol. 68: 251-307.

166. Wyllie, A. H., Morris, R. G., Smith, A. L., and Dunlop, D. 1984. Chromatin cleavage in apoptosis: Association with condensed chromatin morphology and dependence on macromolecular synthesis. J. Pathol. 142: 67-77.

167. Xiang, J., Chao, D. T., and Korsmeyer, S.J. 1986. BAX-induced cell death may not require interleukin 1 beta-converting enzyme-like proteases. Proc. Natl. Acad. Sci. USA 93: 1455-14563.

168. Xiong, Y., Hannon, G. J., Zhang, H., Casso, D., Kobayashi, R., and Beach, D. 1993. p21 is a universal inhibitor of cyclin kinases. Nature 366: 701-704.

169. Xu, Y. and Baltimore, D. 1996. Dual roles of ATM in the cellular response to radiation and in cell growth control. Genes & Dev. 10: 2401-2410.
170. Yang, J., Liu, X., Bhalla, K., Kim C. N., Ibrado, A. M., Cai, J., Peng, T-I., Jones, D. P. and Wang, X. 1997. Prevention of apoptosis by Bcl-2: release of cytochrome C from mitochondria blocked. Science 275: 1129-1132.
171. Yang, T., Kozogas, K. M., and Craig, R. W. 1995. The intracellular distribution and pattern of expression of Mcl-1 overlap with, but not identical to, those of Bcl-2. J. Cell Biol. 126: 1173-1184.
172. Yin, C., Knudson, C. M., Korsmeyer, S. J., and VanDyke, T. 1997. Bax suppresses tumorigenesis and stimulates apoptosis in vivo. Nature 385: 637-640.
173. Yonisch-Rouach E., Rennitzky, D., Lotem, J., Sachs, I., Kimch, A., and Oren, M. 1991. Wild type p53 induces apoptosis of myeloid leukemic cells that is inhibited by interleukin-6. Nature 352: 545-547.
174. Yu, C. E., Oshima, J., Fu, Y. H., Wijsman, E. M., Hisama, F., Alisch, R., Matthews, S., Nakura, J., Miki, T., Ouais, S., Martin, G. M., Mulligan, J., and Schellenberg, G. D. 1996. Positional cloning of the Werner's syndrome gene. Science 272: 258-262.
175. Yuan, J. and Horvitz, H. R. 1990. Genetic mosaic analyses of ced-3 and ced-4, two genes that control programmed cell death in the nematode C. elegans. Dev. Biol. 138: 33-41.
176. Yuan, J. Y., Shaham, S., Ledous, S., Ellis, H. M., and Horvitz, H. R. 1993. The C. Elegans cell death gene ced3 encodes a protein similar to mammlian interleukin-1B-converting enzyme. Cell 75: 641-652.
177. Zha, J., Harads, H., Yang, E., Jockel, J., and Korsmeyer, J. J. 1996. Serine phosphorylation of death agonist Bad in response to survival factor results in binding to 14-3-3 not BCL-X. Cell 87: 619-628.
178. Zou, H., Henzel, W. ., Liu, X., Lutschg, A. and Wang, X. 1997. Apaf-1, a human protein homologous to C. elegans CED-4, participates in cytochrome C-dependent activation of CASPASE-3. Cell 90: 405-413.
179. Zundel, W. and Giaccia, A. 1998. Inhibition of the anti-apoptotic PI(3)K/Akt/Bad pathway by stress. Genes Dev. 12: 1941-1946.

© 2001 Elsevier Science B.V. All rights reserved.
The Role of DNA Damage and Repair in Cell Aging
B.A. Gilchrest and V.A. Bohr, volume editors.

DNA DAMAGE AND ITS PROCESSING WITH AGING:

HUMAN PREMATURE AGING SYNDROMES AS MODEL SYSTEMS

Vilhelm A. Bohr

Laboratory of Molecular Genetics, National Institute on Aging,
NIH, Baltimore, MD

1. DNA Damage, Age Accumulation, and Consequences

Living organisms are constantly exposed to oxidative stress from environmental agents and from endogenous metabolic processes. The resulting oxidative modifications occur in proteins, lipids and DNA. Since proteins and lipids are readily degraded and resynthesized, the most significant consequence of the oxidative stress is thought to be the DNA modifications, which can become permanent via the formation of mutations and other types of genomic instability. Many different DNA base changes have been seen following oxidative stress, and these lesions are widely considered as instigators for the development of cancer, aging and neurological degradation (for reviews, see (Ames, 1989a; Wiseman and Halliwell, 1996). The endogenous attack on DNA by ROS species generates a low steady-state level of DNA adducts that have been detected in the DNA from human cells (Dizdaroglu, 1993). There are over 100 oxidative base modifications in DNA, and one of these is 8-hydroxyguanosine (8-oxoG), which is the lesion that has been most widely studied. Oxidative DNA damage is thought to contribute to carcinogenesis and studies have shown that oxidative DNA damage accumulates in cancerous tissue. For example, higher levels of oxidative base damage were observed in lung cancer tissue compared with surrounding normal tissue (Olinski et al., 1992). Another study reported a 9-fold increase in 8-oxoG, 8-hydroxyadenine and 2,6-diamino-4-hydroxy-5-formamidopyrimidine in DNA from breast cancer tissue compared with normal tissue (Malins and Haimanot, 1991). Further, the cumulative risk of cancer increases dramatically with age in humans (Ames, 1989b) and cancer can in general terms be regarded as a degenerative disease of old age. There is evidence for the accumulation of oxidative DNA damage with age based on studies mainly measuring the increase in 8-oxoG (Sohal et al., 1994).

It has been more than 40 years since the free radical theory of aging was first put forward by Denham Harman (Harman, 1956). He proposed that free radicals would be produced in the utilization of molecular oxygen by animal cells, and that as a consequence of free radical reactions with nucleic acids and other cellular components, the animal would develop mutations and cancer. He also suggested that damage by endogenous free radicals was the fundamental cause of aging. A second theory, proposed three years afterward by Leo Szilard (Szilard, 1959), postulated specifically that time-dependent changes in somatic DNA, rather than other cellular constituents, were the primary cause of senescence. Both authors based their theories in large part on the be-

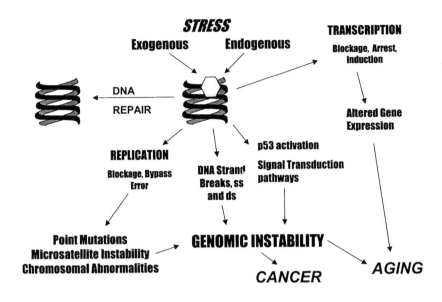

Figure 1. Examples of consequences of DNA damage.

lief, common at the time, that radiation accelerated aging independently of its effects on carcinogenesis.

Fig. 1 shows some of the consequences of DNA damage. It illustrates that DNA damage can be induced by exogenous or endogenous sources. UV irradiation or ionizing irradiation are examples of exogenous sources of stress. Reactive oxygen species generated by the oxidative phosphorylation that occurs in mitochondria and thus via the cellular metabolism is an example of an endogenous type of stress. Mutations in DNA can occur via replication of the damaged DNA whereby they become "fixed". Lesion bypass or replication errors can give rise to other forms of genomic instability, both microscopically and macroscopically. A lesion in DNA can block transcription completely, it may truncate the transcript, or it may cause errors in the transcription. Alternatively, the DNA damage may induce new transcripts, and a number of genes have been shown to be inducible by various forms of cellular stress. Incidentally, whereas induction of DNA repair enzymes in bacteria after stress is a well documented phenomenon, it was thought for many years that DNA repair enzymes were not inducible in mammalian cells. This notion is now under revision, and there is increasing evidence for the induction of DNA repair proteins after stress to mammalian cells (Fornace et al., 1998). These changes in transcription patterns that are affected by DNA damage may be part of the malignant phenotype, as many changes in transcription have been reported in

Table 1.

Pathway	Lesion removed
Base Excision Repair (BER)	Simple, monofunctional lesions, oxidative, alkylating
Mismatch Repair	Mismatches in DNA
Recombinational Repair	Interstrand Crosslinks
Nucleotide excision Repair (NER)	Bulky DNA lesions
Bulk genome repair	4 nitro quinoline, 6-4 photoproducts
Gene Specific Repair	
Transcription Coupled Repair	UV pyrimidine dimers

cancers. It is also likely to explain some of the phenotypes in aging, where reductions or other changes in transcriptional activity are well established (Bohr and Anson, 1995). Lesions in DNA can also lead to the activation of the p53 pathway leading to cell cycle arrest or apoptosis, or they can cause direct strand breaks in DNA, which again appear to be signals for several important enzymes and pathways.

It is estimated that there are several thousand DNA alterations in each cell in our organism per day (Lindahl, 1993) caused by the endogenous and exogenous stresses, and were it not for an efficient DNA repair process, our genetic material would be destroyed by these processes over a normal human lifetime.

DNA repair

Mammalian cells can make use of a variety of DNA repair pathways, and overviews of these are listed in Table 1. The Table also shows some examples of lesions that are removed by particular enzymatic processes. Most studies over the years have been based on the assumption that the DNA repair pathways listed were confined to the removal of specific lesions within certain categories. For example, as shown in Table 1, nucleotide excision repair (NER) was the system that removed bulky lesions in DNA such as pyrimidine dimers, cisplatin adducts and other lesions that dramatically changed the DNA structure. In contrast, base excision repair (BER) would be the process responsible for the removal of simple lesions in DNA, exemplified by alkylation lesions such as methyl guanines that are considered to afflict only small structural changes in DNA and which may not represent major blocks to transcription and replication. This concept has changed somewhat in recent years as it has become evident that many of the DNA repair pathways listed in Table 1 are overlapping and share components. Thus, although it is useful to think of repair pathways as confined to the removal of different types of DNA adducts, that distinction is of limited validity.

It would be much too ambitious to review all the pathways listed in Table 1. I shall refer to some recent reviews for more details and just highlight some aspects that are of importance in this context. Two of the most predominant pathways are NER and BER, and these will be discussed in more detail. The mismatch repair pathway is of particular interest in relation to cancer. There have been a number of recent reviews on the mismatch repair process, and the reader is referred to (Fishel, 1998; Kolodner, 1996; Lindahl et al., 1997; Modrich, 1994; Peltomaki, 1997; Thomas et al., 1996) for further information.

Base Excision Repair (BER)

Base excision repair of oxidative DNA damage is initiated by DNA glycosylases, a class of enzymes that recognize and remove damaged bases from DNA by hydrolytic cleavage of the base-sugar bond leaving an abasic site (AP site) (Lindahl, 1993). Several DNA glycosylases that recognize and process oxidation products of purines and pyrimidines have been identified in human cells. The human homologue of the bacterial endonuclease III (hNTH1) has been cloned and purified to homogeneity. It has a structural and functional homology to the bacterial enzyme and possesses a DNA glycosylase activity against a variety of oxidized pyrimidines (Aspinwall et al., 1997). Two human genes, hOGG1 and hOGG2, have been cloned. The gene products recognize and process 8-oxoG in DNA (Radicella et al., 1997; Roldán-Arjona et al., 1997; Rosenquist et al., 1997). These two proteins differ in their substrate specificity. There are at least two pathways for further processing of the arising AP site (Frosina et al., 1996; Matsumoto et al., 1994). One of these is catalyzed by the AP endonuclease, HAP1 (Ape, Apex, Ref1), which cleaves the phosphodiester bond immediately 5' to the AP site and generates 5'-sugar-phosphate and 3'-OH ends as it nicks DNA (Demple and Harrison, 1994). Removal of the 5'-terminal deoxyribosephosphate residue results in a single-nucleotide gap that is then filled by a DNA polymerase and sealed by DNA ligase (Dianov and Lindahl, 1994; Kubota et al., 1996; Singhal et al., 1995). In mammalian cells DNA polymerase β (pol β) is the major polymerase involved in the single-nucleotide excision patch repair pathway (Singhal et al., 1995; Sobol et al., 1996). It was recently demonstrated that DNA pol β has an intrinsic AP lyase activity that removes 5'-sugar-phosphate by β-elimination, and then pol β fills the single-nucleotide gap (Matsumoto and Kim, 1995; Piersen et al., 1996). This pathway is shown in Fig. 2. The hNTH glycosylase and the glycosylases OGG1 and OGG2 also have intrinsic AP lyase activities which introduces DNA strand breaks 3' to the baseless sugar (Radicella et al., 1997). Purified human OGG1 protein can catalyze a β-elimination reaction and most probably, like its yeast counterpart yOGG1, also catalyzes the removal of a 3'-sugar phosphate from the 3' incised AP site (Sandigurski et al., 1997). This suggests that the hOGG1 alone, or in combination with other enzymes, can generate a one-nucleotide gap even in the absence of pol β.

An alternative, long patch BER pathway has been reported (Frosina et al., 1996; Matsumoto et al., 1994). This pathway is shown in Fig. 2. In addition to a DNA glycosylase and AP endonuclease, this pathway also involves flap endonuclease (FEN1), proliferating cell nuclear antigen (PCNA), DNA polymerase and DNA ligase. Neither of these enzymes can remove a 5' sugar phosphate and generate a one nucleotide gap.

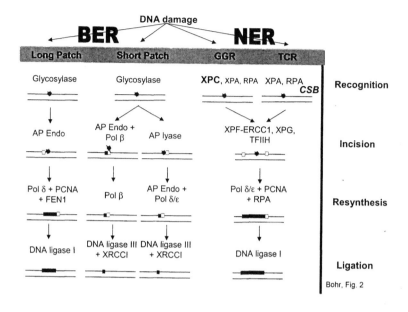

Figure 2. Base excision repair (BER) and Nucleotide excision repair (NER) pathways.

DNA polymerase first adds several nucleotides to the 3' end of the nick and exposes the 5' sugar phosphate as part of a single stranded flap structure. This flap structure is recognized and excised by FEN1 and DNA is finally ligated by DNA ligase (Klungland and Lindahl, 1997). These repair events result in a 2-7 nucleotide long repair patch. In this reaction, PCNA and probably replication factor C (RF-C), assist in loading the DNA polymerase onto the DNA and also stimulate the endonuclease FEN1 (Li et al., 1995). This pathway was recently reconstituted *in vitro* with purified *Xenopus laevis* and human proteins, and resulted in a repair patch of 2-7 nucleotides (Kim et al., 1998; Klungland and Lindahl, 1997).

Currently, there is a lot of research activity in the BER area, and the reader is referred to some recent reviews for further reading (Krokan et al., 1997; Lloyd, 1998; Seeberg et al., 1995; Wilson, 1998).

Nucleotide Excision Repair (NER)

Most of the understanding of the mechanisms of this pathway has come through the study of the human disorder xeroderma pigmentosum (XP). There are seven genetically different complementation groups of this disease, designated A-G. XP proteins are designated after the cell line in which they are mutated, e.g. the XPA protein is the one that would be mutated in XPA cells. The individuals afflicted with this condition suffer from high incidences of skin and internal cancers, hyperpigmentations and premature aging (Cleaver and Kraemer, 1995). Cells from XP patients can not incise their DNA at a site of a UV induced pyrimidine dimer, and are thus in general deficient in the incision process of NER. The enzymatic steps involved in NER are: recognition of the lesion, incision of the DNA, excision of the damaged DNA template, resynthesis of new DNA based on the intact template, and ligation of the newly formed DNA repair patch into reconstructed double helix. These processes are shown in Fig. 2. The recognition step involves the XPC and XPA proteins as well as replication protein A (RPA). The incision involves the structure specific endonucleases XPF-ERRC1 and XPG as well as the basic transcription factor TFIIH that contains the repair proteins XPD and XPB among its nine components. This incision process is a combination of DNA repair and transcription and illustrates the tight linkage between these processes. After incision, there is excision and then resynthesis performed by DNA polymerases followed by ligation performed by ligases. A number of recent reviews have discussed NER in much more detail and the reader is referred to (Friedberg et al., 1995; Friedberg, 1996a; Sancar, 1996; Wood, 1996; Wood, 1997; Wood and Shivji, 1997) for further reading.

Genomic Heterogeneity of NER

A major advance in the study of DNA repair has been the insight that NER and possibly other repair pathways occur with considerable heterogeneity over the mammalian genome. The NER pathway can thus be subdivided into different pathways relating to the functional and structural organization of the genome. Only about 1% of the genome is transcriptionally active, and a repair pathway entitled transcription coupled repair (TCR) operates here. The remainder of the genome, the 99% that is inactive has a separate pathway, general genome repair (GGR). Much work has been vested in the further delineation and clarification of these pathways. These concepts, and the stages of, bulk genome repair, gene repair, and transcription coupled repair are shown in Fig. 3. Most of the steps of NER as described above and seen in Fig. 2 are common to the two pathways, but there are some important differences. Most notably, there is a human condition, Cockayne syndrome (CS) in which the transcription coupled repair is deficient, and there is human condition, XPC in which the general bulk genome repair is deficient.

2. Human Premature Aging Syndromes as Models for Molecular Aging Studies

Human premature aging syndromes are rare disorders where the patients exhibit many of the features of normal aging at an early stage in life. These conditions are also termed segmental progerias, indicating that several, but not all of the symptoms reflect the normal aging process. They are natural human mutants and serve as good model systems for molecular aging research. Two of these conditions are the progeroid disorders Cockayne syndrome (CS) and Werner syndrome (WS), which share several clinical features that are associated with normal aging. WS arises from mutations in a single

Patterns of Nucleotide Excision Repair

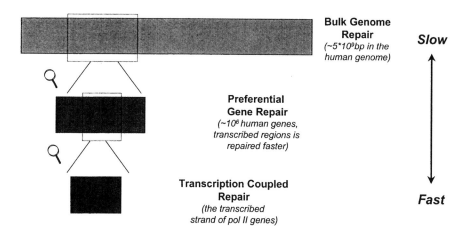

Bulk Genome Repair
(~5*10⁹bp in the human genome)

Slow

Preferential Gene Repair
(~10⁶ human genes, transcribed regions is repaired faster)

Transcription Coupled Repair
(the transcribed strand of pol II genes)

Fast

Figure 3. Stages of DNA repair from bulk to transcription coupling.

gene, *WRN*. The *CSB* and *WRN* gene products contain conserved helicase domains suggesting that these proteins function in diverse DNA metabolic pathways such as transcription, recombination, replication and DNA repair. Helicases disrupt the hydrogen bonds between two strands of a DNA double helix in a nucleoside triphosphate-dependent reaction (West, 1996). Although DNA helicases are ubiquitous in nature, they are specialized to fulfill particular molecular processes in transcription, DNA replication, repair, recombination, and chromosome segregation (Duguet, 1997). In addition, helicases which unwind DNA-RNA or RNA-RNA duplexes have been implicated in transcription and translation, respectively. Most recently, helicases or proteins related to helicases by sequence homology have been directly implicated in a number of human hereditary disorders including xeroderma pigmentosum (XP), CS, trichothiodystrophy, Bloom syndrome, WS, and alpha-thalassemia mental retardation (Ellis, 1997). The diverse clinical symptoms of these diseases suggest that these proteins are involved in distinct molecular genetic pathways of chromosome dynamics.

CS cells are deficient in a special type of DNA repair known as transcription coupled DNA repair (TCR) and they appear to be partially defective in basal transcription. WS cells may have subtle defects in DNA repair, and exhibit reduced transcription as well. The clarification of the precise role of these gene products should provide significant clues into the mechanism of aging.

Cockayne Syndrome (CS)

Patients with CS suffer from cachectic dwarfism, mental retardation, hyperpigmentation and hypersensitivity to sun exposure and various neurological sequelae (Nance and Berry, 1992). Cells from CS patients are sensitive to UV light, exhibit a delay in recovery of DNA and RNA synthesis following irradiation, and are defective in preferential repair and strand-specific repair of active genes (Friedberg, 1996b). Complementation studies demonstrate that there are at least two genes involved in CS, designated *CSA* and *CSB*. CSB protein, by sequence comparison, belongs to the SNF2 family of proteins, which have roles in transcriptional regulation, chromosome stability and DNA repair. The cellular and molecular phenotype of CS include a significantly increased sensitivity to a number of DNA-damaging agents including UV irradiation (Friedberg, 1996b). Studies in CS cells were initially confined to DNA repair in the general, overall genome, where no defect was found (reviewed in (Friedberg et al., 1995)). However, CS cells are defective in the preferential repair of active genes and in the preferential repair of the transcribed strand of such genes (Evans and Bohr, 1994; Venema et al., 1990). This defect in transcription coupled repair (TCR) in CS is not only found after UV exposure but also after exposure to certain forms of oxidative stress (Leadon and Cooper, 1993). Transfection of the *CSB* gene into hamster cells with the CS-B phenotype completely restores TCR and UV resistance to normal levels, demonstrating that the defect in TCR in CS-B is due to mutation in that gene (Orren et al., 1996).

The complex clinical phenotype of CS, however, suggests that DNA repair may not be the primary defect. We have reported a defect in basal transcription in CS both *in vivo* and *in vitro* (Balajee et al., 1997). This transcription defect is seen in CS-B lymphoblastoid cells and fibroblasts without any exposure to stress such as UV light. Other studies support that there is a transcription defect. A previous study found that expression of a metalloprotease was reduced by 50% in CS cells (Millis et al., 1992), and recently it was reported that the purified CSB protein stimulates transcription, presumably as an elongation factor (Selby and Sancar, 1997). It was also shown that the CSB protein binds to RNA polymerase II (vanGool et al., 1997). In all, there are now several indications that the CSB protein plays a role in transcription.

We have used an *in vitro* assay to measure the incision event of the DNA repair process. During the first step of BER, there is an incision in DNA 5' to the lesion. The incision can be quantitated in cell extracts by using oligonucleotide duplexes that contain a single 8-oxoG lesion at a defined site. In primary CS-B cell lines, we observe a deficiency in incision (Dianov et al., 1999). In contrast, when we used a substrate containing other single lesions, such as uracil or thymine glycol, there was no defect in CS-B cells compared to normal. The deficiency in incision at 8-oxoG was seen in four different CS-B primary fibroblast cell lines. Further, we measured incision activity in a CS family. The affected proband had the lowest levels of incision, and the rest of the measurements were compatible with the possibility that the other sibling was a normal homozygote, and that the parents were heterozygotes for the *CSB* mutation (Dianov et al., 1999). The incision deficiency can be complemented by transfection of the CS-B cell line with a plasmid containing the intact *CSB* gene, suggesting a role for CSB in the recognition of 8-oxoG (Dianov et al., 1999). This is the first report of a general, global repair defect in CS-B. This deficiency in incision most likely reflects a decrement in the activity of the 8-oxoG glycosylase, and in support of this we detect lower levels of expression of the human *OGG1* gene in CS-B cells than in normal cells. The expression of *OGG1* is markedly higher in CS-B cells transfected with the wild type *CSB* gene (Dianov et al., 1999). This is a novel and not previously reported property of the *CSB* gene and it leads to the suggestion that the CSB protein is involved in the transcriptional regulation of the *OGG1* and perhaps other DNA repair genes.

The CSB protein apparently functions at the crossroads of DNA repair and transcription. It has been reported to interact with the structure specific incision endonuclease XPG, CSA protein (Friedberg, 1996b), and RNA polymerase II (vanGool et al., 1997). It has considerable homology to the SWI/SNF complex, which in yeast is associated with RNA polymerase II (Van Gool et al., 1997). The SWI/SNF complex is involved in the initiation phase of the transcription process. We have also observed that CS-B cells appear to have a "looser" chromatin structure than normal cells (Balajee et al., 1997), and this would be compatible with a function that involves a role in chromatin structural assembly. It would appear that the CSB protein has more than one function and is most likely involved in a large number of protein-protein interactions in transcription and repair pathways. One or more of these is likely to be very important for the assembly of the DNA repair and transcription factory at the nuclear matrix. This is supported by previous studies showing that CS-B cells are defective in the early, nuclear matrix associated DNA repair (Mullenders et al., 1988)

A functional analysis of the *CSB* gene has been undertaken in our laboratory to better understand the nature of the molecular deficiencies observed in CS. Mutants, generated by site-directed mutagenesis have been tested for genetic complementation of *CSB* null cell lines by cell viability and RNA synthesis recovery upon exposure to UV light and other genotoxic agents. We have also measured another phenotypic alteration in the CSB cells, a marked increase in apoptosis. Point mutations in ATPase motifs I and II of *CSB* dramatically reduce CSB function *in vivo* suggesting that ATP hydrolysis by CSB protein is required for transcription-coupled repair of DNA damage. In contrast to the ATPase point mutations, deletions in the conserved acidic domain do not appear to interfere with the repair capacity of CSB protein. As for the repair studies, we also find that the ATPase activity is essential for resistance to apoptosis. Mutation in this domain leads to dramatically increased apoptosis, and mutations in the acidic, protein binding domain, do not affect the apoptosis. Our results suggests that the acidic domain may be conserved in the SWI-SNF family for some other function than repair and apoptosis. Further studies are in progress to address other important functions of CSB as they relate to the structural domains of the protein.

Werner Syndrome (WS)

Werner Syndrome (WS) is a homozygous recessive disease characterized by early onset of many characteristics of normal aging, such as wrinkling of the skin, graying of the hair, cataracts, diabetes, and osteoporosis. Cancers, particularly sarcomas, have been seen in these patients with increased frequency. The symptoms of WS begin to appear around the age of puberty, and most patients die before age 50.

The premature aging disorder WS is caused by mutations in a gene (*WRN*) belonging to the RecQ family of DNA helicases (Yu et al., 1996). The WRN protein has been demonstrated biochemically to be a DNA unwinding enzyme by several laboratories (Gray et al., 1997; Suzuki et al., 1997) including ours (see later). The enzyme is a DNA-dependent ATPase and catalyzes strand displacement in a nucleotide-dependent reaction. It now appears that the WRN protein (WRNp) has more than one enzymatic activity. It has recently been shown that it also has a 3'-5' exonuclease activity (Huang et al., 1998; Kamath-Loeb et al., 1998; Shen et al., 1998).

The precise molecular deficiencies involved in the clinical phenotypes of WS remain to be defined. The well-documented genomic instability of WS may point to a defect in replication or recombination (Fukuchi et al., 1989). WS cells exhibit replicative senescence and delayed S phase progression in some cell lines (Poot et al., 1992). A defect in recombination is suggested by the hyper-rec phenotype observed in some WS cell lines. Moreover, the *WRN* homologs in yeast, *sgs1* (*S. cerevisiae*) and *rqh1* (*S. pombe*), as well as the bacterial homolog *recQ*, all suppress illegitimate recombination (Hanada et al., 1997). It is conceivable that the helicase function of the WRN protein may also serve as an anti-recombinase, but this hypothesis remains to be directly tested.

Some evidence suggests a role for the WRN protein in DNA repair. This notion is supported by a report that WS cells are sensitive to the carcinogen 4-nitroquinoline 1-oxide (4-NQO) (Ogburn et al., 1997). However, WS cells do not exhibit a hypersensitivity to UV light or other DNA damaging agents suggesting that a repair deficiency in WS is subtle and perhaps not the primary defect in the disease. A fine structure repair defect in lymphoblast but not fibroblast WS cell lines was reported by this laboratory using the gene-specific and strand-specific repair assays (Webb et al., 1996). Also, mismatch repair was found to be defective in some WS cell lines, but more recent data suggests that this is not the case (Bennett et al., 1999).

In addition to repair or replication defects, evidence from this laboratory suggests WS cells may be partially defective in transcription. We have used a variety of WS lymphoblast cell lines that carry homozygous mutations in the *WRN* gene to assess the role of *WRN* gene in transcription. Transcription was measured both *in vivo*, by [³H]-UTP incorporation in the chromatin of permeabilized cells and *in vitro*, by using a plasmid template with a RNA polymerase II specific promoter. The transcription efficiency in different WS cell lines was found to be reduced to 40-60% of the transcription observed in normal cells using both assays. This defect can be complemented by the addition of normal extract to the chromatin of WS cells. Furthermore, the addition of purified WRN protein to the *in vitro* transcription assay stimulates RNA polymerase II driven transcription. These findings are supported by a yeast hybrid-protein reporter assay that reveals that a direct repeat of 27 amino acids, located proximally to the helicase domain, is important for the function of WRN as a transcriptional activator. These observations suggest that the WRN protein may act as a general activator of RNA polymerase II transcription. Similar to the XPB and XPD helicases of the TFIIH particle, the WRN protein may play a direct role in transcription initiation and/or elongation. A future area of work will be to elucidate the protein complexes involving the WRN protein. These studies will improve our understanding of the WRN helicase in cellular-molecular pathways that are deficient in WS.

Recently, we have overexpressed, purified and characterized a recombinant WRN protein. Using the baculovirus system, overexpression of a His-tagged WRN protein in *Sf9* insect cells was achieved. The WRN protein was subsequently purified to near homogeneity using a number of chromatographic steps, including Ni-affinity chromatography. Likewise, insect cells infected with baculovirus either without the WRN cDNA or with mutant WRN constructs were subjected to purification by the same procedure and used as controls in a number of assays. The purified WRN protein was employed in a number of experiments to investigate its biological function. Firstly, we demonstrated that the protein has helicase activity as previously shown by other laboratories (Gray et al., 1997; Suzuki et al., 1997). We have recently observed that the purified protein binds to DNA, with somewhat higher affinity for the single-stranded than the double-stranded form. There is no apparent increased affinity of WRN to damaged DNA. Moreover, extracts from EBV-transformed WS cells are nucleotide excision repair-proficient as shown by an *in vitro* DNA repair synthesis assay using a variety of damaged substrates. The WRN protein is capable of unwinding larger duplex molecules, up to more than two hundred basepairs. For the unwinding of these larger DNA duplexes, however, we find that human replication protein A (RPA) is an essential requirement (Brosh et al., 1999). This suggest a functional interaction between RPA and the WRN protein, and we have recently strengthened the evidence for this notion by showing that there is also a physical interaction between these two molecules (Brosh et al., 1999). This is the first demonstration of a protein interaction with WRN protein, but there are bound to be many more.

3. Transcription Changes in Aging

Our results from both the CS and WS syndromes show a deficiency in transcription as part of the molecular phenotype. This is in some accordance with the general concept in aging where changes and deficiencies in transcription have been reported. The literature on this is voluminous and there are opposing conclusions, but a thorough review on this subject (Van Remmen and Ward, 1995) concluded that more studies showed down regulation than up regulation of transcription with advancing age. Based on our knowledge about the WRN and CSB proteins we speculate that they are involved in the regulation of transcription and thus more or less directly affect a stimulation of transcription. There are other possible explanations for a decline in transcription with age. One of these is based on recent studies on how DNA damage might affect transcription.

DNA damage, which would be expected to accumulate in a repair deficient syndrome, can attract transcription factors including the TBP element of the basal transcription factor TFIIH (Vichi et al., 1997). It was recently shown that DNA damage can lead to a decrease in transcription rate via its "titration" of transcription factors (You et al., 1998). In our lab, we have extended these observations and shown that the addition of external DNA damage to a DNA transcription system *in vitro* greatly affects the transcription by inhibition (Cullinane et al., 1999). Treatment of Adenovirus major late promoter (AdMLP) containing templates with cisplatin resulted in a drug concentration dependent inhibition of RNA pol II transcription initiation. RNA pol II transcription of non-damaged template was subsequently shown to be significantly inhibited by the addition of exogenously damaged DNA, suggesting that cisplatin damage may sequester an essential transcription factor. Constructs containing site-specifically placed cisplatin lesions were subsequently utilized to investigate the effect of individual lesions on RNA pol II elongation. Differential sensitivity of the polymerase to cisplatin damage was observed in which elongating RNA pol II was shown to stall at a GTG intrastrand adduct but bypass a GG adduct.

Thus, transcription rates might be decreased under conditions where DNA damage accumulates, and this may contribute to the CS and WS phenotypes as well as to the aging phenotype in general.

4. Outlook

There have been dramatic advances in the understanding of molecular dysfunctions in the premature aging syndromes. This has been greatly facilitated by the cloning and identification of a number of the genes responsible for these phenotypes. A number of these genes have helicase sequences, and helicases are involved in various important DNA metabolic processes such as transcription, DNA repair and replication. It is not yet clear how important the helicase activity is for the phenotypes, and other functional domains of the genes may turn out to be at least as important. The function of the CSB and WRN proteins appear to be at the crossroads of aging, DNA repair, DNA replication, and transcription, so these studies nicely combine a mechanistic interest in basic DNA related processes with an interest in aging. Some of the points in support of studying the premature aging disorders are shown in Fig. 4.

Acknowledgements

I thank Morten Sunesen for help with the Figures and critical comments. I thank Tanja Frederiksen for comments. I appreciate the interaction with the Danish Center for Molecular Gerontology.

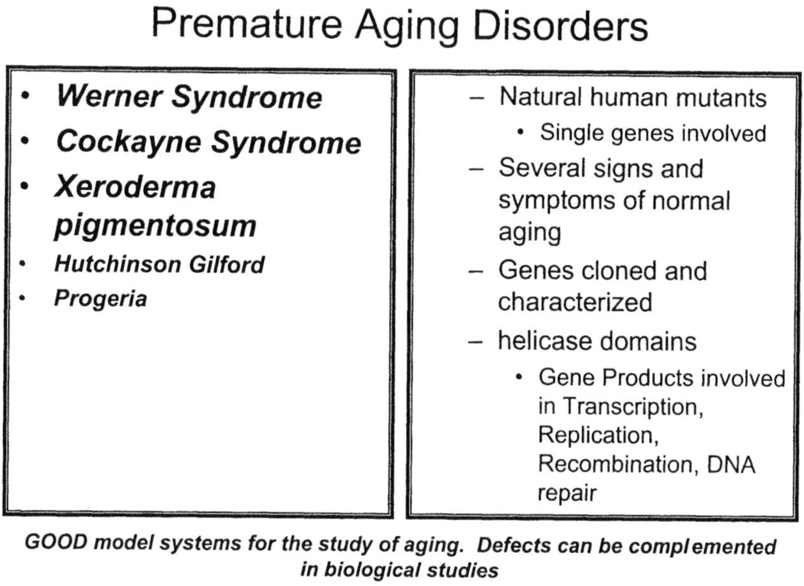

Premature Aging Disorders

• **Werner Syndrome** • **Cockayne Syndrome** • **Xeroderma pigmentosum** • *Hutchinson Gilford* • *Progeria*	– Natural human mutants • Single genes involved – Several signs and symptoms of normal aging – Genes cloned and characterized – helicase domains • Gene Products involved in Transcription, Replication, Recombination, DNA repair

GOOD model systems for the study of aging. Defects can be complemented in biological studies

Figure 4. Premature aging syndromes are good model systems.

References

Ames, B.N. (1989a). Endogenous DNA damage as related to cancer and aging. Mutat Res *214*, 41-6.

Ames, B.N. (1989b). Endogenous oxidative DNA damage aging and cancer. Free Radical Research Communications *7*, 121-128.

Aspinwall, R., Rothwell, D.G., Roldan-Arjona, T., Anselmino, C., Ward, C.J., Cheadle, J.P., Sampson, J.R., Lindahl, T., Harris, P.C., Hickson, I.D. (1997). Cloning and characterization of a functional human homolog of Escherichia coli endonuclease III. Proc.Natl.Acad.Sci.U.S.A. *94*, 109-114.

Balajee, A.S., May, A., Dianov, G.L., Friedberg, E.C., Bohr, V.A. (1997). Reduced RNA polymerase II transcription in intact and permeabilized Cockayne syndrome group B cells. Proc Natl Acad Sci U.S.A *94*, 4306-4311.

Bennett, S.E., Umar, A., Kodama, S., Barrett, J.C., Monnat, R.J., Kunkel, T.A. (1999). Evidence Against a Role for the Werner Syndrome Gene Product in DNA Mismatch Repair. In: Alfred Benzon Symposium 44, Molecular Biology of Aging, ed. V.A.Bohr, B.F.Clark, T.StevnsnerCopenhagen: Denmark, 217-229.

Bohr, V.A., Anson, R.M. (1995). DNA damage, mutation and fine structure DNA repair in aging . Mutat Res *338*, 25-34.

Brosh, R.M.O.D.K., Nehlin, J.O., Ravn, P.H., Kenny, M.K., Machwe, A., Bohr, V.A. (1999). Functional and physical interaction between WRN helicase and human replication protein A. J.Biol Chem. 38, 6204-12.

Cleaver, J.E., Kraemer, K.H. (1995). Xeroderma Pigmentosum and Cockayne Syndrome.

Cullinane, C., Mazur, S., Essigman, J.M., Phillips, D.R., Bohr, V.A. (1999). Inhibition of RNA polymerase II transcription in human cell extracts by cisplatin DNA damage. Biochemistry *In Press.*

Demple, B., Harrison, L. (1994). Repair of oxidative damage to DNA: Enzymology and biology. Annu Rev Biochem *63*, 915-948.

Dianov, G., Bischoff, C., Sunesen, M., Bohr, V.A. (1999). Repair of 8-oxoguanine is deficient in Cockayne syndrome B cells. Nucleic Acid Res. 27, 1365-1368.

Dianov, G., Lindahl, T. (1994). Reconstitution of the DNA base excision-repair pathway. Curr Biol 4, 1069-76.

Dizdaroglu, M. (1993). Chemistry of Free Radical Damage to DNA and Nucleoproteins. London, England: Ellis Horwood, Ltd.

Duguet, M. (1997). When helicase and topoisomerase meet! J Cell Sci. 110, 1345-1350.

Ellis, N.A. (1997). DNA helicases in inherited human disorders. Current Opinion In Genetics & Development 7, 354-363.

Evans, M.K., Bohr, V.A. (1994). Gene-specific DNA repair of UV-induced cyclobutane pyrimidine dimers in some cancer-prone and premature-aging human syndromes. Mutat Res 314, 221-231.

Fishel, R. (1998). Mismatch repair, molecular switches, and signal transduction. Genes Dev. 12, 2096-2101.

Fornace, A.J., Fuks, Z., Weichselbaum, R., Milas, R. (1998). Radiation Therapy., ed. J.Mendelsohn, P.M.Howley, M.A.Israel, L.LiottaPhiladelphia: W.B. Saunders Co..

Friedberg, E.C. (1996b). Cockayne syndrome--a primary defect in DNA repair, transcription, both or neither? Bioessays 18, 731-738.

Friedberg, E.C. (1996a). Relationships between DNA repair and transcription. Annu Rev Biochem 65, 15-42.

Friedberg, E.C., Walker, G.C., Siede, W. (1995). DNA Repair and Mutagenesis. Washington, D.C.: ASM Press.

Frosina, G., Fortini, P., Rossi, O., Carrozzino, F., Raspaglio, G., Cox, L.S., Lane, D.P., Abbondandolo, A., Dogliotti, E. (1996). Two pathways for base excision repair in mammalian cells. J Biol Chem 271, 9573-9578.

Fukuchi, K., Martin, G.M., Monnat, R.J. (1989). Mutator phenotype of Werner syndrome is characterized by extensive deletions. Proc.Nat.Acad.Sci., USA 86, 5893-5897.

Gray, M.D., Shen, J.C., Kamath-Loeb, A.S., Blank, A., Sopher, B.L., Martin, G.M., Oshima, J., Loeb, L.A. (1997). The Werner syndrome protein is a DNA helicase. Nat.Genet. 17, 100-103.

Hanada, K., Ukita, T., Kohno, Y., Saito, K., Kato, J., Ikeda, H. (1997). RecQ DNA helicase is a suppressor of illegitimate recombination in Escherichia coli. Proc.Natl.Acad.Sci.U.S.A. 94, 3860-3865.

Harman, D. (1956). Aging: A theory based on free radical and radiation chemistry. Journal of Gerontology 11, 298-300.

Huang, S., Li, B., Gray, M.D., Oshima, J., Mian, I.S., Campisi, J. (1998). The premature ageing syndrome protein, WRN, is a 3'-->5' exonuclease. Nat.Genet. 20, 114-116.

Kamath-Loeb, A.S., Shen, J.C., Loeb, L.A., Fry, M. (1998). Werner Syndrome Protein. Ii. characterization of the integral 3' --> 5' dna exonuclease. J.Biol.Chem. 273, 34145-34150.

Kim, K., Biade, S., Matsumoto, Y. (1998). Involvement of flap endonuclease 1 in base excision DNA repair. J.Biol.Chem. 273, 8842-8848.

Klungland, A., Lindahl, T. (1997). Second pathway for completion of human DNA base excision-repair: reconstitution with purified proteins and requirement for DNase IV (FEN1). EMBO J 16, 3341-3348.

Kolodner, R. (1996). Biochemistry and genetics of eukaryotic mismatch repair. Genes Dev. 10, 1433-1442.

Krokan, H.E., Standal, R., Slupphaug, G. (1997). DNA glycosylases in the base excision repair of DNA. Biochem J 325, 1-16.

Kubota, Y., Nash, R.A., Klungland, A., Schar, P., Barnes, D.E., Lindahl, T. (1996). Reconstitution of DNA base excision-repair with purified human proteins: interaction between DNA polymerase beta and the XRCC1 protein. EMBO J 15, 6662-6670.

Leadon, S.A., Cooper, P.K. (1993). Preferential repair of ionizing radiation-induced damage in the transcribed strand of an active gene is defective in Cockayne syndrome. Proc.Natl.Acad.Sci.USA 90, 10499-10503.

Li, X., Li, J., Harrington, J., Lieber, M.R., Burgers, P.M. (1995). Lagging strand DNA synthesis at the eukaryotic replication fork involves binding and stimulation of FEN-1 by proliferating cell nuclear antigen. J.Biol.Chem. 270, 22109-22112.

Lindahl, T. (1993). Instability and decay of the primary structure of DNA. [Review]. Nature 362, 709-715.

Lindahl, T., Karran, P., Wood, R.D. (1997). DNA excision repair pathways. Curr.Opin.Genet Dev 7, 158-169.

Lloyd, R.S. (1998). Base excision repair of cyclobutane pyrimidine dimers. Mutat.Res. 408, 159-170.

Malins, D.C., Haimanot, R. (1991). Cancer Res. 51, 5430-5432.

Matsumoto, Y., Kim, K. (1995). Excision of deoxyribose phosphate residues by DNA polymerase beta during DNA repair. Science *269*, 699-702.

Matsumoto, Y., Kim, K., Bogenhagen, D.F. (1994). Proliferating cell nuclear antigen-dependent abasic site repair in Xenopus laevis oocytes: an alternative pathway of base excision DNA repair. Mol Cell Biol *14*, 6187-6197.

Millis, A.J., Hoyle, M., McCue, H.M., Martini, H. (1992). Differential expression of metalloproteinase and tissue inhibitor of metalloproteinase genes in aged human fibroblasts. Expt.Cell.Res. *201*, 373-379.

Modrich, P. (1994). Mismatch Repair, Genetic Stability, and Cancer. Science *266*, 1959-1960.

Mullenders, L.H., van Kesteren van Leeuwen AC, van, Z.A., Natarajan, A.T. (1988). Nuclear matrix associated DNA is preferentially repaired in normal human fibroblasts, exposed to a low dose of ultraviolet light but not in Cockayne's syndrome fibroblasts. Nucleic Acids Res *16*, 10607-10622.

Nance, M., Berry, S. (1992). Cockayne syndrome: review of 140 cases. Am.J.Med.Genet. *42*, 68-84.

Ogburn, C.E., Oshima, J., Poot, M., Chen, R., Gollahon, K.A., Rabinovitch, P.S., Martin, G.M. (1997). An apoptosis-inducing genotoxin differentiates heterozygotic carriers for Werner helicase mutations from wild-type and homozygous mutants. Human Genetics *101*, 121-125.

Olinski, R., Zastawny, T., Budzbin, J., Skokowski, J., Zegarski, W., Dizdaroglu, M. (1992). FEBS Lett.FEBS Letters *309*, 193-198.

Orren, D.K., Dianov, G.L., Bohr, V.A. (1996). The human CSB (ERCC6) gene corrects the transcription-coupled repair defect in the CHO cell mutant UV61. Nucleic Acids Research 24(17):3317-22 3317-222.

Peltomaki, P. (1997). DNA mismatch repair gene mutations in human cancer. Environ Health Perspect. *105 Suppl 4:775-80*, 775-780.

Piersen, C.E., Prasad, R., Wilson, S.H., Lloyd, R.S. (1996). Evidence for an imino intermediate in the DNA polymerase beta deazyribose phosphate excision reaction. J Biol Chem *271*, 17811-17815.

Poot, M., Hoehn, H., Runger, T.M., Martin, G.M. (1992). Impaired S-phase transit of Werner syndrome cells expressed in lymphoblastoid cells. Exp.Cell.Res. *202*, 267-273.

Radicella, J.P., Dherin, C., Desmaze, C., Fox, M.S., Boiteux, S. (1997). Cloning and characterization of *hOGG1* , a human homolog of the *OGG1* gene of *Saccharomyces cerevisiae*. Proc.Natl.Acad.Sci.USA *94*, 8010-8015.

Roldán-Arjona, T., Wei, W.-F., Carter, K.C., Klungland, A., Anselmino, C., Wang, R.-P., Augustus, M., Lindahl, T. (1997). Molecular cloning and functional expression of a human cDNA encoding the antimutator enzyme 8-hydroxyguanine-DNA glycosylase. Proc.Natl.Acad.Sci.USA *94*, 8016-8020.

Rosenquist, T.A., Zharkov, D.O., Grollman, A.P. (1997). Cloning and characterization of a mammalian 8-oxoguanine DNA glycosylase. Proc Natl Acad Sci U.S.A *94*, 7429-7434.

Sancar, A. (1996). DNA excision repair. Ann.Rev.Biochem. *65*, 43-81.

Sandigurski, M., Yacoub, A., Kelley, M., Xu, Y., Franklin, W.A., Deutsch, W.A. (1997). The yeast 8-oxoguanine DNA glycosylase (OGG1) contains a DNA deoxyribophospodiesterase (dRpase) activity. Nucl.Acid Res. *25*, 4557-4561.

Seeberg, E., Eide, L., Bjoras, M. (1995). The base excision repair pathway. Trends Biochem Sci *20*, 391-397.

Selby, C. P. and Sancar, A. Cockayne syndrome group B protein enhances elongation by RNA polymerase II. Proc Natl Acad Sci U S A 94, 11205-11209. 1997.

Shen, J.C., Gray, M.D., Oshima, J., Kamath-Loeb, A.S., Fry, M., Loeb, L.A. (1998). Werner Syndrome Protein. I. dna helicase and dna exonuclease reside on the same polypeptide. J.Biol.Chem. *273*, 34139-34144.

Singhal, R.K., Prasad, R., Wilson, S.H. (1995). DNA polymerase beta conducts the gap-filling step in uracil-initiated base excision repair in a bovine testis nuclear extract. J Biol Chem *270*, 949-957.

Sobol, R.W., Horton, J.K., Kuhn, R., Gu, H., Singhal, R.K., Prasad, R., Rajewsky, K., Wilson, S.H. (1996). Requirement of mammalian DNA polymerase-beta in base-excision repair [published erratum appears in Nature 1996 Oct 3; 383(6599):457]. Nature *379*, 183-186.

Sohal, R.S., Ku, H.H., Agarwal, S., Forster, M.J., Lal, H. (1994). Oxidative damage, mitochondrial oxidant generation and antioxidant defenses during aging and in response to food restriction in the mouse. Mechanisms of Ageing and Development *74*, 121-133.

Suzuki, N., Shimamoto, A., Imamura, O., Kuromitsu, J., Kitao, S., Goto, M., Furuichi, Y. (1997). DNA helicase activity in Werner's syndrome gene product synthesized in a baculovirus system. Nucleic Acids Research 25, 2973-2978.

Szilard, L. (1959). On the nature of the aging process. Proc Natl Acad Sci USA 45, 30-45.

Thomas, D.C., Umar, A., Kunkel, T.A. (1996). Microsatellite instability and mismatch repair defects in cancer cells. Mutat.Res.Fundam.Mol.Mech.Mutagen.Mutation Research: Fundamental and Molecular Mechanisms of Mutagenesis 350, 201-205.

Van Gool, A.J., Citterio, E., Rademakers, S., Van Os, R., Vermeulen, W., Constantinou, A., Egly, J.M., Bootsma, D., Hoeijmakers, J.H. (1997). The Cockayne syndrome B protein, involved in transcription-coupled DNA repair, resides in an RNA polymerase II-containing complex. EMBO J 16, 5955-5965.

Van Remmen, H., Ward, W.F. (1995). Gene Expression and Protein Degradation. In: Handbook of Physiology: Aging, ed. E.J.MasoroNew York: Oxford University Press, 171-234.

vanGool, A.J., vanderHorst, G., Citterio, E., Hoeijmakers, J.J. (1997). Cockayne syndrome: defective repair of transcription? EMBO 16, 4155-4162.

Venema, J., Mullenders, L.H., Natarajan, A.T., van Zeeland, A.A., Mayne, L.V. (1990). The genetic defect in Cockayne syndrome is associated with a defect in repair of UV-induced DNA damage in transcriptionally active DNA. Proc.Natl.Acad.Sci.U.S.A. 87, 4707-4711.

Vichi, P., Coin, F., Renaud, J.P., Vermeulen, W., Hoeijmakers, J.H., Moras, D., Egly, J.M. (1997). Cisplatin- and UV-damaged DNA lure the basal transcription factor TFIID/TBP. EMBO J. 16, 7444-7456.

Webb, D.K., Evans, M.K., Bohr, V.A. (1996). DNA repair fine structure in Werner's syndrome cell lines. Exp.Cell Res 224, 272-278.

West, S.C. (1996). DNA helicases: new breeds of translocating motors and molecular pumps. Cell 177-180.

Wilson, S.H. (1998). Mammalian base excision repair and DNA polymerase beta. Mutat.Res. 407, 203-215.

Wiseman, H., Halliwell, B. (1996). Damage to DNA by reactive oxygen and nitrogen species: role in inflammatory disease and progression to cancer. Biochem J. 313. 17-29.

Wood, R.D. (1996). DNA repair in Eukaryotes. Ann.Rev.Biochem. 65 , 135-167.

Wood, R.D. (1997). Nucleotide excision repair in mammalian cells. J Biol.Chem. 272, 23465-23468.

Wood, R.D., Shivji, M.K.K. (1997). Which DNA polymerases are used for DNA-repair in eukaryotes? Carcinogenesis 18, 605-610.

You, Z., Feaver, W.J., Friedberg, E.C. (1998). Yeast RNA polymerase II transcription in vitro is inhibited in the presence of nucleotide excision repair: complementation of inhibition by Holo-TFIIH and requirement for RAD26. Mol.Cell Biol. 18, 2668-2676.

Yu, C.E., Oshima, J., Fu, Y.H., Wijsman, E.M., Hisama, F., Alisch, R., Matthews, S., Nakura, J., Miki, T., Ouais, S., Martin, G.M., Mulligan, J., Schellenberg, G.D. (1996). Positional cloning of the Werner's syndrome gene. Science 272, 258-262.

© 2001 Elsevier Science B.V. All rights reserved.
The Role of DNA Damage and Repair in Cell Aging
B.A. Gilchrest and V.A. Bohr, volume editors.

GENE ACTION AT THE WERNER HELICASE LOCUS: ITS ROLE IN THE PATHOBIOLOGY OF AGING

Junko Oshima, George M. Martin, Matthew D. Gray, Martin Poot,
and Peter S. Rabinovitch

Department of Pathology, University of Washington, Seattle, WA

1. Introduction

Gerontologists have invoked DNA damage as a mechanism of aging for at least the past forty years (Szillard, 1959). One approach has been to determine the effects of putative DNA damaging agents on life span and on the rates of appearance of various senescent phenotypes in experimental systems (Alexander, 1967). More recently, endogenous reactive oxygen species have been invoked as the usual pathway to DNA damage in aging aerobic organisms (Wallace, 1997). Another approach has been to determine the extent to which aberrations in DNA metabolism might form the basis for various human progeroid syndromes (Martin, 1978). The present Chapter represents an example of this latter approach. We provide a review of the present status of our knowledge of what is arguably the most striking of the segmental progeroid syndromes - a rare autosomal recessive genetic disorder known as the Werner syndrome (WRNs; Progeria of the adult) (entry *27770 in McKusick, 1998). Although the WS locus Werner (*WRN*) was discovered only three years ago (GenBank accession number L76937; Yu et al., 1996a), there has been steady progress towards the elucidation of its structure and function. It is now quite clear that the *WRN* gene product (WRNp) has both helicase and exonuclease functions. There is also good evidence that null mutations of *WRN* are responsible for classical forms of WRNs. But there is much more to be learned. We do not yet know the relative contributions of WRNp to DNA transactions involving replication, repair, recombination, transcription or chromosomal segregation. We also require a great deal more information on the possible impacts upon various aging phenotypes of heterozygosity for null mutations, of heterozygosity and homozygosity for non-null mutations, and of heterozygosity or homozygosity for various polymorphic forms of *WRN*.

A. *A Brief History of the Werner Syndrome and the Discovery of the Werner Helicase Gene*

The first description of the WRNs appeared in a University of Kiel doctoral thesis by Otto Werner in 1904 (Werner, 1904). He described "scleroderma-like" thin, tight skin and bilateral cataracts in a sibship. There was parental consanguinity, leading Werner to suspect a genetic basis for the disease.

An International Registry of Werner Syndrome (http://rice.u.washington.edu/werner) was established in 1984 at the University of Washington, Seattle, WA. It currently includes data for over 80 cases of WRNs from all over the world, confirming widely scattered literature that suggests that *WRN* disease mutations can be found among all ethnic groups that have been evaluated for this syndrome. The prevalence of WRNs has been estimated to vary from around 1-22 per million (Epstein et al., 1966). The higher frequencies of the disorder in certain populations probably reflect both a relatively higher rate of consanguineous mating as well as an increased awareness of practicing physicians. Both of these factors appear to account for the relatively high prevalence of the disorder in various regions of Japan.

In 1992, the *WRN* locus was mapped to the short arm of human chromosome 8 (Goto et al., 1992). This was the result of a productive collaboration between M. Goto of Tokyo, who had collected a large number of nuclear pedigrees, and Dennis Drayna and his colleagues at Genentech. Drayna had to examine over 150 markers before he discovered statistical evidence of linkage. This result was almost immediately confirmed by our Seattle group, working in collaboration with T. Miki and his associates in Osaka (Schellenberg et al., 1992). That latter study was of special interest in that it represented an early example of the utility of homozygosity mapping in human genetics (Lander and Botstein, 1987).

For the next two years, the *WRN* locus was further narrowed down to a region of approximately 400 kb (Oshima et al., 1994; Yu et al., 1994; Nakura et al., 1994). A range of methodologies (cDNA hybridization, exon trapping, database searches and genomic sequencing) (Goddard et al., 1996; Yu et al., 1996b; Hisama et al., 1998) eventually led to the positional cloning of the *WRN* gene. More than a dozen genes in this gene-sparse region of the genome (Yu et al., 1996b; Hisama et al., 1998; Ichikawa et al., 1998) had to be fully sequenced from WRNs patients and control samples before the *WRN* gene was identified as a member of the *RecQ* family of DNA helicases (Yu et al., 1996a).

B. *The Clinical Picture of the Werner Syndrome*

Development of patients with WRNs appears to be normal at least until around the time of puberty (Epstein et al., 1966; Tollefsbol and Cohen, 1984). Subsequent signs and symptoms have been characterized as a segmental progeroid syndrome. The adjective "segmental" was used because not all phenotypes seen in "normal" elderly individuals are seen in WRNs (Martin et al., 1978). The earliest sign, often noted retrospectively, is the lack of growth spurt during early adolescence. An aged appearance can be recognized during the third decade of life. Skin manifestations include thin tight skin of hands and facial wrinkles. Gray hair and thinning of hair may also start during this period. The first clinically detectable degenerative pathology is bilateral ocular cataracts. These typically require surgical treatment about age 30. Other common features include type 2 diabetes mellitus, osteoporosis, atrophy of skin and subcutaneous tissues, atherosclerosis, soft tissue calcifications, and neoplasia. The onsets of these and other phenotypes seem to follow a certain order (Goto et al., 1997). Interestingly, hypertension and Alzheimer type dementia are not common.

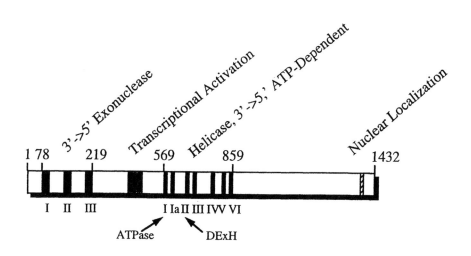

Figure 1. **Functional Domains of human WRN protein.** *Arabic numerals indicate amino acid position. Roman numerals represent the consensus motifs of exonuclease and helicase domains. The approximate location of domains associated with biologically demonstrated functions are shown.*

To date, more than 25 different mutations have been identified in the *WRN* gene (Yu et al., 1996a; Oshima et al., 1997; Yu et al., 1997; Goto et al., 1997; Matsumoto et al., 1997a; Moser et al., 1999). The Web site of the International Registry of Werner Syndrome (http://rice.u.washington.edu/werner) displays the locations of these mutations within the *WRN* gene.

2. The Werner Syndrome Gene Product

A. *The Werner Syndrome Locus and Related Genes*

The Werner syndrome gene (*WRN*) consists of 35 exons. The genomic sequence spans more than 200kb of the short arm of human chromosome 8 (Yu et al., 1997; Matsumoto et al., 1997a). *WRN* encodes a 1,432 amino acid nuclear protein. Investigation of the regional homologies of the *WRN* cDNA and computer-based structural analysis predicted an *E. coli* RecQ type of helicase function associated with the central domain (Gorbalenya et al., 1989; Umezu et al., 1990) and an exonuclease function associated with the N terminal domain (Mushegian et al., 1997; Moser et al., 1997).

Two groups have independently demonstrated that the WRN helicase has ATP-dependent 3' → 5' DNA helicase activity (Gray et al., 1997; Suzuki et al., 1997). A 3' → 5' exonuclease activity was also demonstrated by two groups (Huang et al., 1998; Shen et al., 1998). A C-terminal nuclear localization signal has been demonstrated in experi-

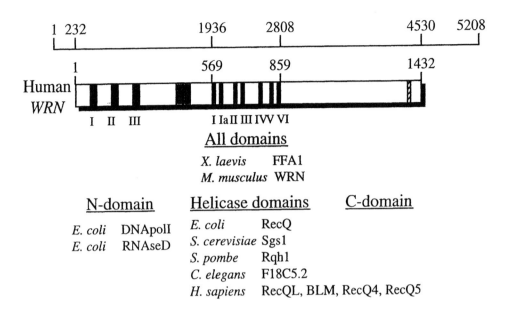

Figure 2. **Homologies of human WRN protein.** *Top line indicates nucleotide numbers, second line, amino acids. Diagram shows WRNp functional domains as in Fig. 1. Bottom table shows the species and homologs of the human WRNp and those for each of the three general regions. All proteins showing homology with the central helicase region of WRNp are members of the RecQ family.*

ments expressing serially truncated *WRN* cDNA in mammalian cells (Matsumoto et al. 1997b; Matsumoto et al., 1998). In addition, a highly acidic region, located between the exonuclease and helicase domains, can function to activate transcription both in yeast and biochemical assays (Ye et al., 1998; Balajee et al., submitted). WRNp functional domains that have been confirmed by biochemical or cell biological studies are shown in Fig. 1. Details of the helicase and exonuclease functions will be discussed later.

Other members of the human RecQ family of helicases include *RecQL* (Puranam et al., 1994), *BLM* (Ellis et al., 1995), *RecQ4* and *RecQ5* (Kitao et al., 1999). Homozygous mutations in *BLM* are responsible for a growth deficiency and cancer prone syndrome, the Bloom syndrome (BLMs) (Ellis and German, 1996). Cells from BLMs subjects exhibit markedly elevated levels of sister chromatid exchanges (German et al. 1995). Homologs of the human *WRN* have been identified in mice and in Xenopus (Imamura et al., 1997; Yan et al., 1998). Among yeast genes, *SGS1* in *S. cerevisiae* (Watt et al., 1995; Watt et al., 1996) and *RQH1* in *S. pombe* (Stewart et al., 1997; Murray et al., 1997) share substantial homologies with *WRN*. Both *WRN* and *BLM* have been shown to partially complement a hyperrecombination phenotype of an sgs1 mutant (Yamagata et al., 1998). These yeast genes, however, are more closely related to *BLM* since, like *BLM*, they lack exonuclease domains. A phylogenetic tree of the structural evolution of these helicases supports this interpretation (Murray et al., 1997; Kusano et al., 1999). A summary of proteins containing sequence homologous to WRNp are presented in Fig. 2.

B. *Expression and Regulation of the Werner Helicase Gene*

The structure of the human *WRN* promoter has been investigated by two independent groups (Wang et al., 1998; Yamabe et al., 1998). The human *WRN* promoter lacks a TATA-box and a CAT-box, features commonly seen in constitutively expressed genes. Western analyses of WRNp in human diploid fibroblasts under various conditions provided no evidence for cell cycle associated alterations in WRNp expression levels, supporting the conclusions of the structural studies that *WRN* may still be a housekeeping gene (Gray et al., 1998).

Other lines of evidence, however, are consistent with some degree of regulation of *WRN* gene expression. These include luciferase reporter assays showing that an active form of the retinoblastoma gene product (pRb) up-regulates *WRN* expression, while p53 appears to down-regulate *WRN* promoter activity (Yamabe et al., 1998). In addition, decreased *WRN* promoter activity is observed in WRNs cells (Wang et al., 1998). Expression of *WRN* mRNAs is reduced to less than half of wildtype levels in cells from *WRN* mutation heterozygotes (Yamabe et al., 1997). These observations suggest that *WRN* expression is subject to positive autoregulation.

WRN mRNA is expressed at relatively low levels in all human and mouse tissue so far examined (Yu et al., 1996a; Imamura et al., 1997). The highest level of expression is in the testes (Fig. 3).

WRN is expressed during fetal development despite the fact that many symptoms of WRNs begin after puberty. Semi-quantitative RT-PCR and Western analysis of human aortic tissues ranging from 49 days of gestation to a 91 year-old adult showed large variations of the *WRN* expression levels, even within the same individual (Oshima, unpublished data). The possible roles of such variations in physiology and pathophysiology should be investigated.

C. *Localization of the Werner Helicase Protein*

The subcellular localization of WRNp has been described by several groups (Matsumoto et al., 1997b; Marciniak et al., 1998; Gray et al., 1998; Shiratori et al., 1999). Human WRNp localizes to the nucleus when overexpressed in Hela cells. In actively replicating populations of normal human diploid cells, it is predominantly localized to nucleoli, where rDNA transcription and ribosome assembly take place. By contrast, mouse WRNp is primarily localized to nucleoplasm. In our study (Gray et al., 1998; Fig. 4), immunodetectable nucleolar WRNp disappeared when cultures were made quiescent by serum starvation or upon treatment with a genotoxic agent, 4-nitroquinoline-1-oxide (4NQO), to which WRNs cells are hypersensitive (Ogburn et al., 1997). Steady state levels of WRNp, determined by Western analysis, did not change, yet virtually all WRNp was in the nuclear fraction regardless of culture conditions. These results are

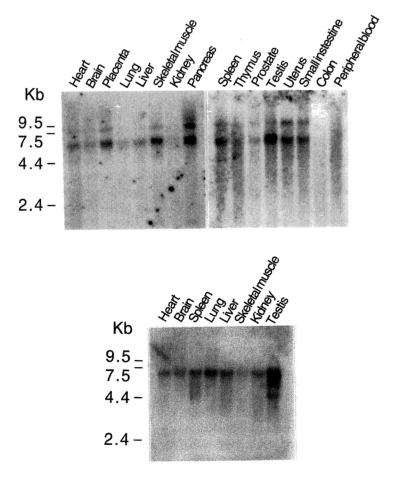

*Figure 3. WRN **mRNA expression levels in human (top) and mouse (bottom).** Multitissue blots (Clonech Inc. Pala Alto, CA) were probed with C-terminal region of the human (top) or mouse (bottom) WRN cDNA.*

consistent with either translocation of WRNp from nucleoli to nucleoplasm, and/or a masking of epitopes (Gray et al., 1998).

We suggest that in exponentially growing cultures, the WRNp has a function or functions within nucleoli - for example, to suppress illegitimate recombination between arrays of rDNA repeats, or to participate in rDNA transcription and/or replication. When DNA damage occurs, however, WRNp disperses throughout the nucleoplasm presumably in order to participate in the repair of recombination or non-ribosomal DNA. Such translocation could also contribute to a diminution of rDNA metabolisms observed dur-

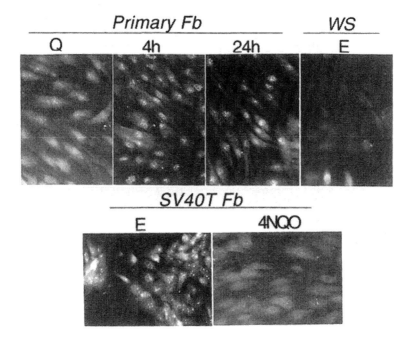

Figure 4. **Indirect immunofluorescence staining of human WRNp.** *Top panels show the WRNp staining in primary skin fibroblasts during serum starvation (Q: 0.2%, 72 hrs) followed by 4 hrs and 24 hrs serum stimulation with 15% FCS. Primary WS fibroblasts are shown as a negative control. Bottom panels show the staining of WRNp in SV40 T fibroblasts during exponential growth (E) and 4NQO treatment (4NQO). The detailed procedures were described previously (Gray et al., 1998).*

ing the establishment of a quiescent state. Our studies with phosphatase inhibitors (Gray et al., 1998) suggest that the shuttling of WRNp between nucleoli and nucleoplasm may be associated with alterations in the state of phosphosphorylation. The accumulation within, or diffusion from, nuclei may be accomplished by associating with a "shuttling protein" which undergoes successive phosphorylation and dephosphorylation, as has been demonstrated for nucleophosmin/B23 (Peter et al., 1990).

3. The Functions of the Werner Helicase Protein

A. *Biochemistry of the Werner Helicase Protein*

Human and mouse WRNp contain the canonical sequences of the seven motifs that define the RecQ class of helicases, named after the prototypic helicase of this type first

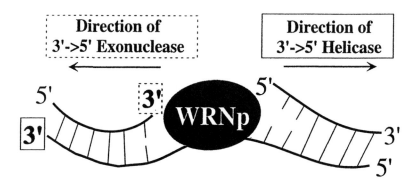

Figure 5. **Directionalities of WRN helicase and exonuclease.** *The direction of a helicase is defined by its progression on the longer strand whose overhang area initially binds helicase. The direction of exonuclease is defined by the direction of exonuclease activity. It should be noted that, by definition, the directionalities of helicase and exonuclease are opposite.*

described in *E. coli* (Nakayama et al., 1985) (Figs. 1 and 2). Motif I corresponds to the ATPase domain, and motif II contains the DHxA sequence, the distinguishing feature of a much lager family of helicases. An eighty amino acids region of unknown function c-terminal to the helicase domains is also conserved in WRNp, BLMp, Sgs-1p, RecQp and RecQLp (Morozov et al., 1997). ATP-dependent 3' → 5' helicase activity was first demonstrated with the E. coli RecQ helicase using a DNA strand displacement assay (Umezu et al., 1990). Directionality is defined by the longer strand of the duplex DNA substrate to which these helicases bind (Fig. 5). Comparable results have now been demonstrated using recombinant human WRNp (Gray et al., 1997; Suzuki et al., 1997). This type of helicase activity also has been biochemically confirmed with related gene products such as SGS-1p (Lu et al., 1996; Bennett et al., 1998), human RecQL (Tada et al., 1996), and human BLMp (Karow et al., 1997).

Proof that the helicase activity of WRNp requires ATPase activity was provided by the co-purification of helicase and ATPase activitities and the loss of unwinding activities in displacement assays when utilizing a recombinant protein bearing an inactivating mutation in the ATPase domain (K577M) (Gray et al., 1997; Shen et al., 1998).

Recombinant WRNp efficiently displaces 20-24mer oligonucleotides. It is less efficient at displacing longer nucleotides (40-53mers). The displacement of longer nucleotides can be improved, however, by the addition of *E. coli* single strand binding protein (SSB), T4 gene 32 product, or, more effectively, by human replication protein A (RPA) (Gray et al., 1997; Suzuki et al., 1997; Shen et al., 1998). The enhancement of unwinding by these three single strand binding proteins is thought to be occur via the stabilization of displaced ssDNA. The displacement was shown to become more efficient by the addition of 10 extra unpaired nucleotides to both ends of a 53mer (Suzuki et a., 1997).

WRNp can also displace a 18mer RNA in a DNA-RNA heteroduplex, though not as effectively as DNA-DNA duplex (Suzuki et al., 1997). This observation raises the question of a potential role of WRNp in the synthesis of the lagging strand and the displacement during transcription.

A 3' → 5' proofreading exonuclease activity was predicted by a structural analysis revealing homology of motif I, II and III of bacterial RNaseD (Mushegian et al., 1997; Moser et al., 1997). Huang et al., (1998) provided biochemical evidence of exonuclease activity with both a full length recombinant WRNp and an N-terminal fragment (amino acid 1-333). The exonuclease degraded the recessed ends of DNA-DNA duplexes. E84A or D82A mutations in the N-terminal fragment essentially abolished 3' → 5' exonuclease activity. An important result was that exonuclease activity could be demonstrated with recombinant mutant K577M, which eliminates the ATPse activity of the helicase. This established that helicase activity and exonuclease activity can be uncoupled.

Additional biochemical studies revealed that exonuclease activity can be stimulated by the presence of ATP, dATP or CTP, and that WRNp can hydrolyze both 3'-OH and 3'-PO4 ends of DNA oligomers from the recessed end of DNA duplexes to the similar extents (Shen et al., 1998; Kamath-Loeb et al., 1998).

Further biochemical studies of WRNp should be focused on three aspects. The first is the search for the proteins which modulate the helicase and exonuclease activities. The second is the search for the physiological DNA substrates. BLMp has been shown to be able to unwind a G4 structure of DNA, which can be formed during recombination and replication (Sun et al., 1998). Comparable studies with the WRNp would be of great interest. A third avenue for new research would involve assays for other enzymatic activities.

B. *Cell biology of the Werner helicase protein and Genomic Instability of Somatic Cells from Werner Syndrome Patients*

A particularly well documented cellular phenotype of WRNs is the striking reduction of the replicative life span of cultured skin fibroblast-like cells (Martin et al., 1970; Salk et al., 1985; Kill et al., 1994; reviewed by Tollefsbol and Cohen, 1984). The cell cycle time, particularly the S phase, is elongated in primary WRNs fibroblast cultures (Takeuchi et al., 1982b). In WRNs LCLs, the mean duration of the S phase was shown to be 51% longer than that of control cells, while the mean durations of the G1 and G2 phases were similar (Poot et al., 1992). A decreased number of replication forks (Takeuchi et al., 1982b) or a retarded rate of elongation of replication forks (Fujiwara et al., 1977) are possible explanations for these findings.

Associated with the phenotype of accelerated replicative senescence is a propensity of WRNs cells to generate a variety of different stable chromosomal mutations, including deletions, inversions and reciprocal translocations. This phenotype has been given the name of variegated translocation mosaicism (Hoehn et al., 1975). In addition to this evidence of chromosomal instability, there is evidence for enhanced gene mutation. Spontaneous forward mutation rates in cultures of SV-40 transformed WRNs fibroblasts and controls have been determined for an X-linked recessive locus, *HPRT* (hypoxanthine guanine phosphoribosyl transferase) using 6-thioguanine (6TG) as a selective agent. 6TG is metabolized to cytotoxic 6-thio-GMP by *HPRT* enzyme activity. Only those cells in which *HPRT* has been inactivated can survive in 6TG. The mutation rates in WS cells were some 10-100 times those of controls (Fukuchi et al., 1989). Evidence that hypermutation at that locus also occurs *in vivo* was developed from single cell assays for putative loss of enzyme activity using freshly isolated cells from the peripheral blood lymphocytes (Fukuchi et al., 1990). The frequencies of *HPRT* mutations in cells from human subjects increase with the age of the donor. This has been shown for both peripheral blood (Trainor et al., 1984) and renal tubular epithelial cells (Martin et al., 1996). Fukuchi et al. (1989) showed that the most common *HPRT* mutation in cultured SV40-transformed fibroblasts from WS patients is deletion within or spanning this locus. Large scale lesions are also common causes of spontaneous mutations in peripheral lymphocytes from normal human donors (Turner et al., 1984). A high proportion of deletions has also been reported for the error-prone DNA ligation of transfected linearized plasmids observed in WRNs LCLs (Runger et al., 1994). Increased *in vivo* mutation frequencies at the two other loci, the glycophorin A (GPA) in erythrocytes and the TCR/CD3 in CD4+ T cells were observed in WRNs patients (Kyoizumi et al., 1998). Although studies of DNA repair in WRNs cells were initiated more than twenty years ago, the subject is still at a very early stage of development, as knowledge in the field as a whole has greatly outstripped the pace of research that has been applied to the problem of WRNs. Simple measures of DNA repair, such as the disappearance of X-ray-induced single strand breaks and UV damaged DNA, and sensitivity to X-ray and UV killing, appear to be normal in WRNs fibroblasts (Fujiwara et al., 1977). More recently, some evidence has been developed supporting a deficiency of transcription-coupled DNA repair in WRNs lymphoblastoid cell lines (Webb et al., 1996), while the possibility of mismatch repair in WRNs SV40 transformed cell lines seems unlikely (Bennett et al., 1997; Bennett et al., 1999).

LCLs from WRNs patients are more sensitive to the genotoxic agent, 4-nitroquinoline-1-oxide (4NQO) than controls (Ogburn et al., 1997). Furthermore, intermediate sensitivities were found in heterozygous carriers of the *WRN* mutation. Hypersensitivity of WRNs cells to 4NQO raises the possibility that an inadequate response to oxidative DNA damage, an important cause of aging according to one theory, may be responsible for the accelerated aging in WRNs. WRNs LCLs are also hypersensitive to topoisomerase I inhibitor camptothecin, but not to topoisomerase II inhibitors (Okada et al., 1998). We found that LCLs from WRNs patients showed lower cell survival and more dead cells when exposed to camptothecin in the rage from 10 to 50nM (Poot et al., 1999). Differences in cell survival levels were statistically significant, but a much clearer differentiation between WRNs patients and wild type individuals was seen by examining cell death by apoptosis. Although wild type and *WRN* mutated cells both show accumulation in the S phase (due to slowing and/or arrest), only *WRN* mutated cells underwent apoptosis, all of which occurred in S phase. We hypothesize that WRNp function is required for resolution of topoisomerase I-DNA adducts. Conflict with DNA replication may lead to apoptosis, increased mutation rates and cancer in WRNs.

The challenge for the future is to discover to what extent the underlying mechanisms that cause the increased genomic instability in WRNs and in other genomic instability syndromes (Weirich-Schwaiger et al., 1994) are also operative during aging in most people.

C. *Mechanism of DNA Transactions Mediated by the Werner Helicase Protein*

WRN could be potentially involved in virtually all DNA transactions: DNA repair, recombination, DNA replication, transcription and chromosomal segregation. It remains to be determined which of these transactions are crucial functions of the WRNp and how these relate to the pathogenesis of WRNs and the normal aging process. It is reasonable to assume that those pathways whose defects are manifested in WRNs subjects are crucial, while those which do not appear to be defective in WRNs cells are auxiliary functions. Examples of the former would include double-strand (ds) break repair/recombination, and replication. Examples of the latter would include transcriptional activation. To derive information upon which to build a model of how WRNp works, it is obviously important to review experimental findings with other members of the RecQ family of homologues, especially those with more complete WRNp homologies.

The RecQ of *E. coli* is a 3' → 5' helicase that is required for homologous recombination by the RecF pathway. RecF is a minor recombination pathway in *E. coli*. It is invoked when the major pathway, RecBCD, is defective (Nakayama et al., 1984). *E.coli* RecQ mutants exhibit 20-300 fold increases in illegitimate recombination (Hanada, et al., 1997). RecQ has been shown to initiate or to disrupt recombinational intermediates (Harmon and Kowalczykowski, 1998). These findings suggest a model in which RecQ functions in homologous recombination and in suppression of illegitimate recombination by modulating the structures of recombination intermediates.

It is conceivable that WRNp can both suppress and enhance recombination, depending on the target DNA structures and/or the state of the microenvironment, as may be the case for *E. coli* RecQ helicase (Harmon and Kowalczykowski, 1998). We, however, speculate that WRNp acts more as a suppresser of recombination than an enhancer because of observations of increased homologous recombination in WRNs fibroblasts (Cheng et al., 1990) and the ability of *WRN* to suppress hyperrecombination in yeast sgs1 mutant (Yamagata et al., 1998).

In *E.coli*, 3' → 5' exonuclease activity is also associated with the regulation of recombination (Phillips et al., 1988; Myers et al., 1995). While WRNp contains both 3' → 5' helicase and 3' → 5' exonuclease domains (Fig. 2), these activities are associated with different genes in *E.coli*. 5' → 3' is the direction of nucleotide synthesis while 3' → 5' is the direction of proofreading. *WRN* might therefore have a role in DNA repair.

The *E. coli* RecF pathway has been shown to be required for the resumption of replication following the disruption of replication (Courcelle et al., 1997). Recently, a Xenopus homolog of *WRN*, FFA-1 (focus forming activity 1) was isolated from replication foci. Its helicase activity has been shown to be required for the formation of the replication fork (Yan et al., 1998). FFA-1 co-localized with a single strand binding pro-

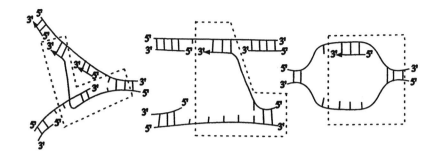

Figure 6. Potential Target Structure of WRNp. Note the similarity in three structures.

tein, RPA, (replication protein A), which has been shown to enhance helicase activity in association with recombinant human WRNp (Shen et al., 1998).

Analogous observations with other other members of the RecQ family help to clarify many, but not all, of the phenomena observed in WRNs cells. For example, they help elucidate the findings of reduced DNA chain elongation rate (Fujiwara et al., 1977), reduced replication initiation (Takeuchi et al., 1982b), and elongated S phase (Takeuchi et al., 1982a; Poot et al., 1992).

There are several possible mechanisms by which defects of WRNp might affect multiple areas of DNA metabolism. First, WRNp or a WRNp complex may play a role in an intermediate step common to different DNA transactions. Second, different sets of functional domains of WRNp might be utilized in different steps of DNA metabolisms. We prefer the former possibility because of the mechanistic similarity of DNA strand processing in DNA replication, recombination/double-strand break repair, and transcription (Kodadek, 1998). WRNp might have been evolved primarily for a single specific purpose, for example, recombinational DNA repair during replication. Once evolved, however, it might also have been utilized for other DNA transactions.

Fig. 6 shows several potential intermediate structures for WRNp complexes during replication (left: repair of the lagging strand), recombination (center) and during the initiation of transcription (right).

A potential role of *WRN* in the maintenance of rDNA has been postulated based on studies in yeast (Guarente, 1997). The yeast homolog of WRNp, SGS1p (Watt et al., 1995), is also localized to nucleoli (Sinclear et al., 1997). Nucleolar fragmentation and extrachromosomal rDNA circles are observed in aged sgs1 mutant strains (Sinclear and Guarente, 1997). This led to the hypothesis that WRNp may be involved in the suppression of recombination in highly repetitive rDNAs. However, more recent studies do not support the hypothesis that WRNp plays a major role in rDNA maintenance in mammalian cells. First, nucleolar fragmentation or rDNA circles have not been observed in higher eukaryotes. Second, the mouse homolog of WRNp has been localized to the nucleoplasm (Marciniak et al., 1998), indicating that nucleolar localization of WRNp ho-

Table 1. Phenotype of Hosts and Deletion Mutants

RecQ Homolog	BLM	SGS-1	FFA-1	mWRN	hWRN
Exonuclease domain	Absent	Absent	Present	Present	Present
Host	Human	Yeast	Xenopus	Mouse	Human
Rate of homologous recombination in the host	Low	High	Low	Low	Low
Subcellular	Nuclear localization	Nucleoli	N/A	Nucleoplasm	Nucleoili/ Nucleoplasm
Phenotype Sister of deletion mutant	Chromatid Exchange	Nucleolar Fragment	N/A	N/A	Variegated Transmosaicism

mologs is not evolutionarily conserved. Third, another human RecQ helicase, BLMp, has been shown to be more similar to SGS1p than WRNp. This conclusion derives from comparative structural studies (both BLMp and SGSp lack exonuclease domains) and from functional studies (BLM cDNA can complement more phenotypes of sgs1 mutant than WRN cDNA) (Yamagata et al., 1998). Table 1 summarizes the differences of RecQ type helicase deletion mutants.

It is likely that SGS1p has a distinct role in yeast to suppress homologous recombination in nucleoli, where more than 200 copies of rDNA are tandemly located. Since yeast has very few repeat sequences other than rDNA, other parts of the yeast genome would not strongly require SGS1p or other mechanisms to maintain an intact genome. On the other hand, higher eukaryotes have numerous repeat sequences in their genome. They would therefore need to evolve lower homologous recombination rates for all such repeats, including rDNA. The generic functions of SGS1p and WRNp may be similar: maintenance of an intact genome. However, characteristics of host genomes might make the primary sites of action different.

4. Future Research Directions

A. *Mouse Models of the Werner Syndrome*

A major problem of studies of WRNs has been the lack of availability of patient materials. In order to further investigate the pathogenesis of WRNs, mouse models of WRNs have been generated by several laboratories.

One model is a deletion mutant of exon 18 and 19 of mouse *WRN* gene, which corresponds to human *WRN* exon 19 and 20, generated by the method of targeted mutagenesis (Label and Leder, 1998). This mutation results in the in-frame deletion of motif III and IV of the helicase region. The other motifs are retained, including the exonuclease domain and the nuclear localization signal. These mutated WRNp are expected to lack helicase activities. ES cell lines with such homozygous mutations showed hypersensitivity to DNA topoisomerase I and II inhibitors (Label and Leder, 1998). Fibroblast cultures derived from homozygous mutant mouse embryos fibroblasts exhibited decreased saturation densities. Thus, deletion of helicase domains alone appears to simulate some of the cellular phenotypes of WRNs.

Another model, developed in our laboratory, is a transgenic mouse line carrying human *WRN* cDNA with a dominant negative mutation, K577M. Primary tail fibroblasts from three different mouse lines showed three cellular phenotypes consistent with WS: hypersensitivity to 4NQO, reduced replicative lifespan determined by the colony size distribution, and reduced endogenous mouse WRNp. K577M mutation in recombinant human WRNp abolishes the helicase activity (Gray et al., 1997), but not exonuclease activity (Huang et al., 1998), indicating that abolishment of helicase alone may be sufficient to create some, if not all, of the cellular characteristics of WRNs. It has not yet determined whether or how much other RecQ helicases are suppressed in this model.

B. *Phenotypic Impacts of a Spectrum of WRN Alleles*

Since the identification of *WRN*, a major question has been to what extent the functions of WRNp are involved in the pathogenesis of age-related disorders and "normal" aging process. We have identified four missense polymorphisms and four conservative polymorphisms in *WRN* (Castro et al., 1999; Fig. 7). Polymorphisms, by definition, are common variants. *WRN* polymorphisms with more subtle effects on age-related phenotypes might therefore constitute important public health issues.

Our initial investigations have utilized a Cys/Arg polymorphism at amino acid 1367, which is four amino acids N-terminal to the nuclear localization signal (Castro et al., 1999). A cell biological study provided some evidence that LCLs with the Arg/Arg genotype are more sensitive to 4NQO than those from donors with a Cys/Cys genotype (Castro et al., unpublished). Moreover, a genetic association study in a Japanese population indicated that homozygosity for a cysteine at amino acid 1367 (the most prevalent genotype) was associated with a 2.78 times greater risk of myocardial infarction (95% confidence intervals: 1.23 to 6.86) (Ye et al., 1997). We therefore have speculated that cells with the WRNp from individuals with the 1367 Arg/Arg genotype might be capa-

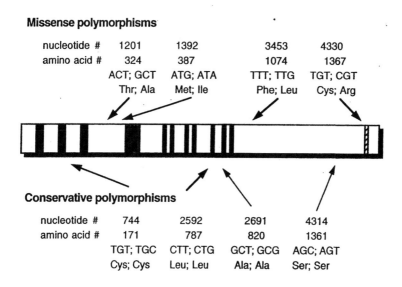

Figure 7. **Human** WRN **Polymorphisms.** *The nucleotide numbers and amino acid numbers of the polymorphisms correspond to the sequence of GenBank accession number, L76937.*

ble of more efficient transport to nucleoli (since basic amino acids are critical for efficient nuclear localization signals) or that they could have slightly greater affinity for critical targets. Efforts are underway to evaluate the role of this polymorphism in modulating the risk of myocardial infarction in other populations. Protection from myocardial infarction might be expected to contribute to unusual longevities, particularly in populations in which this is highly prevalent. Finland has such a population. We therefore determined the relative prevalence of the Arg/Arg polymorphism in Finnish centenarians and Finnish newborns. No significant differences were observed. However, the prevalence of the Arg allele was approximately three times that of the Japanese population, raising the possibility that the allele had some protective effect in that population over some undetermined period of time. An alternative interpretation is that the high frequency of the Arg allele in Finland could have resulted from a genetic founder effect. Arg allele frequencies are also relatively high in the North American Caucasian population.

5. Summary

The positional cloning of the *WRN* gene in 1996 has led to a great acceleration in the pace of research on the nature of the Werner syndrome and its relationship to usual aging. We have learned that the WRNp has both helicase and exonuclease domains and, as such is a member of a new sub-family of the RecQ helicases. There are many possibilities for how this gene product functions (replication, repair, recombination, transcription, chromosome segregation) but as of 1999, there is no definitive evidence implicating one of these as the critical function that, when defective, leads to the progeroid features that characterize WRNs. Moreover, the jury is out as to what extent an understanding of the pathogenesis of WRNs will give us an understanding of how aging occurs in most of us. The association of a helicase mutation with genomic instability and cancer is not surprising. What is surprising is that this class of mutation can lead to such diverse phenotypes as coronary artery atherosclerosis, ocular cataracts and osteoporosis. Thus, work on the molecular biology of WRN, has led to a renaissance of interest in learning more of the details of how gene action protects the integrity of the genetic material during the early phases of the life course and how it may fail during the senescent phases of the life course.

References

Alexander, P (1967) The role of DNA lesions in the processes leading to aging in mice. Symp. Soc. Exp. Biol. 21, 29-50.

Balajee, A.S., Machwe, A., May, A., Gray, M.D., Oshima, J., Matrin, G.M., Nehlin, J.O., Brosh R, Orren D, & Bohr VA. The Werner protein is involved in RNA polymerase II transcription. submitted.

Bennett, R.J., Sharp, J.A., & Wang, J.C. (1998). Purification and characterization of the Sgs1 DNA helicase activity of Saccharomyces cerevisiae. J. Biol. Chem. 73, 9644-9650

Bennett, S.E., Umar, A., Oshima, J., Monnat, R.J. Jr., & Kunkel, T.A. (1997). Mismatch repair in extracts of Werner syndrome cell lines. Cancer Res. 57, 2956-2960.

Bennett, S.E., Umar, A., Kodama. S.,. Bennett, J.C., Monnat, R.J., & Kunkel, T.A. (1999). Evidence against a role for the Werner syndrome gene product in DNA mismatch repair. Alfred Benson Symposium Prodeedings, 44, in press.

Castro, E., Ogburn, C.E., Hunt, K.E., Tilvis, R, Louhija, J., Penttinen, R., Erkkola, R., Panduro, A., Riestra, R., Piussan, C., Deeb, S.S., Wang, L., Edland, S.D., & Martin, G.M., Oshima, J. (1999). Polymorphisms at the Werner locus: I. Newly identified polymorphisms, ethnic variability of 1367Cy/Arg, and its stability in a population of Finnish centenarians. Am. J. Med. Genet. 82:399-403

Cheng, R.Z., Murano, S., Kurz, B., & Shmookler Reis, R.J. (1990). Homologous recombination is elevated in some Werner-like syndromes but not during normal in vitro or in vivo senescence of mammalian cells. Mutat. Res. 237, 259-269

Courcelle, J., Carswell-Crumpton, C., & Hanawalt, P.C. (1997). recF and recR are required for the resumption of replication at DNA replication forks in Escherichia coli. Proc. Natl. Acad. Sci. U. S. A. 94, 3714-3719.

Ellis, N.A., & German, J. (1996). Molecular genetics of Bloom's syndrome. Hum. Mol. Genet. 5, 1457-1463.

Ellis, N.A., Groden, J., Ye, T.Z., Straughen, J., Lennon, D.J., Ciocci, S., Proytcheva, M., & German, J. (1995). The Bloom's syndrome gene product is homologous to RecQ helicases. Cell 83, 655-666

Epstein, C.J., Martin, G.M., Schultz, A.L., Motulsky, A.G. (1966). Werner's syndrome: a review of its symptomatology, natural history, pathologic features, genetics and relationships to the natural aging process. Medicine 45, 172-221

Fujiwara, Y., Higashikawa, T., & Tatsumi, M. (1977). A retarded rate of DNA replication and normal level of DNA repair in Werner's syndrome fibroblasts in culture. J. Cell. Physiol. 92, 365-374.

Fukuchi, K., Martin, G.M., & Monnat, R.J. Jr. (1989). Mutator phenotype of Werner syndrome is characterized by extensive deletions. Proc. Natl. Acad. Sci. U. S. A. 86, 5893-5897

Fukuchi, K., Tanaka, K., Kumahara, Y., Marumo, K., Pride, M.B., Martin, G.M., Monnat, R.J. Jr. (1990). Increased frequency of 6-thioguanine-resistant peripheral blood lymphocytes in Werner syndrome patients. Hum. Genet. 84, 249-252.

German, J. (1995). Bloom's syndrome. Dermatol. Clin. 13, 7-18.

Goddard, K.A.D., Yu, C.E., Oshima, J., Martin, G.M., Schellenberg, G.D., & Wijsman, E.M. (1996). Towards localization of the Werner syndrome gene by linkage disequilibrium and ancestral haplotyping: lessons learned from analysis of 35 chromosome 8p11.1-21.1 markers. Am. J. Hum. Genet. 58, 1286-1302.

Gorbalenya, A.E., Koonin, E.V., Donchenko, A.P., Blinov, V.M. (1989). Two related superfamilies of putative helicases involved in replication, recombination, repair, and expression of DNA and RNA genomes. Nucl. Acids. Res. 17, 4713-4730

Goto, M. (1997). Hierarchical deterioration of body systems in Werneris syndrome: Implications for normal aging. Mechanisms Aging. Develop. 98, 239-254.

Goto, M., Imamura, O., Kuromitsu, J., Matsumoto, T., Yamabe, Y., Tokutake, Y., Suzuki, N., Mason, B., Drayna, D., Sugawara, M., Sugimoto, M., & Furuichi, Y. (1997). Analysis of helicase gene mutations in Japanese Werner's syndrome patients. Hum. Genet. 99, 191-193.

Goto, M., Rubenstein, M., Weber, J., Woods, K., & Drayna, D. (1992). Genetic linkage of Werner's syndrome to five markers on chromosome 8. Nature 355, 735-758

Gray, M.D., Shen, J.C., Kamath-Loeb, A.S., Blank, A., Martin, G.M., Oshima, J., & Loeb, L.A. (1997). The Werner syndrome protein is A DNA helicase. Nat. Genet. 17, 100-103.

Gray, M.D., Wang, L., Youssoufian, H., Martin, G.M., & Oshima, J. (1998). Werner helicase is localized to transcriptionally active nucleoli of cycling cells. Exp. Cell. Res. 242, 487-494.

Guarente, L. (1997). Chromatin and ageing in yeast and in mammals. Ciba. Found. Symp. 211:104-107.

Hanada, K., Ukita, T., Kohno, Y., Saito, K., Kato, J., & Ikeda, H. (1997). RecQ DNA helicase is a suppressor of illegitimate recombination in Escherichia coli. Proc. Natl. Acad. Sci. U. S. A. 94, 3860-3865.

Harmon, F.G., & Kowalczykowski, S.C. (1998). RecQ helicase, in concert with RecA and SSB proteins, initiates and disrupts DNA recombination. Genes Dev. 12, 1134-1144.

Hisama, F.M., Oshima, J., Yu, C.E., Fu, Y.H., Mulligan. J, Weissman, E.M., & Schellenberg, G.D. (1998). Comparison of methods for identifying transcription units and transcription map of the werner syndrome gene region. Gemonics 52, 352-357.

Hoehn, H., Bryant, E.M., Au, K., Norwood, T.H., Boman, H., & Martin, G.M. (1975). Variegated translocation mosaicism in human skin fibroblast cultures. Cytogenet. Cell. Genet. 15, 282-298.

Huang, S., Li, B., Gray, M.D., Oshima, J., Mian, I.S., & Campisi, J .(1998). The premature aging syndrome protein, WRN, is a 3í to 5í exonuclease. Nat. Genet. 20, 114-116.

Ichikawa, K., Shimamoto, A., Imamura, O., Tokutake, Y., Yamabe, Y., Kitao, S., Suzuki, N., Sugawara, K., Matsumoto, T., Thomas, W., Drayna, D., Goto, M., Sugimoto, M., Sugawara, M., & Furuichi, Y. (1998). Physical map of the human chromosome 8p12-p21 encompassing tumor suppressor and Werner's syndrome gene loci. DNA Res. 5, 103-113.

Imamura, O., Ichikawa, K., Yamabe, Y., Goto, M., Sugawara, M., & Furuichi, Y. (1997). Cloning of a mouse homologue of the human Werner syndrome gene and assignment to 8A4 by fluorescence in situ hybridization. Genomics 41, 298-300.

Kamath-Loeb, A.S., Shen, J.C., Loeb, L.A., & Fry, M. (1998) Werner syndrome protein. II. Characterization of the integral 3' --> 5' DNA exonuclease. J. Biol. Chem. 273, 34145-34150.

Karow, J.K., Chakraverty, R.K., Hickson, I.D. (1997). The Bloom's syndrome gene product is a 3'-5' DNA helicase. J. Biol. Chem. 272, 30611-30614.

Kill, I.R., Faragher, R.G., Lawrence, K., & Shall, S. (1994). The expression of proliferation-dependent antigens during the lifespan of normal and progeroid human fibroblasts in culture. J. Cell. Sci. 107, 571-579.

Kitao, S., Ohsugi, I., Ichikawa, K., Goto, M., Furuichi, Y., & Shimamoto, A. (1998). Cloning of two new human helicase genes of the RecQ family: biological significance of multiple species in higher eukaryotes. Genomics 54, 443-452.

Kodadek, T. (1998). Mechanistic parallels between DNA replication, recombination and transcription. Trends. Biochem. Sci. 23, 79-83.

Kusano, K., Berres, M.E., & Engels, W.R. (1999). Evolution of the RECQ family of helicases. A drosophila homolog, dmblm, is similar to the human Bloom syndrome gene. Genetics 151, 1027-1039.

Kyoizumi, S., Kusunoki, Y., Seyama, T., Hatamochi, A., & Goto, M. (1998). In vivo somatic mutations in Werner's syndrome. Hum. Genet. 103, 405-410.

Lander, E.S., & Botstein, D. (1987). Homozygosity mapping: a way to map human recessive traits with the DNA of inbred children. Science 236, 1567-1570.

Lebel, M., & Leder, P. (1998). A deletion within the murine Werner syndrome helicase induces sensitivity to inhibitors of topoisomerase and loss of cellular proliferative capacity. Proc. Natl. Acad. Sci. U. S. A. 95, 13097-13102.

Lu, J., Mullen, J.R., Brill, S.J., Kleff, S., Romeo, A.M., & Sternglanz, R. (1996). Human homologues of yeast helicase. Nature 383, 678-679.

Marciniak, R.A., Lombard, D.B., Johnson, F.B., & Guarente, L. (1998). Nucleolar localization of the Werner syndrome protein in human cells. Proc. Natl. Acad. Sci. U.S.A. 95, 6887-6892.

Martin, G.M. (1978) Genetic syndromes in man with potential relevance to the pathology og aging. Birth Defects 14, 5-39.

Martin, G.M., Ogburn, C.E., Colgin, L.M., Gown, A.M., Edland, S.D., & Monnat, R.J. Jr. (1996). Somatic mutations are frequent and increase with age in human kidney epithelial cells. Hum. Mol. Genet. 5, 215-221.

Martin, G.M., Sprague, C.A., & Epstein, C.J. (1970). Replicative life-span of cultivated human cells. Effects of donor's age, tissue, and genotype. Lab. Invest. 23, 86-92.

Matsumoto, T., Imamura, O., Yamabe, Y., Kuromitsu, J., Tokutake, Y., Shimamoto, A., Suzuki, N., Satoh, M., Kitao, S., Ichikawa, K., Kataoka, H., Sugawara, K., Thomas, W., Mason, B., Tsuchihashi, Z., Drayna, D., Sugawara, M., Sugimoto, M., Furuichi, Y., & Goto, M. (1997a). Mutation and haplotype analyses of the Werner's syndrome gene based on its genomic structure: genetic epidemiology in the Japanese population. Hum. Genet. 100, 123-30

Matsumoto, T., Imamura, O., Goto, M., & Furuichi, Y. (1998). Characterization of the nuclear localization signal in the DNA helicase involved in Werner's syndrome. Int. J. Mol. Med. 1, 71-76.

Matsumoto, T., Shimatoto, A., Goto, M., & Furuichi, Y. (1997b). Impaired nuclear loclization of defective DNA helicases in Werneris syndrome. Nat. Genet. 16,335-336.

McKusick, V.A. (1998) Mendelian inheritance in man, 12th Edition. The Johns Hopkins University Press, Baltimore, MD

Morozov, V., Mushegian, A.R., Koonin, E.V., & Bork, P. (1997). A putative nucleic acid-binding domain in Bloom's and Werner's syndrome helicases. Trends. Biochem. Sci. 22, 417-418.

Murray, J.M., Lindsay, H.D., Munday, C.A., & Carr, A.M. (1997). Role of Schizosaccharomyces pombe RecQ homolog, recombination, and checkpoint genes in UV damage tolerance. Mol. Cell. Biol. 17, 6868-6875.

Moser MJ, Holley WR, Chatterjee A, SI Mian (1997). The proofreading domain of Escherichia coli DNA polymerase I and other DNA and/or RNA exonuclease domains. Nucleic. Acids. Res. 25, 5110-5118.

Moser, M.J., Oshima, J., & Monnat, R.J.(1999). Mutation update: WRN mutations in Werner symdrome. Hum. Mut. in press.

Mushegian, A.R., Bassett,D.E., Boguski, M.S., Bork, P., Konin, E.V. (1997). Positionally cloned human disease genes: Pattern of evolutionary conservation and functional motifs. Proc. Natl. Acad. Sci. U.S.A. 94, 5831-5836.

Myers, R.S., Kuzminov, A., & Stahl, F.W. (1995). The recombination hot spot chi activates RecBCD recombination by converting Escherichia coli to a recD mutant phenocopy. Proc. Natl. Acad. Sci. U. S. A. 92, 6244-6248.

Nakayama, K., Irino, N., & Nakayama, H. (1985). The recQ gene of Escherichia coli K12: molecular cloning and isolation of insertion mutants. Mol. Gen. Genet. 200, 266-271.

Nakayama, H., Nakayama, K., Nakayama, R., Irino, N., Nakayama, Y., & Hanawalt, P.C. (1984). Isolation and genetic characterization of a thymineless death-resistant mutant of Escherichia coli K12: identification of a new mutation (recQ1) that blocks the RecF recombination pathway. Mol. Gen. Genet. 195, 474-480.

Nakura, J., Wijsman, E.J., Miki, T., Kamino, K., Yu, C.E., Oshima, J., Fukuchi, K.I., Weber, J.L., Piussan, C., Melaragno, M.I., Epstein, C.J., Scappaticci, S., Fraccaro, M., Matsumura, T., Murano, S., Yoshida, S., Fujiwara, Y., Saida, T., Ogihara, T., Martin, G.M., & Schellenberg, G.D. (1994). Homozygosity mapping of the Werner syndrome (WRN). Genomics 23, 600-608.

Ogburn, C.E., Oshima, J., Poot, M., Chen, R., Hunt, K.E., Gollahon, K.A., Rabinovitch, P.S., & Martin, G.M. (1997). An apoptosis-inducing genotoxin differentiates heterozygotic carriers for Werner helicase mutations from wild-type and homozygous mutants. Hum. Genet. 101, 121-125.

Okada, M., Goto, M., Furuichi, Y., & Sugimoto, M. (1998). Differential effects of cytotoxic drugs on mortal and immortalized B-lymphoblastoid cell lines from normal and Werner's syndrome patients. Biol. Pharm. Bull. 21, 235-239.

Oshima, J., Yu, C.E., Boehnke, M., Weber, J.L., Edelhoff, S., Wagner, M.J., Wells, D.E., Wood, S., Disteche, C.M., Martin, G.M., & Schellenberg, G.D (1994). Integrated mapping analysis of the Werner syndrome region of chromosome 8. Genomics 23, 100-113.

Oshima, J., Yu, C.E., Piussan, C., Klein, G., Jabkowski, J., Balci, S., Miki, T., Nakura, J., Ogihara, T., Ells, J., Smith, M., Melaragno, M.I., Fraccaro, M., Scappaticci, S., Matthews, J., Ouais, S., Jarzebowicz, A., Schellenberg, G.D., & Martin, G.M. (1996). Homozygous and compound heterozygous mutations at the Werner syndrome locus. Hum. Mol. Genet. 5, 1909-1913.

Peter, M., Nakagawa, J., Doree, M., Labbe, J. C., & Nigg, E. A. (1990). Identification of major nucleolar proteins as candidate mitotic substrates of cdc2 kinase. Cell 60, 791-801.

Phillips, G.J., Prasher, D.C., & Kushner, S.R. (1988). Physical and biochemical characterization of cloned sbcB and xonA mutations from Escherichia coli K-12. J. Bacteriol. 170, 2089-2094.

Poot, M, Gollahon, K.A., & Rabinovitch, P.S. (1999). Werner syndrome lymphoblastoid cells are sensitive to camptothecin-induced apoptosis in S-phase. Hum. Genet. 104, 10-14.

Poot, M., Hoehn, H., Runger, T.M., & Martin, G.M. (1992). Impaired S-phase transit of Werner syndrome cells expressed in lymphoblastoid cell lines. Exp. Cell. Res. 202, 267-273.

Puranam, K.L., & Blackshear, P.J. (1994). Cloning and characterization of RECQL, a potential human homologue of the Escherichia coli DNA helicase RecQ. J. Biol. Chem. 269, 29838-29845.

Runger, T.M., Bauer, C., Dekant, B., Moller, K., Sobotta, P., Czerny, C., Poot, M., & Martin, G.M. (1994). Hypermutable ligation of plasmid DNA ends in cells from patients with Werner syndrome. J. Invest. Dermatol. 102, 45-48.

Salk, D., Bryant, E., Hoehn, H., Johnston, P., & Martin, G.M. (1985). Growth characteristics of Werner syndrome cells in vitro. Adv. Exp. Med. Biol. 190, 305-311.

Schellenberg, G.D., Martin G.M., Wijsman, E.M., Nakura, J., Miki, T., & Ogihara, T. (1992). Homozygosity mapping and Werneris syndrome. Lancet 339, 1002.

Shen, J.C., Gray, M.D., Kamath-Loeb, A.S., Fry, M., Oshima, J., & Loeb, L.A. (1998). Werner syndrome protein. I. DNA helicase and DNA exonuclease reside on the same polypeptide. J. Biol. Chem. 273, 34139-34144.

Shiratori, M., Sakamoto, S., Suzuki, N., Tokutake, Y., Kawabe, Y., Enomoto, T., Sugimoto, M., Goto, M., Matsumoto, T., & Furuichi, Y. (1999). Detection by epitope-defined monoclonal antibodies of Werner DNA helicases in the nucleoplasm and their upregulation by cell transformation and immortalization. J. Cell. Biol. 144, 1-9.

Sinclair, D.A., & Guarente, L. (1997). Extrachromosomal rDNA circles—a cause of aging in yeast. Cell 91, 1033-1042.

Sinclair, D.A., Mills, K., & Guarente, L. (1997). Accelerated aging and nucleolar fragmentation in yeast sgs1 mutants. Science 277, 1313-1316.

Stewart, E., Chapman, C.R., Al-Khodairy, F., Carr, A.M., & Enoch, T. (1997). rqh1+, a fission yeast gene related to the Bloom's and Werner's syndrome genes, is required for reversible S phase arrest. EMBO J. 16, 2682-2692.

Sun, H., Karow, J.K., Hickson, I.D., & Maizels, N. (1998). The Bloom's syndrome helicase unwinds G4 DNA. J. Biol. Chem. 273:27587-27592.

Suzuki, N., Shimamoto, A., Imamura, O., Kuromitsu, J., Kitao, S., Goto, M., Furuichi, Y. (1997) DNA helicase activity in Werner's syndrome gene product synthesized in a baculovirus system. Nucl. Acids. Res. 25:2973-2979

Szilard, L. (1959). On the nature of the aging process. Proc. Natl. Acad. Sci. U.S.A. 45, 30-45.

Tada, S., Yanagisawa, J., Sonoyama, T., Miyajima, A., Seki, M., Ui, M., & Enomoto, T . (1996). Characterization of the properties of a human homologue of Escherichia coli RecQ from xeroderma pigmentosum group C and from HeLa cells. Cell. Struct. Funct. 21, 123-132

Takeuchi, F., Hanaoka, F., Goto, M., Akaoka, I., Hori, T., Yamada, M., & Miyamoto, T. (1982a). Altered frequency of initiation sites of DNA replication in Werner's syndrome cells. Hum. Genet. 60, 365-368.

Takeuchi, F., Hanaoka, F., Goto, M., Yamada, M., & Miyamoto, T. (1982b). Prolongation of S phase and whole cell cycle in Werner's syndrome fibroblasts. Exp. Gerontol. 17, 473-480.

Tollefsbol, T.O., & Cohen, H.J. (1984). Werner's syndrome: An underdiagnosed disorder resembling premature aging. Age 7, 75-88

Trainor, K.J., Wigmore, D.J., Chrysostomou, A., Dempsey, J.L., Seshadri, R., & Morley, A.A. (1984). Mutation frequency in human lymphocytes increases with age. Mech. Ageing Dev. 27, 83-86.

Turner, D.R., Morley, A.A., Haliandros, M., Kutlaca, R., &Sanderson, B.J. (1985) In vivo somatic mutations in human lymphocytes frequently result from major gene alterations. Nature 315, 343-345

Umezu, K., Nakayama, K., & Nakayama, H. (1990). Escherichia coli RecQ protein is a DNA helicase. Proc. Natl. Acad. Sci. U. S. A. 87, 5363-5367.

Wallace, D.C. (1997). Mitochondrial DNA in aging and disease. Sci. Am. 277, 40-47.

Wang, L., Hunt, K.E., Martin, G.M., & Oshima, J. (1998). Structure and function of the human Werner syndrome gene promoter: evidence for transcriptional modulation. Nucl. Acids. Res. 26, 3480-3485.

Watt, P.M., Hickson, I.D., Borts, R.H., Louis, E.J. (1996). SGS1, a homologue of the Bloom's and Werner's syndrome genes, is required for maintenance of genome stability in Saccharomyces cerevisiae. Genetics 144, 935-945.

Watt, P.M., Louis, E.J., Borts, R.H., & Hickson, I.D. (1995). Sgs1: a eukaryotic homolog of E. coli RecQ that interacts with topoisomerase II in vivo and is required for faithful chromosome segregation. Cell 81, 253-260.

Webb, D.K., Evans, M.K., & Bohr, V.A. (1996). DNA repair fine structure in Werner's syndrome cell lines. Exp. Cell. Res. 224, 272-278.

Weirich-Schwaiger, H., Weirich, H.G., Gruber, B., Schweiger, M., & Hirsch-Kauffmann, M. (1994). Correlation between senescence and DNA repair in cells from young and old individuals and in premature aging syndromes. Mutat. Res. 316, 37-48.

Werner, O. (1904). On cataract associated in conjunction with scleroderma (doctoral dissertation, Kiel University), Schmidt and Klaunig, Kiel.

Yamabe, Y., Shimamoto, A., Goto, M., Yokota, J., Sugawara, M., Furuichi, Y. (1998). Sp1-mediated transcription of the Werner helicase gene is modulated by Rb and p53. Mol. Cell. Biol. 18, 6191-200,

Yamabe, Y., Sugimoto, M., Satoh, M., Suzuki, N., Sugawara, M., Goto, M., & Furuichi, Y. (1997). Downregulation of the defective transcripts of the Werner's syndrome gene in the cells of patients. Biochem. Biophys. Res. Commun. 236, 151-154.

Yamagata, K., Kato, J., Shimamoto, A., Goto, M., Furuichi, Y., & Ikeda, H. (1998). Bloom's and Werner's syndrome genes suppress hyperrecombination in yeast sgs1 mutant: implication for genomic instability in human diseases. Proc. Natl. Acad. Sci. U. S. A. 95. 8733-8738.

Yan, H., Chen, C.Y., Kobayashi, R., & Newport, J. (1998). Replication focus-forming activity 1 and the Werner syndrome gene product. Nat. Genet. 19, 375-378.

Ye, L., Miki, T., Nakura, J., Oshima, J., Kamino, K., Rakugi, H., Ikegami, H., Higaki, J., Edland, S.D., Martin, G.M., & Ogihara, T. (1997). Association of a polymorphic variant of the Werner helicase gene with myocardial infarction in a Japanese population. Am. J. Med. Genet. 68:494-498

Ye, L., Nakura, J., Morishima, A., Miki, T. (1998). Transcriptional activation by the Werner syndromegene product in yeast. Exp. Gerontol. 33:805-812.

Yu, C.E., Oshima, J., Fu, Y.H., Hisama, F., Wijsman, E.M., Alisch, R., Matthews, S., Nakura, J., Miki, T., Ouais, S., Martin, G.M., Mulligan, J., & Schellenberg, G.D. (1996a). Positional cloning of the Werneris syndrome gene. Science 272:258-262.

Yu, C.E., Oshima, J., Goddard, K.A.B., Miki, T., Nakura, J., Ogihara, T., Fraccaro, M., Piussan, C., Martin, G.M., Schellenberg, G.D., Wijsman, E.M. (1994). Linkage disequilibrium and haplotype analysis of chromosome 8p11.1-21.1 markers and Werner syndrome. Am. J. Hum. Genet. 55, 356-364.

Yu, C.E., Oshima, J., Hisama, F., Matthews, S., Trask, B., & Schellenberg, G.D. (1996b). A Yac. P1 and cosmid contig and 16 new polymorphic markers for the Werneris syndrome region at 8p12-21. Genomics 35, 431-440.

Yu, C.E., Oshima, J., Wijsman, E.M., Nakura, J., Miki, T., Piussan, C., Matthews, S., Fu, Y.H., Mulligan, J., Martin, G.M., Schellenberg, G.D., & the Werneris Syndrome Collaborative Group. (1997). Mutations in the consensus helicase domains of the Werneris syndrome gene. Am. J. Hum. Genet. 60, 330-341.

© 2001 Elsevier Science B.V. All rights reserved.
The Role of DNA Damage and Repair in Cell Aging
B.A. Gilchrest and V.A. Bohr, volume editors.

INDUCIBLE PHOTOPROTECTION IN SKIN:
EVIDENCE FOR A EUKARYOTIC SOS RESPONSE

Mark S. Eller and Barbara A. Gilchrest

Department of Dermatology, Boston University School of Medicine, Boston, MA

1. Aging and DNA Damage

Aging is a complex process in which both genetic (intrinsic) and environmental (extrinsic) factors play an essential role in determining cell and organism viability over time. Damage and repair of DNA plays a key role in this process. DNA damage arises from exogenous agents such as ultraviolet radiation or carcinogenic chemicals as well as from free radical-generating physiologic metabolic processes (Wei et al., 1993; Agarwal and Sohal, 1994; Szillard, 1995; Bohr and Anson, 1995).

Because organisms are constantly subjected to environmental DNA damage, they must possess mechanisms to repair this damage in order to protect functioning of the organism as a whole. DNA repair is accomplished in prokaryotes and eukaryotes largely through the nucleotide excision repair (NER) pathway. NER is a versatile yet complicated repair mechanism effective not only in repair of UV-induced photoproducts, but also in repair of chemically modified DNA formed by agents such as 4-nitroquinoline oxide and photoactivated psoralen (Sancar and Sancar, 1988). The enzymology of excision repair is fairly well understood for prokaryotes, particularly *E. coli*, but much less so for eukaryotic cells. However, recent work with mutant cells deficient in excision repair has begun to explore this process in yeast, higher eukaryotic cells, and even human cells.

Excision repair, in concept, is simple and can be divided into several distinct steps (Eller, 1995). Briefly, damaged DNA is first recognized by the repair complex. The DNA is then incised both 5' and 3' to the damaged nucleotides, the damaged oligonucleotide is removed, and the gap created is resynthesized by DNA polymerase. Finally, the newly synthesized DNA strand is ligated to the pre-existing DNA to form a covalently linked, double-stranded molecule. Most experimental methods used to detect excision repair measure the DNA incision step or repair DNA synthesis.

The balance between DNA damage and repair greatly influences mutation frequency and, as would be expected, has a major impact on aging. Indeed, studies looking at specific gene loci in both human and animal cells have confirmed that accumulation of mutations in somatic cells is associated with aging (Alexander, 1967; Trainor et al., 1984; Dutkowski et al., 1985; Grist et al., 1992; Wei et al., 1993; King et al., 1994). Also, Moriwaki et al. (1996) reported a decrease in DNA repair capacity and an increase in mutation frequency in cultured skin fibroblasts and blood cells from donors of increasing age. Furthermore there is an age-associated increase in skin cancer in the general population (Scotto et al., 1974; Scotto et al., 1981) and many of these cancers contain "signature" mutations from UV-induced DNA photoproducts (Brash et al., 1991).

2. The Prokaryotic SOS Response

In prokaryotes such as Escherichia coli, DNA damage and replication blocks induce genes involved in DNA repair, mutagenesis and cell survival, termed the "SOS response" (Walker, 1984). Early indications of this coordinated response came from studies of DNA repair and prophage lambda induction and mutagenesis in DNA-damaged bacteria (Weigle, 1953; Radman, 1974; Radman 1975). In 1953, Weigle found that UV-irradiation of host E. coli before infection with UV-damaged bacteriophage enhanced the repair and survival of these damaged viruses. Furthermore, if the UV-induced thymine dimers were removed from the bacterial DNA by activation of the repair enzyme photolyase, this enhanced survival of the bacteriophage did not occur. It was also found that there was a high mutation frequency in these surviving phage (Weigle, 1953) as well as in the chromosomal DNA of the irradiated bacteria (Witkin, 1969). Radman (1974, 1975) termed this phenomenon of heightened, albeit less accurate, DNA repair after an initial DNA injury the SOS response, as it conferred increased probability of survival on affected bacteria.

The bacterial genes induced as part of the SOS response are normally under repression by the LexA protein (Fig. 1). Single-stranded DNA generated by excision repair of DNA damage or at stalled replication forks activates the RecA protease which in turn cleaves the LexA repressor protein, de-repressing the SOS genes (Walker, 1984). Among the genes induced as part of the SOS response are the *uvrD* gene which codes for an excision repair-associated helicase (Pang and Walker, 1983; Siegel, 1983) and the *umuC* and *umuD* genes, coding for proteins which facilitate translesional DNA synthesis (Reuven et al., 1998; Tang et al., 1998). This induced ability to synthesize DNA across a damaged template decreases lethal frameshift mutations while increasing the frequency of non-lethal point mutations from base substitution (Reuven et al., 1998; Tang et al., 1998). The resulting mutagenesis is viewed as evolutionarily benefcial in that it enables populations of microorganisms to adapt to environmental stresses such as UV irradiation.

In addition to its role as a main regulator of gene expression, the RecA protein also has a more direct function in SOS-induced DNA repair, particularly in recombinational repair. By virtue of its ability to bind DNA, especially single-stranded DNA, the RecA protein facilitates the pairing of homologous strands (Radding, 1991; Roca and Cox, 1990) facilitating DNA strand exchange between two molecules. The RecA protein also promotes the cleavage of the UmuD protein to the active form, designated UmuD', facilitating mutagenic translesional DNA synthesis (Burckhardt et al., 1988; Nohmi et al., 1988; Shinagawa et al., 1988).

Although most extensively studied in E. coli, many other bacteria also possess a coordinately regulated SOS response. These bacteria include *Salmonella typhimurium* (Orrego and Eisenstadt, 1987), *Haemophilus influenza* (Notani and Setlow, 1980),

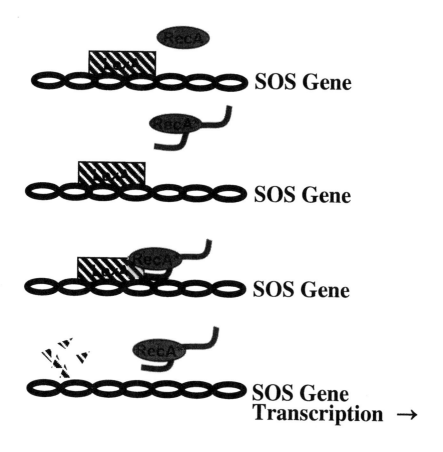

Figure 1. Single stranded DNA induces expression of SOS genes in bacteria. The LexA protein normally represses genes expressed as part of the SOS response. After DNA damage, single-stranded DNA is generated by the excision repair process and at stalled replication forks. This single-stranded DNA interacts with and activates the RecA protease (designated as RecA*). This protease then cleaves the LexA protein, derepressing and inducing the expression of the SOS response genes.

Pseudomonas aeruginosa (Miller and Kokjohn, 1988), and *Bacillus subtilis* (Lovett et al., 1988). Furthermore, many species of bacteria were found to constitutively repress a heterologous gene construct consisting of a promoter containing a LexA-binding region and a lacZ reporter sequence, and to induce the expression of this fusion gene after DNA damage (de Henestrosa et al., 1991). Therefore, many physiologic and mechanistic aspects of the SOS system appear common in prokaryotic cells.

3. Responses to DNA Damage in Eukaryotic Cells

Eukaryotic responses to DNA damage include cell cycle arrest and induction of specific genes. Although many such inducible genes have been identified (Fornace et al., 1988; Sebastian and Sancar, 1991; Luethy and Holbrook, 1992; Hollander et al., 1993), the existence of a coordinately regulated SOS response in eukaryotic cells has not been demonstrated. In yeasts, cell cycle arrest and regulation of transcription in response to DNA damage and replication blocks is thought to be mediated largely through the Rad53/SAD1/MEC2 protein (Allen et al., 1994; Kato and Ogawa, 1994; Sanchez et al., 1996; Weinert et al., 1994). This protein kinase activates the DUN1 protein, potentially by phosphorylation, and activated DUN1 controls a broad transcriptional response to DNA damage (Zhou and Elledge, 1993; Sanchez et al., 1996). Rad53 likely regulates other proteins that are involved in the yeast cell cycle arrest in response to genotoxic stress, although this pathway downstream of Rad53 is poorly understood.

In higher eukaryotes, many of the cell cycle and transcriptional responses to DNA damage from ionizing and UV-irradiation as well as genotoxic chemicals are mediated through the p53 transcription factor and tumor suppressor protein. In response to DNA damage, most likely DNA strand breaks (Lu and Lane, 1993; Nelson and Kastan, 1994), there is a rapid increase in nuclear p53 levels (Kastan et al., 1991; Lu et al., 1992; Fritsche et al., 1993; Hall et al., 1993). Although the mechanism(s) by which p53 levels are regulated are poorly understood, data suggest that changes in the phosphorylation state in response to genotoxic stress may decrease the rate of ubiquitin-mediated degradation of this protein (Milne et al., 1992; Fiscella et al., 1993; Chowdary et al., 1994), accounting for the often striking increase in p53 protein level after DNA injury (Fritsche et al., 1993). Furthermore, there is evidence that p53 transcriptional and DNA binding activity is regulated independently of protein levels (Hupp et al., 1995; Lu et al., 1996). The activation of p53 is likely affected by covalent modification such as phosphorylation (Milne et al., 1992; Fiscella et al., 1993; Hupp et al., 1995; Lu et al., 1996; Banin et al., 1998; Canman et al., 1998) and/or dephosphorylation (Waterman et al., 1998), or interaction with peptides (Hupp et al., 1995; Waterman et al., 1998) or single-stranded DNAs (Jayaraman and Prives, 1995; Bayle et al., 1995). Activation of p53 subsequently leads to the increased transcription of genes involved in cell cycle arrest, such as p21 (El-Deiry et al., 1993; Di Leonardo et al., 1994) and GADD45 (Kastan et al., 1992; Smith et al., 1994). In addition, GADD45 protein has been shown to have a direct role in NER (Smith et al., 1994), although these findings are controversial and highly dependent on the in vitro assay systems used (Kazantsev and Sancar, 1995, Kearsey et al., 1995A). That p53 plays a role in NER is further supported by the findings of Ford and Hanawalt (1995) that human fibroblasts homozygous for p53 mutations are deficient in overall (global) DNA repair, yet retain the DNA repair pathway coupled to transcription. Furthermore, Tron et al. (1998) recently showed that repair of UV-induced DNA damage was reduced in p53 null (-/-) mouse skin compared to wild-type animals.

Alternatively, under conditions of severe damage, p53 can direct not only the repair and survival of the cell, but its death by apoptosis. Although the mechanism(s) of p53-

induced apoptosis are not fully understood, transcriptional modulation of the apoptosis-inducing genes *Bax* and *Fas/APO*-1 by p53, at least in some cell types, has been demonstrated (Miyashita et al., 1994; Owen-Schaub et al., 1995). In contrast, p53 may also affect apoptosis independently of transcription because cells transiently expressing mutant p53 that is incapable of trans-activating genes have also been shown to undergo apoptosis (Haupt et al., 1995), albeit at a slower rate than cells expressing transactivation-competent wild type p53.

4. Inducible DNA Repair in Eukaryotes

A coordinately regulated eukaryotic SOS response has not been conclusively demonstrated, but there is growing evidence that many mammalian cells can induce their DNA repair capacity. As early as 1976, Lytle et al. (1976) reported that human fibroblasts UV-irradiated four days prior to infection with UV-irradiated herpes virus supported the growth of this virus about twice as well as control, unirradiated cells, implying that after irradiation the human cells repaired the viral DNA more efficiently. Furthermore, even fibroblasts from excision repair deficient xeroderma pigmentosum (XP) patients exhibited this enhanced reactivation of UV-damaged virus, suggesting the enhanced DNA repair was not attributable to any single gene product whose mutation characterized these cells. These findings were later confirmed in CV-1 monkey kidney cells using UV-irradiated SV40 virus (Taylor et al., 1982). Furthermore, this enhanced repair of damaged virus was found not to be associated with an increased mutation frequency as is characteristic of the prokaryotic SOS response (Taylor et al., 1982). Jeeves and Rainbow (1983A & 1983B) expanded these findings using normal and repair-deficient human fibroblasts infected with adenovirus type 2 (Ad2). They found that the enhanced repair response could be stimulated not only by UV irradiation, but also by gamma-irradiation of the cells and was effective in enhancing the survival of both UV- and gamma-irradiated virus (Jeeves and Rainbow, 1983A, Jeeves and Rainbow, 1983B). Normal fibroblasts and repair deficient fibroblasts from XP and Cockayne's syndrome patients also exhibited the inducible repair of UV- and gamma-irradiated virus, although to different degrees (Jeeves and Rainbow, 1983A; Jeeves and Rainbow, 1983C). Interestingly, cells from ataxia telangiectasia (AT) patients, who are hypersensitive to ionizing radiation, failed to exhibit gamma irradiation-enhanced Ad2 reactivation (Jeeves and Rainbow, 1986).

The gene mutated in AT (ATM, for AT-mutated) has now been identified and cloned and shown to share extensive homology with a family of phosphatidylinositol (PI) kinases (Savitsky et al., 1995). AT cells show a diminished and/or delayed induction of p53 in response to ionizing radiation (Khanna and Lavin, 1993; Lu and Lane, 1993), and the ATM kinase has been shown to phosphorylate p53 in vitro in response to ionizing radiation (Banin et al., 1998; Canman et al., 1998). In addition, ATM has been shown to mediate the dephosphorylation of serine 376 on p53 and subsequent activation of p53 through interactions with specific binding proteins (Waterman et al., 1998). Interestingly, the enhanced DNA repair response that is induced by UV irradiation and heat shock in normal cells cannot be induced in p53-deficient tumor cells (Rainbow et al., 1995; McKay and Rainbow, 1996) or Li-Fraumeni syndrome fibroblasts that are heterozygous for a dominant negative p53 mutation (McKay et al., 1997). Together, these data suggest a role for p53 in the DNA damage-induced enhancement of viral reactivation and, at least in the case of enhanced reactivation following gamma irradiation, ATM may mediate this activity of p53.

In addition to radiation, mammalian cells also exhibit an enhanced DNA repair capacity after treatment with the chemical carcinogen mitomycin C, as shown by Protic et al. (1988). In their experiments, an SV40-transformed human XP fibroblast line did not show enhanced reactivation of their transfected UV-damaged chloramphenicol acetyltransferase (CAT) reporter vector, but this possibly reflected inactivation of p53 by SV40 transformation.

Although the vast majority of studies on enhanced repair responses have focused on inducible NER, low doses of ionizing radiation have been shown to "prime" human lung carcinoma cells to enhance the removal of oxidative base damage (Le et al., 1998), thought to be repaired predominantly by the base excision repair pathway. Le et al. found that these cells showed an initial enhanced repair of thymine glycol after exposure to ionizing radiation if they were exposed to a much smaller dose four hours prior.

5. Photoprotective Responses in Human Skin

A. *Tanning*

Human skin exhibits protective responses to UV irradiation such as repair of the DNA damage and increased melanogenesis (tanning) to help protect against subsequent exposures. Epidermal melanin is widely acknowledged as being photoprotective, and many studies have demonstrated an inverse relationship between melanin content and the subsequent UV-induction of DNA photoproducts. Kobayashi et al. (1993) found that highly pigmented melanoma cells were more resistant to UV exposure and formed fewer cyclobutane dimers and (6-4) photoproducts than did lightly pigmented cells. Similarly, Hill and Setlow (1982) found that lightly pigmented mouse melanoma cells formed fewer dimers after UV irradiation than did nonpigmented mouse mammary carcinoma cells. Interestingly, they found the protective effect to be more pronounced at UVC wavelengths (200-290 nm) compared to wavelengths greater than 290 nm. A full understanding of melanin's protective qualities must also take into account such features as the type of melanin present and the subcellular localization of this pigment.

Exposure of skin to UV radiation results in increased pigmentation which occurs in two stages, an immediate darkening and a delayed tanning reaction. Immediate pigment darkening is thought to result from oxidation of pre-existing melanin (Pathak and Stratton, 1968) and redistribution of melanosomes from a perinuclear position in the melanocyte to a peripheral, dendritic distribution, but is minimally photoprotective

(McGregor et al., 1999). In contrast, the delayed tanning response is photoprotective against subsequent UV injury, begins as the immediate pigmentation reaction fades and progresses for at least 3-5 days after UV exposure. Appearance of tanning clinically parallels increased melanocyte tyrosinase activity (Rosdahl et al., 1979; Friedmann and Gilchrest, 1987). The delayed tanning response can be induced by both UVA (320-400 nm) and UVB (290-320 nm), although the response induced by UVA alone is 2 to 3 orders of magnitude less efficient and differs mechanistically from that induced by UVB. The UVA-induced delayed tan requires oxygen at the time of irradiation, whereas the UVB response does not (Auletta et al., 1986). Also, the UVA delayed tanning response occurs fairly early, often directly after the immediate pigmentation phase, whereas tanning from UVB irradiation begins only several days after exposure (Lavker and Kaidbey, 1982; Kochevar, 1992).

Although the regulation of melanin synthesis is complex and incompletely understood, evidence suggests that UV-induced DNA photodamage and/or its repair is at least one of the initial signals that stimulates melanogenesis in response to UV irradiation (Gilchrest and Eller, 1999). First, the action spectrum for the tanning response in human skin is the same as that for the induction of the major DNA photoproducts (Parrish et al., 1982; Freeman et al., 1989). Second, enhancing DNA repair by treatment of melanocytic cells post-irradiation with the prokaryotic DNA repair enzyme T4 endonuclease V approximately doubles the melanin content of these cells compared to irradiated cells treated with either heat-inactivated enzyme or diluent alone (Gilchrest et al., 1993).

Melanogenesis and DNA repair are photoprotective to skin, similar to the photoprotective prokaryotic SOS response initiated by single stranded DNA (ssDNA). In addition, the data discussed above provide evidence that these responses are initiated, at least in part, by DNA damage. This suggests that melanogenesis and an enhanced DNA repair capacity are part of a eukaryotic "SOS-like" response similarly stimulated by ssDNA. Because skin is the primary target for damage by UV light, we chose intact skin and cultured skin cells to study the effects of ssDNA on the induction of these photoprotective responses. Although the exact sequence of the ssDNA fragments released during excision repair varies from site to site, most such fragments contain pyrimidine dimer photoproducts, particularly thymine dimers. Therefore, we used the substrate for these photoproducts, thymidine dinucleotide, pTpT, to examine the induction of photoprotective responses in skin. First, we determined the effect of pTpT on pigmentation (Eller et al., 1994) using the melanotic Cloudman S91 mouse melanoma cell line which is known to pigment in response to UV irradiation (Chakraborty et al., 1991). Melanoma cells treated for 4 days with 50 μM pTpT showed a 7-fold increase in melanin content compared to control, diluent-treated cells (Eller et al., 1994) (Figure 2). Cells similarly treated with deoxyadenine dinucleotide, pdApdA, a dinucleotide rarely involved in photoproduct formation under physiologic conditions (Gasparro and Fresco, 1986), showed only a minimal 20-30% increase in melanin content. In agreement with an increase in melanin content, pTpT-treated cells showed a 2 to 3-fold elevation in the mRNA for tyrosinase, the rate-limiting enzyme for melanin biosynthesis, 3 to 4 days after addition of the dinucleotide to the cell cultures. The

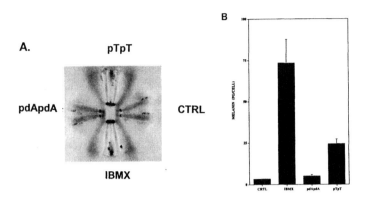

Figure 2. Effect of dinucleotides on melanogenesis. (A) Photograph of paired S91 cell cultures treated as indicated. (B) Values of melanin content are the mean ±S.D. of the paired cultures depicted in A. Both pTpT and pdApdA were resuspended in water at a concentration of 1mM and filter-sterilized. Cloudman S91 murine melanoma cells were grown in DME +10% calf serum (control) or supplemented with 100 μM isobutylmethylxanthine (IBMX) as a positive control, 50 μM pTpT or 50 μM pdApdA for 4 days. The cells were then washed, collected by trypsinization and counted by Coulter counter. An equal number of cells were then pelleted by microcentrifugation and photographed (A) Because of the angle of centrifugation, only the thickest portion of the lightly-pigmented cell pellets are clearly visible, making these pellets appear smaller. NaOH (1 M; 1ml) was then added to each pellet and vortexed until the pellet was dissolved. Melanin concentration was calculated by absorbance at 475 nm compared with a standard curve of melanin in 1 M NaOH (B). (From Eller et al., Nature 372:413-414, 1994 with permission.)

increase in tyrosinase mRNA in pdApdA-treated cells was much more modest. Human melanocytes also increased their pigment content in response to pTpT, but less dramatically than S91 cells. pTpT induced a tanning response not only in cells in culture, but also in intact guinea pig skin (Eller et al., 1994). Aliquots of pTpT (100 μM–300 μM) or diluent were applied to shaved guinea pig skin twice daily for 5 days. A tanning response was noticed on the pTpT-treated sites approximately one week after final application, with maximal response 1-2 weeks later. The sites were biopsied and sections were stained with Fontana-Masson to highlight melanin. Histologic evaluation showed an increase in melanin mainly in the basal epidermal layer, but also in the suprabasal cells, particularly in "caps" over the cell nucleus, closely resembling a UV-induced tan.

   The melanogenic activity of pTpT was confirmed and further examined by Pedeux et al. (1998), who reported that human melanocytes and melanoma cells pigment in response to this dinucleotide, that the monomeric form pT is not melanogenic, and that pTpT at the concentrations examined is non-toxic to these cells and does not induce

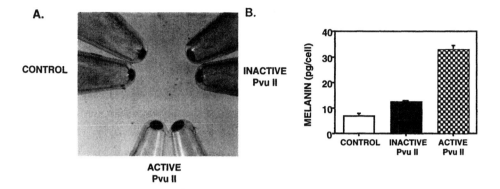

Figure 3. Effect of DNA damaging agents on pigmentation in S91 cells. Duplicate cultures of S91 cells were treated with the restriction enzyme Pvu II after poration with streptolysin O. After treatment, cells were cultured for 6 days, collected, and counted and an equal number of cells were pelleted for analysis. The cell pellets were photographed and melanin/cell calculated for each pellet as described. Averaged values of the duplicate cultures of each treatment are graphed in (B). The small melanogenic effect of "inactive" Pvu II is attributed to incomplete heat denaturation of the enzyme in this protocol. (From Eller et al., PNAS 93:1087-1092, 1996 with permission.)

apoptosis. Cell cycle analysis of pTpT-treated melanoma cells revealed that they were temporarily blocked in the S phase of the cell cycle 24 hours after addition of pTpT, yet resumed cycling within 48 hours. Although no mechanism for cell cycle regulation by pTpT was proposed, similar inhibition was not observed with equimolar amounts of thymidine monophosphate, pT.

Other evidence that DNA damage stimulates pigmentation in the S91 mouse melanoma cell line is that cutting the genomic DNA with the restriction enzyme Pvu II stimulates pigmentation to 4 to 5 times control levels (Eller et al., 1996) (Figure 3). Similarly, treatment of S91 cells and normal human melanocytes with methyl methanesulfonate (MMS) or 4-nitroquinoline 1-oxide (4NQO), chemicals known to damage DNA, increases their melanin content, at least in part through up-regulation of tyrosinase mRNA and protein levels, in a similar time course and to a similar degree as following UV-irradiation or pTpT-treatment of these cells (Eller et al., 1996). In addition, pre-treatment of cells with pTpT or MMS enhances the response to alpha melanocyte stimulating hormone (α-MSH), known to cause pigmentation in these cells (Fuller et al., 1987; Bolognia et al., 1989; Hoganson et al., 1989; Rungta et al., 1996), at least in part through increased binding of α-MSH to its cell surface receptor, the mechanism proposed by Bolognia et al. (1989) for the enhanced pigmentation in S91 cells and intact mouse skin after UV irradiation.

Although pTpT was originally chosen for study because it is the precursor of the major UV photoproduct, it was subsequently found that other, but not all, DNA

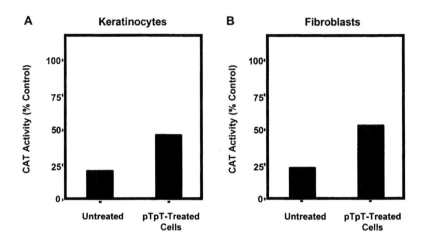

Figure 4. pTpT treatment enhances the repair of UV-induced DNA damage. Newborn keratino-cytes and fibroblasts were established from dermal explants. Cells were treated with either 100 μM pTpT or an equal volume of diluent for 5 days prior to transfection. Duplicate cultures of each condition were then transfected with 5 μg reporter DNA, pCAT-Control vector. This non-replicating vector contains the chloramphenicol acetyltransferase gene under control of SV40 promoter and enhancer sequences. Before transfection, the vector DNA was either sham irradiated or exposed to solar-simulated UVB radiation. Dose response experiments showed that irradiation of the pCAT-control vector with 70, 100 and 140 mJ/cm² UVB resulted in 65, 75 and 88% inhibition, respectively, of CAT activity in transfected fibroblasts and keratinocytes. For the repair experiments, 100 mJ/cm² UVB irradiation was used to give an effective but non-saturating degree of DNA damage to the pCAT expression vector. Cells were collected 24 hours after transfection and CAT enzyme activity was determined. CAT activity was expressed as c.p.m./100 μg protein and is represented as percent activity of cells transfected with sham-irradiated plasmid. Actual CAT values for duplicate keratinocyte cultures were all within 8% of the mean. CAT values for duplicate fibroblast cultures were generally within 6% of the mean, with one pair differing by 20%. (From Eller et al., PNAS 94:12627-12632, 1997.)

fragments also stimulate pigmentation. For example, two 9-mer oligonucleotides pGpApGpTpApTpGpApG and pTpApGpGpApGpGpApT, as well as the 7-mer pApGpTpApTpGpA, enhance melanin content in S91 cells up to 800% that of control cells while the 5-mer oligonucleotide pCpApTpApC has no effect (Eller et al., 1998). Among the several oligonucleotides tested to date, the melanogenic activity increases with increasing oligo length, but there is also an apparent influence of base composition and/or sequence. To illustrate this point, pCpApTpApC is inactive in experiments in which pGpTpApTpG increases melanin content to over 600% of control levels (Eller et al., 1998). Determining the relationship between oligonucleotide size and sequence and melanogenic activity should prove instructive in dissecting the mechanisms of this and other UV responses.

B. *Inducible DNA Repair*

The ability of pTpT to stimulate tanning raises the question whether pTpT can also induce other photoprotective, SOS-like responses, such as enhanced DNA repair. Host cell restoration of expression of a transfected UV-damaged chloramphenicol acetyl-transferase (CAT) expression vector was selected as a measure of DNA repair (Wei et al., 1993). Because this reporter plasmid is non-replicating, CAT activity is a measure of host cell enzyme-mediated DNA repair and restoration of biological activity. This assay has been used previously to detect a decrease in DNA repair capacity in human lymphocytes from patients presenting with an early onset of skin cancer compared to healthy controls, and in cells from older (50-60 years old) donors compared to younger (20-40 years old) donors (Wei et al., 1993; Moriwaki et al., 1996). In addition, this technique has been used to measure DNA repair capacity in lymphocytes from workers exposed to benzene (Hallberg et al., 1996) or 1,3 butadiene (Hallberg et al., 1997).

When compared to reporter plasmid sham-irradiated before transfection, plasmid irradiated with 100 mJ/cm^2 UVB radiation generated 23% and 24% as much CAT activity, respectively, in human keratinocytes and fibroblasts under control conditions (Eller et al., 1997) (Figure 4). However, if the cells were pre-treated with 100 μM pTpT for 5 days before transfection, the CAT activity in the cell lysate after 24 hours was approximately twice that for diluent pre-treated keratinocyte and fibroblast cultures. These data demonstrate that pTpT pre-treatment increases the capacity of normal human skin-derived cells to repair UV-induced DNA damage in a transfected plasmid.

In order to directly measure the effect of pTpT on the removal rate for UV-induced photoproducts from genomic DNA, human fibroblasts were pre-treated with either diluent or pTpT for 5 days before irradiation with 50 mJ/cm^2 UVB, using a solar-simulator. Cells were collected immediately after irradiation and at various time-points through 24 hours, the genomic DNA was isolated, blotted onto membranes and probed with monoclonal antibodies specific for (6-4) photoproducts or thymine dimers. Diluent-treated cells showed no decrease in dimers at 12 hr and had removed only 25% of dimers by 24 hr (Eller et al., 1997), a repair rate consistent with previous reports for mouse skin as well as cultured mouse and human fibroblasts (Mitchell et al., 1990; Mitchell and Karentz, 1993). As also previously reported (Mitchell et al., 1990; Mitchell and Karentz, 1993), the rate of repair of (6-4) photoproducts was considerably faster, with an approximately 10-15% removal after 20 min and 50% removal after 1 hr for controls. In contrast, pTpT-treated cells removed 40% and 65% of dimers at 12 and 24 hours post irradiation, respectively, and approximately 50% of (6-4) photoproducts within 20 min, statistically more rapid rates of removal for both types of UV-induced DNA damage (Eller et al., 1997). An accelerated removal of photoproducts can also be detected in UV-irradiated guinea pig skin in vivo following 5 days of topical application of pTpT versus diluent alone (Gilchrest and Eller, 1999).

To determine whether the enhanced DNA repair is truly photoprotective and hence improves cell survival after UV irradiation, the effect of pTpT pre-treatment on cell survival and yield after UV irradiation was also examined. Cultures pre-treated for 5 days with diluent alone before irradiation with UVB doses of 20, 30 and 40 mJ/cm^2

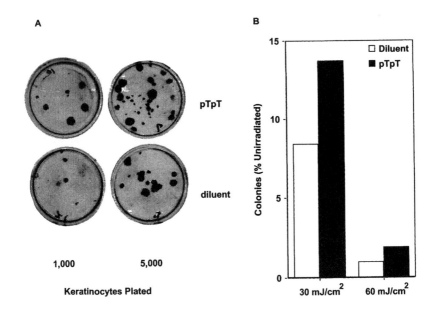

Figure 5. Effect of pTpT on colony-forming efficiency of keratinocytes after UV irradiation. First passage cultures were supplemented with either 100 μM pTpT or an equal amount of diluent for 5 days before irradiation. Cells were either sham-irradiated or exposed to 30 mJ/cm² or 60 mJ/cm² UVB, trypsinized, and immediately replated at various densities onto a 3T3 feeder layer to permit clonal growth. (A) Representative dishes of diluent or pTpT-treated cells exposed to 30 mJ/cm² UVB irradiation are presented. Each colony represents the progeny of a single previously irradiated cell. (B) Colony yield is presented as the percent of colonies formed from unirradiated keratinocytes. Nonpaired T-test analysis found that the number of colonies formed between diluent and pTpT-treated cells was statistically significant with P < 0.01 for 30 mJ/cm² and P < 0.03 for 60 mJ/cm². (From Eller et al., PNAS 94:12627-12632, 1997.)

showed a progressive decrease in cell number over the following 3 days due to cell death and detachment (Eller et al., 1997). However, cells pre-treated with pTpT continued a slow net growth even after irradiation at 40 mJ/cm² and, in contrast to control irradiated cells, appeared healthy after irradiation. In addition, pTpT pre-treatment increased the proportion of human keratinocytes that undergo mitosis and form colonies after UV irradiation, compared to diluent-pre-treated and irradiated cells, approximately doubling the yield, expressed as the percent of colonies formed from unirradiated cells (Eller et al., 1997). Assays of the colony forming efficiency of unirradiated pTpT-treated or diluent-treated keratinocytes confirmed that pTpT exposure is not toxic and does not compromise cells: keratinocytes treated for up to 3 days with 100 μM pTpT or diluent alone had identical colony-forming efficiencies (Figure 5).

Table 1. Genes involved in DNA repair and cell cycle regulation in response to genotoxic stress in mammalian cells

Gene	Comments/Function	Reference
Growth Arrest and DNA Damage Induced Gene 45 (GA□DD45)	Induced by genotoxic stress p53-regulated	Zhan et al., 1994
		Smith et al., 1994
	May participate in NER	Karantsev et al., 1995
		Kearsey et al., 1995B
Proliferating Cell Nuclear Antigen (PCNA)	Essential function in DNA repair and replication	Prelich et al., 1987
		Shiviji et al., 1992
	p53-regulated	Morris et al., 1996
Excision Repair Cross-Complementing Gene 3 (ERCC3)	Helicase activity	Weeda etal., 1990A
	Participates in NER and transcription	Weeda et al., 1990B
		Schaffer et al., 1993
p21	p53-regulated	Harper et al., 1993
	Represses PCNA function in replication	Li et al., 1994
	Inhibits cyclin-dependent kinases and G1 → S transition in cell cycle	Waga et al., 1994

To determine whether pTpT pre-treatment enhances cellular ability to repair DNA damaged by agents other than UV irradiation, experiments were performed using host cell reactivation of a CAT vector reacted with the chemical carcinogen benzo(a)pyrene (BP) (Maeda et al., 1999). BP is a powerful mutagen found in many environmental pollutants such as cigarette smoke and automobile exhaust (Phillips, 1983). The diol epoxide metabolite of BP reacts strongly with guanine bases in DNA to form a bulky adduct (Osborne et al., 1981) that is believed to be removed by NER based on the induction of unscheduled DNA synthesis, a hallmark of NER in BP-treated cells (Celotti et al., 1993). Compared to diluent-treated cells, pTpT-treated human keratinocytes showed an approximate 3-fold increase of CAT expression when transfected with a vector containing 25 BP adducts/plasmid (Maeda et al., 1999). Using CAT vectors with a higher level of damage (~50 adducts/plasmid) a doubling of CAT activity was found in pTpT-treated cells. This increase of CAT activity in pTpT-treated cells in these experiments is similar to the results obtained with a UV-damaged CAT vector, described above. Such a doubling of DNA repair might be expected to have profound biological effects, given that a decrease in DNA repair capacity for UV-induced photoproducts of only 5-8% characterizes patients with early onset of basal cell carcinoma, a malignancy strongly associated with sun exposure, compared to age-matched controls and using the same assay; and DNA repair capacity declines only 15%

during adulthood for normal subjects, a loss proposed to account for the well-known age-associated increased cancer risk (Wei et al., 1993).

In order to determine the mechanism(s) by which pTpT affects DNA repair, northern and western blot analysis were performed for gene products involved in NER and in the cellular response to DNA damage as described in Table 1. In addition, another protein vital to NER, the XPA-correcting (XPAC) protein (Li et al., 1995), was examined. The XPAC gene is mutated at one of several sites (Tanaka et al., 1990; Satokata et al., 1992) in xeroderma pigmentosum group A (XPA) patients, who are characterized clinically by an increased sun sensitivity, a high incidence of skin cancer (Takebe et al., 1987) and neurological abnormalities (Mimaki et al., 1986; Maeda et al., 1995). The XPAC protein plays an important role in recognition of DNA damage (Kuraoka et al., 1996), one of the first steps of NER. Indeed, Park et al. (1995) showed direct binding of the XPA protein to transcription factor IIH (TFIIH), which exhibits helicase activity and partially unwinds the DNA. The human and mouse XPAC genes have been cloned, and like genes encoding many other NER proteins, the XPAC gene lacks common promoter elements including CAAT and TATA boxes in the 5' flanking region (Satokata et al., 1993; van Ostrom et al., 1994), and instead both mouse and human XPAC promoters contain a pyrimidine-rich sequence element. Whether this sequence mediates baseline or inducible expression of *XPAC* or other NER genes, and if so through which transcription factors, is unknown.

Northern blot analysis showed that the steady state level of the mRNAs for *GADD45*, *ERCC3* and *p21* are up-regulated in pTpT-treated human keratinocytes within 24 hours after addition of the dinucleotide and remain elevated at least through 72 hours (Eller et al., 1997). Furthermore, the protein levels of p53, PCNA and XPAC are increased in keratinocytes treated with pTpT for 48 hours compared to diluent-treated cells (Maeda et al., 1999). Together, these data suggest that the enhanced DNA repair capacity of pTpT-treated cells derives, at least in part, from increased expression of NER genes. The increase in XPAC expression by pTpT is particularly interesting with respect to the earlier data showing an enhanced DNA repair response in cells from XP group A patients induced by DNA damage (Jeeves and Rainbow, 1983A; Jeeves and Rainbow, 1983C). Perhaps pTpT induces higher levels of a partially inactive repairprotein in some XP cells to compensate for the lower activity. Alternatively, pTpT as well as DNA damage may induce a protein or proteins that compensate for the mutated protein and at least partially restore DNA repair capacity in these compromised cells.

6. Cell Cycle Regulation by DNA Oligonucleotides

Because p53, p21, GADD45 and PCNA are known to cooperate in regulating cell cycle arrest in response to DNA damage (El-Deiry et al., 1993; Harper et al., 1993; Di Leonardo et al., 1994; Li et al., 1994; Smith et al., 1994; Waga et al., 1994; Zhan et al., 1994) and all were found to be up-regulated by pTpT in skin cells, the effect of pTpT on skin cell proliferation was examined. The dinucleotide was found to inhibit the

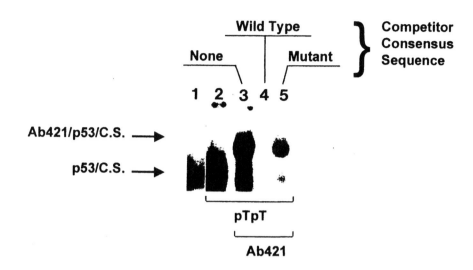

Fig. 6. Activation of p53 by pTpT. Cultures of p53 null H1299 cells were transfected with either control vector or a p53 expression vector, treated with pTpT and processed for the electromobility shift assay. Lane 1, extract from cells transfected with the p53 expression vector. Lane 2, cells were transfected with the p53 expression vector and incubated for 6 hours with 100 :M pTpT. Lane 3, same as lane 2 but the binding reaction contained 0.1 :g mAb421. Lane 4, same as lane 3 but with a 100-fold excess (10 pmol) unlabeled wild-type p53 consensus sequence. Lane 5, same as lane 3 but with a 100-fold excess (10 pmol) of labeled mutant p53 consensus sequence. The consensus sequence DNA/p53 complex (C.S. DNA/p53) and the supershifted complex (C.S. DNA/p53/Ab421) are indicated by arrows. (From Eller et al., PNAS 94:12627-12632, 1997.)

proliferation of normal human keratinocytes and fibroblasts as well as a squamous cell carcinoma line (Eller et al., 1997). These effects on cell growth were rapidly and completely reversible upon removal of pTpT from the medium, in a timeframe consistent with resumption of epidermal cell proliferation after UV irradiation (McGregor et al., 1999). Although the mechanism of this cell cycle arrest has not been studied in detail, Pedeux et al. (1998) found that pTpT, but not pT, induced an S-phase arrest in human melanoma cells, perhaps surprisingly, given that a block in the G_1 to S transition of the cell cycle is predicted for an agent acting through p53 and p21 induction (Kastan et al., 1992; El-Deiry et al., 1993; Di Leonardo et al., 1994).

7. The Role of p53 in the DNA Damage Responses

A. *DNA Repair and Cell Cycle*

Extensive evidence demonstrates that many of the effects of pTpT on the cell cycle and on DNA repair are mediated through p53. pTpT treatment increases the level of p53 protein in normal human keratinocytes (Maeda et al., 1999) and fibroblasts (Goukassian et al., 1999) and in the mRNA level of the p53-regulated genes *p21* and *GADD45* in keratinocytes and fibroblasts (Eller et al., 1997). Furthermore, p21 up-regulation and consequent growth arrest by pTpT is dependent on p53 and can be seen in the p53-null cells only after the cells are transfected with a p53-expression vector (Eller et al., 1997).

In addition to increasing p53 protein levels, pTpT treatment activates the protein as demonstrated by the electromobility shift assay (Eller et al., 1997) (Figure 6) or by increased read-out of a reporter gene under control of a promoter with a p53 consensus sequence (Maeda et al., 1999). Jayaraman and Prives (1995) have shown that the sequence-specific DNA binding activity of p53 is stimulated by the non-specific binding of short single-stranded DNA fragments to a C-terminal domain of p53, and pTpT may similarly interact with p53, directly stimulating p53-mediated transcription. Alternatively, pTpT may act indirectly through upstream effectors of p53 function, such as the ATM gene product (Sanchez et al., 1996; Walworth and Bernards, 1996). Of note, such increases in p53 activity may occur independent of p53 protein levels, as have been reported after exposure of cells to low levels of UV radiation (Hupp et al., 1995).

To further confirm that pTpT enhances DNA repair through p53 activation, repair of the BP-damaged CAT vector was measured in p53-null H1299 cells that were first transfected with either the p53 expression vector or an empty control vector. p53 protein expression was confirmed by Western blot analysis 48 hours after transfection. Parallel cultures were then treated with either 100 μM pTpT or an equal volume of diluent, and transfected with either BP-damaged or undamaged CAT vector. The plasmid containing a low level of BP damage was repaired nearly twice as efficiently in pTpT-treated cells than in diluent-pre-treated cells ($53 \pm 9\%$ vs. $27 \pm 2\%$ of controls, p=0.014, non-paired T-test), and the plasmid containing a high level of BP damage was repaired about 50% more efficiently in pTpT treated cells ($13.5 \pm 2\%$ vs. $8.0 \pm 2.0\%$ of controls) (Figure 7) (Maeda et al., 1999), pTpT-induced increases very similar to those for the repair of UV-induced photoproducts (Eller et al., 1997). In p53-null cells, however, the repair capacity was the same in both pTpT and diluent treatment groups and comparable to that in the p53-transfected cells in which the p53 had not been activated by pTpT. These data demonstrate that enhanced repair of BP-DNA adducts by pTpT is accomplished largely, if not entirely, through p53 activation and also suggest that there is a quantitatively important p53-independent repair pathway in these cells. Because these cells have a repair capacity approximately 50% that of pTpT-treated p53+ cells, this pathway appears to account for up to half the total repair capacity even in cells containing activated p53. These results are consistent with the report of McKay et al. (1997) that host cell reactivation of UV-induced DNA damage is enhanced by prior

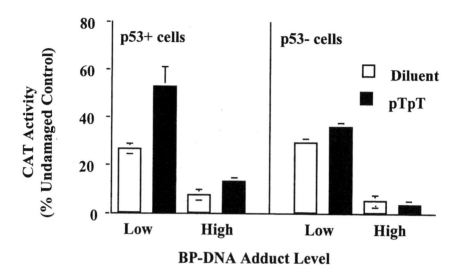

Figure 7. Enhancement of BP-DNA adduct repair by pTpT requires p53. Cultures of p53-null H1299 cells were transfected with either a p53 expression vector (p53 + cells) or a control vector (p53 – cells). One day after transfection, cells were incubated with either 100 μM pTpT or diluent alone for 2 days. Paired cultures were then transfected with either undamaged CAT vector or CAT vector continuing a low level (25 adducts/plasmid) or high level (50 adducts/plasmid) of BP damage. All cultures were cotransfected with the β-gal normalization vector. Each CAT transfection condition was performed in duplicate. Normalized CAT activity from cells expressing the undamaged CAT vector was set at 100%. (From Maeda et al. Mutat Res 433:137-145, 1999 with permission.)

UVC irradiation in normal human fibroblasts, but not in Li-Fraumeni syndrome cells, which contain one allele with a dominant negative p53 mutation, while DNA repair capacity is similar in unirradiated LFS cells and normal human fibroblasts.

B. *Melanogenesis*

If tanning is indeed part of a mammalian SOS response, the above findings suggest that it should also be p53-regulated. However, until recently, little evidence supported a role for p53 in the regulation of melanogenesis. Kichina et al. (1996), who transfected a wild type p53-expressing plasmid into pigmenting human melanoma cells, reported that p53 overexpression decreased tyrosinase mRNA levels and activity and reduced the melanin content of these cells. Furthermore, they determined that p53 negatively affected the expression of a reporter gene driven by the tyrosinase promoter, suggesting that p53 modulated transcription of the rate limiting enzyme for melanogenesis.

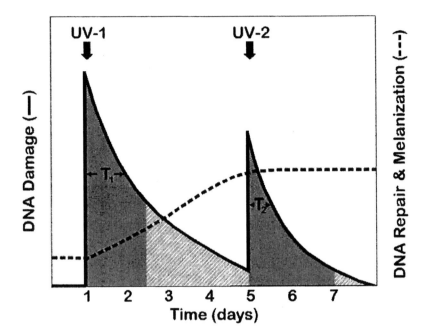

Figure 8. Consequences of the SOS response in human skin. In skin, under basal conditions of relatively low DNA repair capacity and melanin content, an initial UV exposure (UV-1) results in substantial immediate DNA damage (solid line). The repair, as measured by removal of DNA photoproducts, has been shown to occur exponentially, and the time for 50% of total repair (T_1), has been determined to be approximately 24 hr for thymine dimers and 6 hr for (6-4) photoproducts after irradiation of cultured newborn fibroblasts with a UV dose experienced during a modest physiologic exposure in vivo. Concomitantly, over approximately 5 days following the UV exposure, melanin content of skin progressively increases, leading to a visible tan, and rate limiting DNA repair proteins are progressively induced 2-3 fold (dashed line), as determined experimentally. Thereafter, a second equal UV exposure (UV-2) results in less initial damage (due primarily to greater absorption of UV photons by melanin) and the time of repair (T_2) is approximately half, as determined experimentally even in cells lacking protective melanin. After each UV exposure, there is a dose-dependent p53-mediated growth arrest, on the order of 1-2 days, during which DNA repair occurs and mutation (the consequence of DNA replication past an unrepaired photoproduct) cannot. The length of this growth arrest, which depends of the extent of p53 activation and resulting level and duration of induction for p21 and other cell cycle inhibitors, is probably greater after the second UV exposure than the first, but has not been

cont'd on next page

determined experimentally and is therefore graphed as equal in length following both exposures (shaded areas). In combination, these phenomena result in a period of time after the first UV exposure when considerable DNA damage persists, cells have resumed DNA synthesis, and DNA repair capacity is submaximal (striped area below the curve). During this period, there is a relatively high risk of mutation. In contrast, after the second UV exposure that occurs during the period of induced photoprotection, growth arrest persists for essentially the entire period required for DNA repair, greatly reducing the risk of UV-induced mutation. Eventually, in the absence of further UV exposures, melanin content and repair proficiency return to baseline; and any subsequent exposure is handled like UV-1. However, if exposures are repeated during the period encompassed by the SOS response (assumed to be 1-2 weeks, depending on the inciting UV exposure), increased melanin content and repair proficiency are retained or further induced; and each exposure is handled like UV-2. For these reasons, the pattern of UV exposures, in addition to the total cumulative UV dose, has major long-term impacts on skin.

p53 is known to positively regulate gene transcription through binding specific consensus sequences in promotors of target genes with its DNA-binding domain and activating gene expression through its transactivating domain (El-Deiry et al., 1992; Kern et al., 1992A; Kern et al., 1992B). Alternatively, p53 can also negatively regulate gene transcription, apparently through interacting with proteins that are necessary for maintaining basal transcription rates from some promoters. For example, p53 has been shown to interact with TATA box-binding proteins and to interfere with the initiation of transcription of genes dependent on the TATA box element (Seto et al., 1992; Mack et al., 1993). Also, p53 has been shown to negatively regulate the transcription of the human HSP70 gene, potentially through direct interaction with proteins that bind the CCAAT promoter element (Agoff et al., 1993). Whether either of these mechanisms is applicable to the regulation of tyrosinase transcription by p53 is unclear, but the human tyrosinase gene promoter does contain both a TATA-box and CCAAT-box (Ponnazhagan et al., 1994), and the data of Kichina et al. (1996) demonstrate a decrease in the basal transcription rate of the tyrosinase gene with increasing levels of wild type p53. Similarly, it was shown that tyrosinase mRNA levels initially decrease in S91 mouse melanoma cells after UV-irradiation (Eller et al., 1996), coincident with increased p53 protein levels in response to UV-induced DNA damage (Goukassian et al., 1999). Interestingly, in both S91 cells and in a human melanoma cell line known to express wild type p53, after initial down-regulation of tyrosinase mRNA levels, there is a subsequent increase 24 to 48 hours after irradiation (Eller et al., 1996; Khlgatian et al., 1999A). In human melanoma cells stably transfected with a vector over-expressing a dominant negative p53, initial down-regulation of tyrosinase message after UV irradiation is observed, as in normal pigment cells, but the subsequent up-regulation does not occur (Khlgatian et al., 1999A). These data suggest that p53 protein levels transcriptionally down-regulate tyrosinase mRNA levels while increased p53 activity directly or indirectly increases transcription of this message leading to increased steady-state tyrosinase mRNA levels after UV irradiation, compatible with the observed delay in visible tanning.

8. Summary

Extensive data suggest that human skin and skin-derived cells exhibit an SOS response initiated by DNA damage. This response includes increased pigmentation (tanning), transient cell cycle arrest, and enhanced DNA repair capacity. It appears highly homologous to the SOS response in bacteria, which acts through single stranded DNA fragments to transcriptionally upregulate genes relevant to protecting against subsequent DNA damage. Available data pertain primarily to damage resulting from UV irradiation but chemical carcinogens and X-irradiation also appear effective. The consequence of this SOS response is represented schematically in Figure 8, for which many but not all of the quantitative aspects derive from experimental data, as presented in the preceeding sections.

This hypothesis predicts a lower UV-induced mutation frequency in previously (recently) irradiated skin than in previously unirradiated skin, as well as a reduced mutation frequency in pTpT-treated skin in which the dinucleotides have presumably induced the SOS response by mimicking DNA damage or its repair. Indeed, in preliminary experiments, we have recently measured mutation frequency in a reporter vector integrated into the genome of transgenic mice (Boerrigter et al., 1995) and demonstrated that pTpT-pre-treatment, before UV irradiation, reduces UV-induced mutations by approximately half in cultured mouse fibroblasts and 20-30% in intact mouse skin (Khlgatian et al., 1999B), a lesser effect presumably due to poor oligonucleotide absorption through the stratum corneum. Whether this reduction in UV-induced mutations is accomplished solely through increased melanin pigmentation and up-regulation of DNA repair or perhaps also through affecting the duration of the UV-induced cell cycle arrest remains to be determined. Regardless, these results suggest that the SOS hypothesis is correct and that topical application of DNA fragments may induce protective functions in human skin without exposure to genotoxins such as UV light. Such reduced formation and hastened removal of potentially mutagenic DNA lesions would be expected to reduce the risk of photoaging and photocarcinogenesis in sun-exposed human skin.

References

Agarwal, S., and Sohal, R.S. (1994). DNA oxidative damage and life expectancy in houseflies. Proc. Natl. Acad. Sci. USA 91, 12332-12335.

Agoff, S.N., Hou, J., Linzer, D.I.H., and Wu, B. (1993). Regulation of the human hsp70 promoter by p53. Science 259, 84-87.

Alexander, P. (1967). The role of DNA lesions in processes leading to aging in mice. Symp. Soc. Exp. Biol. 21, 29-50.

Allen, J.B., Zhou, Z., Siede, W., Friedberg, E.C., and Elledge, S.J. (1994). The SAD1/RAD53 protein kinase controls multiple checkpoints and DNA damage-induced transcription in yeast. Gen. Dev. 8, 2401-2415.

Athas, W.F., Hedayati, M.A., Matanoski, G.M., Farmer, E.R., and Grossman, L. (1991). Developments and field-test validation of an assay for DNA repair in circulating human lymphocytes. Cancer Res. 51, 5786-5793.

Auletta, M., Gange, W., Tan, O., and Matzinger, E. (1986). Effect of cutaneous hypoxia upon erythema and pigment responses to UVA, UVB, and PUVA (8-MOP + UVA) in human skin. J. Invest. Dermatol. 86, 649-652.

Banin, S., Moyal, L., Shieh, S.-Y., Taya, Y., Anderson, C.W., Chessa, L., Smorodinsky, N.I., Prives, C., Reiss, Y., Shiloh, Y., and Ziv, Y. (1998) Enhanced phosphorylation of p53 by ATM in response to DNA damage. Science 281, 1674-1677.

Bayle, J.H., Elenbaas, B., and Levine, A.J. (1995). The carboxyl-terminal domain of the p53 protein regulates sequence-specific DNA binding through its nonspecific nucleic acid-binding activity. Proc. Natl. Acad. Sci. USA 92, 5729-5733.

Boerrigter, M.E.T.I., Dolle, M.E.T., Martus, H., Gossen, J.A., and Vijg, J. (1995). Plasmid-based transgenic mouse model for studying in vivo mutations. Nature 377, 657-659.

Bohr, V.A., and Anson, M.R. (1995). DNA damage, mutation and fine structure DNA repair in aging. Mutat. Res. 338, 25-34.

Bolognia, J, Murray, M., and Pawelek, J. (1989). UVB-induced melanogenesis may be mediated through the MSH-receptor system. J. Invest. Dermatol. 92, 651-656.

Brash, D.E., Rudolph, J.A., Simon, J.A., Lin, A., McKenna, G.J., Baden, H.P., Halperin, A.J., and Ponten J. (1991). A role for sunlight in skin cancer: UV-induced p53 mutations in squamous cell carcinoma. Proc. Natl. Acad. Sci. USA 88, 10124-10128.

Burckhardt, S.E., Woodgate, R., Scheuermann, R.H., and Echols, H. (1988). UmuD mutagenesis protein of Escherichia coli: overproduction, purification, and cleavage by RecA. Proc. Natl. Acad. Sci. USA 85, 1811-1815.

Canman, C.E., Lim, D.-S., Cimprich, K.A., Taya, Y., Tamai, K., Sakaguchi, K., Appella, E., Kastan, M.B., and Siliciano, J.D. (1998). Activation of ATM kinase by ionizing radiation and phosphorylation of p53. Science 281, 1677-1679.

Celotti, L., Ferraro, P., Furian, D., Zanesi, N., Pavanello, S. (1993). DNA repair in human lymphoctyes treated in vitro with (+) and (-) syn-benzo(a)pyrene diolepoxide. Mutat. Res. 294, 117-126.

Chakraborty, A.K., Orlow, S.J., Bologna, J.L. and Pawelek, J.M. (1991) Structural/functional relationships between internal and external MSH receptors: modulation of expression in Cloudman melanoma cells by UVB radiation. J Cellular Physiol. 147:1-6.

Chowdary, D.R., Dermody, J.J., Jha, K.K., and Ozer H.L. (1994). Accumulation of p53 in a mutant cell line defective in the ubiquitin pathway. Mol. Cell. Biol. 14, 1997-2003.

de Henestrosa, A.F., Calero, S., and Barbe, J. (1991). Expression of the recA gene in Escherichia coli in several species of gram-negative bacteria. Mol. Gen. Genet. 226, 503-506.

Di Leonardo, A., Linke, S.P., Clarkin, K., and Wahl G.M. (1994). DNA damage triggers a prolonged p53-dependent G1 arrest and long-term induction of Cip1 in normal human fibroblasts. Genes Dev. 8, 2540-2551.

Dutkowski, R.T., Lesh, R., Staiano-Coico, L., Thaler, H., Darlington, G.J., and Weksler M.E. (1985). Increased chromosomal instability in lymphocytes from elderly humans. Mutat. Res. 149, 505-512.

El-Deiry, W.S., Kern, S.E., Pietenpol, J.A., Kinzler, K.W., and Vogelstein, B. (1992). Definition of a consensus binding site for p53. Nature Genet. 1, 45-49.

El-Deiry, W.S., Tokino, T., Velculescu, V.E., Levy, D.B., Parsons, R., Trent, J.M., Lin, D., Mercer, E., Kinzler, K.W., and Vogelstein, B. (1993). WAF1, a potential mediator of p53 tumor suppression. Cell 75, 817-825.

Eller, M.S., Yaar, M., and Gilchrest, B.A. (1994). DNA damage and melanogenesis. Nature 372, 413-414.

Eller, M.S. (1995). Repair of DNA Photodamage in Human Skin. In: Photodamage, Gilchrest, B.A. (ed) Blackwell Science, Inc., NY pp. 26-50.

Eller, M.S., Ostrom, K., and Gilchrest, B.A. (1996). DNA damage enhances melanogenesis. Proc. Natl. Acad. Sci, USA 93, 1087-1092.

Eller, M.S., Maeda, T., Magnoni, C., Atwal, D., and Gilchrest, B.A. (1997). Enhancement of DNA repair in human skin cells by thymidine dinucleotides: Evidence for a p53-mediated mammalian SOS response. Proc. Natl. Acad. Sci. USA. 94, 12627-12632.

Eller, M.S., Gasparro, F.P., Amato, P.E., and Gilchrest, B.A. (1998). Induction of melanogenesis by DNA oligonucleotides: effect of oligo size and sequence. J. Invest. Dermatol. 110, 474.

Fiscella, M., Ullrich S.J., Zambrano, N., Shields, M.T., Lin, D., Lees- Miller, D.P., Anderson, C.W., Mercer, W.E., and Appella, E. (1993). Mutation of the serine 15 phosphorylation site of human p53 reduced the ability of p53 to inhibit cell cycle progression. Oncogene 8, 1519-1528.

Ford, J.M., and Hanawalt, P.C. (1995). Li-Fraumeni syndrome fibroblasts homozygous for p53 mutations are deficient in global DNA repair but exhibit normal transcription-coupled repair and enhanced UV resistance. Proc. Natl. Acad. Sci. USA 92, 8876-8880.

Fornace, A.J., Jr., Alamo, I., Jr., and Hollander M.C. (1988). DNA damage-inducible transcripts in mammalian cells. Proc. Natl. Acad. Sci. USA 85, 8800-8804.

Freeman, S.E., Hacham, H., Gange, R.W., Maytum, D.J., Sutherland, J.C., and Sutherland, B.M. (1989). Wavelength dependence of pyrimidine dimer formation in DNA of human skin irradiated in situ with ultraviolet light. Proc. Natl. Acad. Sci. USA 86, 5605-5609.

Friedmann, P.S., and Gilchrest, B.A. (1987). Ultraviolet radiation directly induces pigment production by cultured human melanocytes. J. Cell Phys. 133, 88-94.

Fritsche, M., Haessler, C., and Brandner, G. (1993). Induction of nuclear accumulation of the tumor-suppressor protein p53 by DNA-damaging agents. Oncogene 8, 307-318.

Fuller, B.B., Lunsford, J.B., and Iman, D.S. (1987). α-Melanocyte-stimulating hormone regulation of tyrosinase in Cloudman S-91 mouse melanoma cell cultures. J. Biol. Chem. 9, 4024-4033.

Gasparro, F.P., and Fresco, J.R. (1986) Ultraviolet-induced 8.8-adenine dehydrodimers in oligo- and polynucleotides. Nucleic Acids Res. 63, 1-10.

Gilchrest, B.A., Zhai, S., Eller, M.S., Yarosh, D.B., and Yaar, M. (1993). Treatment of human melanocytes and S91 melanoma cells with the DNA repair enzyme T4 endonuclease V enhances melanogenesis after ultraviolet irradiation. J. Invest. Dermatol. 101, 666-672.

Gilchrest, B.A. and Eller, M.S. (1999) DNA photodamage stimulates melnogenesis and other photoprotective responses. J. Invest. Dermatol. Symposium Proceedings 4, 35-40.

Grist, S.A., McCarron, M., Kutlaca, A., Turner, D.R., and Morley, A.A. (1992). In vivo somatic mutation: frequency and spectrum with age. Mutat. Res. 266, 189-196.

Goukassian, D.A., Eller, M.S., Yaar, M., and Gilchrest, B.A. (1999). Thymidine dinucleotide mimics the effect of solar simulated irradiation on p53 and p53-regulated proteins. J. Invest. Dermatol. 112, 25-31.

Hall, P., McKee, P.H., Menage, H.D., Dover, R., and Lane, D.P. (1993). High levels of p53 protein in UV-irradiated normal human skin. Oncogene 8, 203-207.

Hallberg, L.M., El Zein, R., Grossman, L., and Au, W.W. (1996). Measurement of DNA repair deficiency in workers exposed to benzene. Environ. Health Perspect. 104, 529-534.

Hallberg, L.M., Bechtold, W.E., Grody, J., Legator, M.D., and Au, W.W. (1997). Abnormal DNA repair activities in lymphocytes of workers exposed to 1, 3-butadiene. Mutat. Res. 303, 213-221.

Harper, J.W., Adami, G.R., Wei, N., Keyomarsi, K., and Elledge, S.J. (1993). The p21 Cdk-interacting protein Cip1 is a potent inhibitor of G1 cyclin-dependent kinases. Cell 75, 805-816.

Haupt, Y., Rowan, S., Shaulian, E., Vousden, K.H., and Oren, M. (1995). Induction of apoptosis in HeLa cells by trans-activation-deficient p53. Genes Dev. 9, 2170-2183.

Hill, H.Z., and Setlow, R.B. (1982). Comparative action spectra for pyrimidine dimer formation in Cloudman S91 mouse melanoma and EMT6 mouse mammary carcinoma cells. Photochem. Photobiol. 35, 681-684.

Hoganson, G.E., Ledwitz-Rigby, F., Davidson, R.L., and Fuller, B.B. (1989). Regulation of tyrosinase mRNA levels in mouse melanoma cell clones by melanocyte-stimulating hormone and cyclic AMP. Som. Cell Mol. Gene. 15, 255-263.

Hollander, M.C., Alamo, I., Jackman, J., Wang, M.G., McBride, O.W., and Fornace, A.J., Jr. (1993). Analysis of the mammalian gadd45 gene and its response to DNA damage. J. Biol. Chem. 268, 24385-24393.

Hupp, T.R., Sparks, A., and Lane, D.P. (1995). Small peptides activate the latent sequence-specific DNA binding function of p53. Cell 83, 237-245.

Jayaraman, J., and Prives, C. (1995). Activation of p53 sequence-specific DNA binding by short single strands of DNA requires the p53 C-terminus. Cell 81, 1021-1029.

Jeeves, W.P., and Rainbow, A.J. (1983A). Gamma ray-enhanced reactivation of irradiated adenovirus in Xeroderma pigmentosum and Cockayne syndrome fibroblasts. Radiat. Res. 94, 480-498.

Jeeves, W.P., and Rainbow, A.J. (1983B). U.V. enhanced reactivation of U.V.- and gamma-irradiated adenovirus in normal human fibroblasts. Int. J. Radiat. Biol. Relat. Stud. Phys. Chem. Med. 43, 599-623.

Jeeves, W.P., and Rainbow, A.J. (1983C). U.V. enhanced reactivation of U.V.- and gamma-irradiated adenovirus in Cockayne syndrome and Xeroderma pigmentosum fibroblasts. Int. J. Radiat. Biol. Relat. Stud. Phys. Chem. Med. 43, 625-647.

Jeeves, W.P., and Rainbow, A.J., (1986). An aberration in gamma-ray-enhanced reactivation of irradiated adenovirus in ataxia telangiectasia fibroblasts. Carcinogenesis 7, 381-387.

Kastan, M.B. Onyekwere, O., Sidransky, D., Vogelstein B., and Craig R.W. (1991). Participation of p53 protein in the cellular response to DNA damage. Cancer Res. 51, 6304-6311.

Kastan, M.B., Zhan, Q., El-Deiry, W.S., Carrier, F., Jacks, T., Walsh, W.V., Plunkett, B.S., Vogelstein, B., and Fornace, A.J. A mammalian cell cycle checkpoint pathway utilizing p53 and GADD45 is defective in ataxia-telangiectasia. Cell 71, 587-597.

Kato, R., and Ogawa, H. (1994). An essential gene, ESR1, is required for mitotic cell growth, DNA repair and meiotic recombination in Saccharomyces cerevisiae. Nucleic Acids Res. 22, 3104-3112.

Kazantsev, A., and Sancar, A. (1995) Does the p53 up-regulated Gadd45 protein have a role in excision repair? Science 270, 1003-1004.

Kearsey, J.M., Shivji, M.K.K., Hall, P.A., and Wood, R.D. (1995A). Does the p53 up-regulated Gadd45 protein have a role in excision repair? Science 270, 1004-1005.

Kearsey, J.M., Shivji, M.K.K., Hall, P.A., and Wood, R.D. (1995B). Technical comments: does the p-53 regulator Gadd 45 protein have a role in excision repair? Science. 270, 1005-1006.

Kern, S., Kinzler, K., Bruskin, A. Jarosz, D. Friedman, P., Prives, C., and Vogelstein, B. (1991). Identification of p53 as a sequence-specific DNA-binding protein. Science 252, 1708-1711.

Kern, S.E., Pietenpol, J.A., Thiagalingam, S., Seymour, A., Kinzler, K.W., and Vogelstein, B. (1992). Oncogenic forms of p53 inhibit p53 regulated gene expression. Science 256, 827-830.

Khanna, K.K., and Lavin, M.F. (1993). Ionizing radiation and UV induction of p53 protein by different pathways in ataxia-telangiectasia cells. Oncogene 8, 3307-3312.

Khlgatian, M., Asawanonda, P., Eller, M.S., Yaar, M., Fujita, M., Norris, D.A., and Gilchrest, B.A. (1999A). Tyrosinase expression is regulated by p53. J. Invest. Dermatol. 112, 548.

Khlgatian, M., Hadshiew, E., Eller, M., Giese, H., Vijg, J., and Gilchrest, B.A. (1999B). Thymidine dinucleotide pre-treatment reduces DNA mutation frequency. J. Invest. Dermatol. 112, 557.

Kichina, J., Green, A., and Rauth, S. (1996). Tumor suppressor p53 down-regulates tissue-specific expression of tyrosinase gene in human melanoma cell lines. Pig. Cell. Res. 9, 85-91.

King, C.M., Gillespie, E.S., McKenna, P.G., and Barnett, Y.A. (1994). An investigation of mutation as a function of age in humans. Mutat. Res. 316, 79-90.

Kobayashi, N., Muramatsu, T., Yamashine, Y., Shirai, T., Ohnishi, T., and Mori, T. (1993). Melanin reduces ultraviolet-induced DNA damage formation and killing rate in cultured human melanoma cells. J. Invest. Dermatol. 101, 685-689.

Kochevar, I.E. (1992). Acute effects of ultraviolet radiation on skin. In: Holick, M.F., and Kligman, A.M., eds. Biological Effects of Light. Berlin: Walter de Gruyter, pp. 3-10.

Kuraoka, I., Morita, E.H., Saijo, M., Matsuda, T., Morikawa, K., Shirakawa, M., and Tanaka, K. (1996). Identification of a damaged-DNA binding domain of the XPA protein. Mutat. Res. 362, 87-95.

Lavker, R.M., and Kaidbey, K.H. (1982). Redistribution of melanosomal complexes within keratinocytes following UVA irradiation: a possible mechanism for cutaneous darkening in man. Arch. Dermatol. Res. 272, 215-228.

Le, X.C., Xing, J.Z., Lee, J., Leadon, S.A., and Weinfeld, M. (1998). Inducible repair of thymine glycol detected by an ultrasensitive assay for DNA damage. Science 280, 1066-1069.

Li, L., Peterson, C.A., Lu. X., and Legerski, R.J. (1995). Mutations in XPA that prevent association with ERCC1 are defective in nucleotide excision repair. Mol. Cell Biol. 15, 1993-1998.

Li, R., Waga, S., Hannon, G.J., Beach, D., and Stillman, B. (1994). Differential effects by the p21 CDK inhibitor on PCNA-dependent DNA replication and repair. Nature 371, 534-537.

Lovett, C.M., Jr., Love, P.E., Yasbin, R.E., and Roberts, J.W. (1988). SOS-like induction in Bacillus subtilis: induction of the RecA protein analog and a damage-inducible operon by DNA damage in Rec[+] and DNA repair-deficient strains. J. Bacteriol. 170, 1467-1474.

Lu, X., Park, S.H., Thompson T.C., and Lane D.P. (1992). Ras-induced hyperplasia occurs with mutation of p53, but activated ras and myc together can induce carcinoma without p53 mutation. Cell 70, 153-161.

Lu, X., and Lane D.P. (1993). Differential induction of transcriptionally active p53 following UV or ionizing radiation: defects in chromosome instability syndromes. Cell 75, 765-778.

Lu, X., Burbidge, S.A., Griffin, S., and Smith H.M. (1996). Discordance between accumulated p53 protein level and its transcriptional activity in response to U.V. radiation. Oncogene 13, 413-418.

Luethy, J.D., and Holbrook N.J. (1992). Activation of the Gadd53 promoter by genotoxic agents: a rapid and specific response to DNA damage. Can. Res. 52, 5-10.

Lytle, C.D., Day, R.S., III, Hellman, K.B., and Bockstahler, L.E. (1976). Infection of UV-irradiated xeroderma pigmentosum fibroblasts by herpes simplex virus: study of capacity and Weigle reactivation. Mutat. Res. 36, 257-264.

Mack, D.H., Vartikar, J., Pipas, J.M., and Laimins, L.A. (1993). Specific repression of TATA-mediated but not initiator-mediated transcription by wild-type p53. Nature, 363, 281-283.

Maeda, T., Eller, M.S., Hedayati, M., Grossman, L., and Gilchrest, B.A. (1999). Enhanced repair of benzo(a)pyrene-induced DNA damage in human cells treated with thymidine dinucleotides. Mutat. Res. 433:137-145.

Maeda, T. Sato, K., Minarni, Tagochi, H., and Yoshikawa, K. (1995). Chronological difference in walking impairment among Japanese group A xeroderma pigmentosum (XP-A) patients with various combinations of mutations. Clin. Genet. 48, 225-231.

McGregor, J.M., and Hawk, J.L.M. (1999). Acute effects of ultraviolet radiation on the skin. In: Fitzpatrick, Dermatology in General Medicine Freedberg, I.M., Eisen, A.Z., Wolff, K., Austen, K.F., Goldsmith, L.A., Katz, S.I., and Fitzpatrick, T.B. eds., 5th ed. McGraw-Hill, Inc., New York, pp. 1555-1561.

McKay, B.C., and Rainbow, A.J. (1996). Heat-shock enhanced reactivation of a UV-damaged reporter gene in human cells involves the transcription coupled DNA repair pathway. Mut. Res. 363, 125-135.

McKay, B.C., Francis, M.A., and Rainbow, A.J. (1997). Wildtype p53 is required for heat shock and ultraviolet light enhanced repair of a UV-damaged reporter gene. Carcinogenesis 18, 245-249.

Miller, R.V., and Kokjohn T.A. (1988). Expression of the recA gene in Pseudomonas aeruginosa PAO is inducible by DNA-damaging agents. J. Bacteriol. 170, 2384-2387.

Milne, D.M., Palmer, R.H., and Meek, D.W. (1992). Mutation of the casein kinase II phosphorylation site abolishes the anti-proliferative activity of p53. Nuc Acids Res 21, 565-5570.

Mimaki, T., Itoh, N., Abe, J., Tagawa, T., Sato, K., Yabuuchi, H., and Takebe, H. (1986). Neurological manifestations in xeroderma pigmentosum, Annu. Neurol. 20, 70-75.

Mitchell, D.L., Cleaver, J.E., and Epstein, J.H. (1990). Repair of pyrimidine (6-4) pyrimidone photoproducts in mouse skin. J. Invest. Dermatol. 95, 55-59.

Mitchell, D.L., and Karentz, D. (1993). The induction and repair of DNA photodamage in the environment. In: Environmental UV Photobiology. Young, A.R., Bjorn, L.O., Moan, J., and Nultsch, W., eds. (Plenum, New York.). pp. 345-377.

Miyashita, T., Krajewski, S., Krajewska, M., Wang, H.G., Lin, H.K., Liebermann, D.A., Hoffman, B., and Reed, J.C. (1994). Tumor suppressor p53 is a regulator of bcl-2 and bax gene expression in vitro and in vivo. Oncogene 9, 1799-1805.

Moriwaki, S.I., Ray, S., Tarone, R.E., Kraemer, K.H., and Grossman, L. (1996). The effect of donor age on the processing of UV-damaged DNA by cultured human cells: reduced DNA repair capacity and increased DNA mutability. Mutat. Res. 364, 117-123.

Morris, G.F., Bischuff, J.R., and Matthews, M.B. (1996). Transcriptional activation of the human proliferating cell nuclear antigen promoter by p53. Proc. Natl. Acad. Sci. USA 93, 895-899.

Nelson, W.G., and Kastan, M.B. (1994). DNA strand breaks: the DNA template alterations that trigger p53-dependent DNA damage response pathways. Mol. Cell. Biol. 14, 1815-1823.

Nohmi, T., Battista, J.R., Dodson L.A., and Walker, G.C. (1988). RecA-mediated cleavage activates UmuD for mutagenesis: mechanistic relationship between transcriptional derepression and post-translational activation. Proc. Natl. Acad. Sci. USA 85, 1816-1820.

Notani, N.K., and Setlow J.E. (1980). Inducible repair system in Haemophilus influenzae. J. Bacteriol. 143, 516-519.

Orrego, C., and Eisenstadt E. (1987). An inducible pathway is required for mutagenesis in Salmonella typhimurium LT2. J. Bacteriol. 169, 2885-2888.

Osborne, M.R., Jacobs, S., Harvey, R.G., B.P. (1981). Minor products from the reaction of (+) and (-) benzo(a) pyrene anti-diolepoxide with DNA. Carcinogenesis 2, 553-558.

Owen-Schaub, L., Zhang, W., Cusack, J.C., Angelo, L.S., Santee, S.M., Fujiwara, T., Roth, J.A., Deisseroth, A.B., Zhang, W.W., Kruzel, E., et al. (1995). Wild-type human p53 and a temperature-sensitive mutant induce Fas/APO-1 expression. Mol. Cell. Biol. 15, 3032-3040.

Pang, P.P., and Walker, G.C. (1983). The Salmonella typhimurium LT2 uvrD gene is regulated by the lexA gene product. J. Bacteriol. 154, 1502-1504.

Park, C.H., Mu. D., Reardon, J.T., and Sancar, A. (1995). The general transcription-repair factor TFIIH is recruited to the excision repair complex by the XPA protein independent of the TFII E transcription factor. J. Biol. Chem. 270, 4896-4902.

Parrish, J.A., Jaenicke, K.F., and Anderson, R.R. (1982). Erythema and melanogenesis action spectra of normal human skin. Photochem. Photobiol. 36, 187-191.

Pathak, M.A., and Stratton, K. (1968). Free radicals in human skin before and after exposure to light. Arch. Biochem. Biophys. 123, 468-476.

Pedeux, R., Al-Irani, N., Marteau, C., Pellicier, F., Branche, R., Ozturk, M., Ranchi, J., and Dore, J.F. (1998). Thymidine dinucleotides induce S phase cell cycle arrest in addition to increased melanogenesis in human melanocytes. J. Invest. Dermatol. 111, 472-477.

Phillips, D.H. (1983). Fifty years of benzo(a)pyrene. Nature 303, 468-472.

Ponnazhagan, S., Hou, L., and Kwon, B.S. (1994). Structural organization of the human tyrosinase gene and sequence analysis and characterization of its promoter region. J. Invest. Dermatol. 102, 744-748.

Prelich, G., Tan, C.K., Kostura, M., Mathews, M.B., So, A.G., Downey, K.M., and Stillman, B. (1987). The cell-cycle regulated proliferating cell nuclear antigen is required for SV4O DNA replication in vivo. Nature 215,471-475.

Protic, M., Roilides, E., Sevine, A.S., and Dixon, K. (1988). Enhancement of DNA repair capacity of mammalian cells by carcinogen treatment. Som. Cell Mol. Gene. 14, 351-357.

Radding, C.M. (1991). Helical interactions in homologous pairing and strand exchange driven by RecA protein. J. Biol. Chem. 266, 5355-5358.

Radman, M. (1974). Phenomenology of an inducible mutagenic DNA repair pathway in Escherichia coli: SOS repair hypothesis. In: Molecular and Environmental Aspects of Mutagenesis (Prakash, L., Sherman, F., Miller, M., Lawrence, C., and Tabor H.W. eds.) Charles C Thomas, Springfield, IL, pp. 128-142.

Radman, M. (1975). SOS repair hypothesis: phenomenology of an inducible DNA repair which is accompanied by mutagenesis. In: Molecular Mechanisms for Repair of DNA, Part A (Hanawalt, P., and Setlow, R.B. eds.) Plenum Publishing Corp., New York, pp. 355-367.

Rainbow, A.J., Francis, M.A., McKay, B.C., and Hill, C.A. (1995). Involvement of the p53 protein in UV and heat shock enhanced DNA repair in an actively transcribed reporter gene in human cells. Mutagenesis 10, 572-573.

Reuven, N.B., Tomer, G., and Livneh, Z. (1998). The mutagenesis proteins UmuD and UmuC prevent lethal frameshifts while increasing base substitution mutations. Mol. Cell 2, 191-199.

Roca, A., and Cox, M. (1990). The RecA protein: structure and function. Crit. Rev. Biochem. Mol. Biol. 25, 415-456.

Rosdahl, I.D. (1979). Local and systemic effects of the epidermal melanocyte population in UV-irradiated mouse skin. J. Invest. Dermatol. 73, 306-309.

Rungta, D., Corn, T.D., and Fuller, B.B. (1996). Regulation of tyrosinase mRNA in mouse melanoma cells by α-melanocyte-stimulating hormone. J. Invest. Dermatol. 107, 689-693.

Sancar, A. and Sancar, G.B. (1988) DNA repair enzymes. Ann Rev. Biochem. 57, 29-67

Sanchez, Y., Desany, B.A., Jones, W.J., Liu, Q., Wand, B., and Elledge, S.J. (1996). Regulation of RAD53 by the ATM-like kinases MEC1 and TEL1 in yeast cell cycle checkpoint pathways. Science 271, 357-360.

Satokata, I., Tanaka, K., Minura, N., Narita, M., Mimaki, T., Satoh, Y., Kondo, S., and Okada, Y. (1992). Three nonsense mutations responsible for group A xeroderma pigmentosum. Mutat. Res. 273, 193-202.

Satokata, K., Iwai, T, Matsuda, Y., Okada, K. Tanaka K. (1993). Genomic characterization of the human DNA excision repair-controlling gene XPAC. Gene 136, 345-348.

Savitsky, K., Bar-Shira, A., Gilad, S., Rotman, G., Ziv, Y., Vanagaite, L., Tagle, D.A., Smith, S., Uziel, T., Sfez, S., Ashkenazi, M., Pecker, I., Frydman, M., Harnik, R., Patanjali, S.R., Simmons, A., Clines, G.A.,

Sartiel, A., Gaiti, R.A., Chessa, L., Sanal, O., Lavin, M.F., Jaspers, N.G.J., Taylor, A.M.R., Arlett, C.F., Miki, T., Weissman, S.M., Lovett, M., Collins F.S., and Shiloh, Y. (1995). A single ataxia telangiectasia gene with a product similar to PI-3 kinase. Science 268, 1749-1753.

Schaeffer, L., Roy, R., Humbert, S., Moncollin, V., Vermeulen, W., Hoeijmakers, H.J., Chambon, P, and Egly, J.M. (1993). DNA repair helicase: a component of BTF (TFIIH) basic transcription factor. Science 260, 58-63.

Scotto, J., Kopf, A., and Urbach, F. (1974) Non-melanoma skin cancer among Caucasians in four areas of the United States. Cancer 34, 1333-1338.

Scotto, J., Fears, T.R., and Fraumeni, J.F., Jr. (1981). Incidence of nonmelanoma skin cancer in the United States. U.S. Dept. of Health and Human Services Publication No. (NIH) 82-2443.

Sebastian J., and Sancar G.B. (1991). A damage-responsive DNA binding protein regulates transcription of the yeast DNA repair gene PHR1. Proc. Natl. Acad. Sci. USA 88, 11251-11255.

Seto, E., Usheva, A., Zambetti, G.P., Momand, J., Horikoshi, N., Weinmann, R., Levine, A.J., and Shenk, T. (1992). Wild-type p53 binds to the TATA-binding protein and represses transcription. Proc. Natl. Acad. Sci. USA 89, 12028-12032.

Shinagawa, H., Iwasaki, H., Kato, T., and Nakata, A. (1988). RecA protein-dependent cleavage of UmuD protein and SOS mutagenesis. Proc. Natl. Acad. Sci. USA 85, 1806-1810.

Shiviji, K.K. Kenny, M.K., and Wood, R.D. (1992). Proliferating cell nuclear antigen is required for DNA excision repair. Cell 69, 367-374.

Siegel, E.C. (1983). The Escherichia coli uvrD gene is inducible by DNA damage. Mol. Gen. Genet. 191, 397-400.

Smith, M.L., Chen, I.-T., Zhan, Q., Bae, I., Chen C.-Y., Gilmer, T.M., Kastan, M.B., O'Connor, P.M., and Fornace, A.J. (1994). Interaction of the p53-regulated protein GADD45 with proliferating cell nuclear antigen. Science 266, 1376-1379.

Szillard, L. (1995). On the nature of the aging process. Proc. Natl. Acad. Sci. USA 45, 30-45.

Takebe, H, Nishigori, C., Satoh, Y. (1987). Genetics and skin cancer of xeroderma pigmentosum in Japan. Jpn. J. Cancer Res. 78, 1135-1143.

Tanaka, K., Minura, N., Satokata, I., Miyamoto, I., Yoshida, M.C., Satoh, Y., Kondo, S., Yasui, A., Okayama, H., and Okada, Y. (1990). Analysis of human DNA excision repair gene involved in group A xeroderma pigmentosum and containing a zinc finger domain. Nature 348, 73-76.

Tang, M., Bruck, I., Eritja, R., Turner, J., Frank E.G., Woodgate, R., O'Donnell, M., and Goodman, M.F. (1998). Biochemical basis of SOS-induced mutagenesis in Escherichia coli: reconstitution of in vitro lesion bypass dependent on the UmuD2C mutagenic complex and RecA protein. Proc. Natl. Acad. Sci. USA 95, 9755-9760.

Taylor, W.D., Bockstahler, L.E., Montes, J., Babich, M.A., and Lytle, C.D. (1982). Further evidence that ultraviolet radiation-enhanced reactivation of simian virus 40 in monkey kidney cells is not accompanied by mutagenesis. Mutat. Res. 105, 291-298.

Trainor, K.J., Wigmore, D.J., Chrysostomou, A., Dempsey, J.L., Seshadri, R., and Morley, A.A. (1984). Mutation frequency in human lymphocytes increases with age. Mech. Ageing. Dev. 27, 83-86.

Tron, V.A., Trotter, M.J., Ishikawa, T., Ho, V.C., and Li, G. (1998). p53-dependent regulation of nucleotide excision repair in murine epidermis in vivo. J. Cutan. Med. Surg. 3, 16-20.

Van Ostrom, C.T.M., deVries, A., Verbeek, S.J., van Kreijl, C.F., and van Steeg, H. (1994). Cloning and characterization of the mouse XPAC gene. Nucleic Acids Res. 22, 11-14.

Waga, S., Hannon, G.J., Beach, D., and Stillman, B. (1994). The p21 inhibitor of cyclin-dependent kinases controls DNA replication by interaction with PCNA. Nature 369, 574-578.

Walker, G.C. (1984). Mutagenesis and inducible responses to deoxyribonucleic acid damage in Escherichia coli. Microbiol. Rev. 48, 60-80.

Walworth, N.C., and Bernards, R. (1996). Rad-dependent response of the chk1-encoded protein kinase at the DNA damage checkpoint. Science 271, 353-356.

Waterman, M.J.F., Stavridi, E.S., Waterman, J.L.F., and Halazonetis, T.D. (1998). ATM-dependent activation of p53 involves dephosphorylation and association with 14-3-3 proteins. Nature Genet 19, 175-178.

Weeda, G., Van Ham, R.C.A., Vermeuler, W., Bootsma, D., van der Eb, A.J., and Hoeijmakers, H.J. (1990A) A presumed DNA helicase encoded by ERCC-3 is involved in the human repair disorders xeroderma pigmentosum and Cockayne's syndrome. Cell 62, 777-791.

Weeda, G., Van Ham, R.C.A., Masurel, R., Westerveld, A., Odijk, H., de Wit, J., Bootsma, D., van der Eb, A.J., and Hoeijmakers, H.J. (1990B) Molecular cloning and biological characterization of the human excision repair gene ERCC-3. Mol. Cell. Biol. 10, 2570-2581.

Wei, Q., Matanoski, G.M., Farmer, E.R., Hedayati M.A., and Grossman L. (1993). DNA repair and aging in basal cell carcinoma: a molecular epidemiology study. Proc. Natl. Acad. Sci. USA 90, 1614-1618.

Weigle, J.J. (1953). Induction of mutation in a bacterial virus. Proc. Natl. Acad. Sci. USA 39, 628-636.

Weinert, T.A., Kiser, G.L., and Hartwell, L.H. (1994). Mitotic checkpoint genes in budding yeast and the dependence of mitosis on DNA replication and repair. Gen. Dev. 8, 652-665.

Witkin, E.M. (1969). The mutability towards ultraviolet light of recombination-deficient strains of Escherichia coli. Mutat. Res. 8, 9-14.

Zhan, Q., Lord, K.A., Alamo, I., Hollander, M.C., Carrier, F., Ron, D., Kohn, K.W., Hoffman, B., Lieberman, D.A., and Fornace, A. (1994). The Gadd and MyD genes define a novel set of mammalian genes encoding acidic proteins that synergistically suppress cell growth. Mol. Cell. Biol. 14, 2361-2371.

Zhou, Z., and Elledge, S.J. (1993). DUN1 encodes a protein kinase that controls the DNA damage response in yeast. Cell 75, 1119-1127.

Barbara A. Gilchrest, M.D.
Boston University School of Medicine
Department of Dermatology
609 Albany Street, J-507
Boston, MA 02118
Phone: (617)638-5538
Fax: (617) 638-5550
E-Mail: bgilchrest@bu.edu

Judith Campisi, Ph.D.
Lawrence Berkeley National Laboratory
Building 70A, Room 1118
1 Cyclotron Road
Berkeley, CA 94720
Phone: (510)486 4416
Fax: (510)486 4475
E-Mail: jcampisi@lbl.gov

Lawrence Grossman, PhD.
Johns Hopkins University
School of Hygiene and Public Health
615 N. Wolfe Street
Baltimore, MD 21205
Phone: (410)614-4226
Fax: (410) 955-2926
E-Mail: lgrossman@jhsph.edu

James Gaubatz, Ph D.
Dept of Biochemistry and Molecular Biology
University of South Alabama
College of Medicine
Mobile, Alabama 36688
Phone: (334)4606852
Fax: (334) 460-6127
E-Mail: gaubatzj@suncg.ushouthal.edu

Arlan Richardson, Ph.D.
GRECC (182)
Audie Murphy VA Hospital
7400 Merton Minter Boulevard
San Antonio, IX 78284
Phone: (210) 617-5300 ext. 6670
Fax: (210) 617-5312
E-Mail: Richardson@uthscsa.edu

Vilhelm A. Bohr, MID., Ph.D.
Laboratory of Molecular Genetics
National Institute on Aging, NIH
5600 Nathan Shock Drive
Baltimore, MD 21224b823
Phone: (410)558-8162
Fax: (410) 558-8157
E-Mail: vbohr@nih.gov

Huber Warner, Ph.D.
Gateway Building, Suite 2C231
Biology of Aging Program
National Institute on Aging
Bethesda, MD 20892
Phone: (301)496-6402
Fax: (301) 402-0010
E-Mail: warner@exmur.nia.nih.gov

Ronald Hart, PhD.
National Center for Toxicologic Research
3900 NCTR Road
Jefferson, AR 72079
Phone: (870) 543-7116
Fax: (870) 543-7332
E-Mail: Rhart@nctr.fda.gov

Jan Vijg, PhD.
University of Texas Health Science Center and
CTR Institute for Drug Development
8122 Datapoint Drive, Suite 700
San Antonio, TX 78229
Phone: (210)616-5850
Fax: (210) 692-7502
E-Mail: jvijg@saci.org

ZhongMao Guo, PhD.
Department of Physiology
University of Texas Health Science Center
   at San Antonio
San Antonio, TX 78284
Phone: (210) 617-5300 ext. 5556
Fax: (210) 617-5312
E-Mail: Guo@uthscsa.edu

Dean Rosenthal, Ph.D.
Basic Science Building, Room 347
Dept of Biochemistry and Molecular Biology
Georgetown University School of Medicine
3900 Reservoir Road NW
Washington, DC 20007
Phone: (202)687-1056
Fax: (202) 687-7186
E-Mail: drosen@bc.georgetown.edu

Wen Fang Liu
Basic Science Building, Room 348
Dept of Biochemistry and Molecular Biology
Georgetown University School of Medicine
3900 Reservoir Road NW
Washington, DC 20007
Phone: (202)687-1708
Fax: (202) 687-7186

255

Robert Shmookler Reis, PhD.
Depts of Geriatrics, Biochemistry and Molecular
    Biology, and Medicine
University of Arkansas for Medical Sciences and
McClellan Veterans Medical Ctr, Research 151
4300 West 7th Street
Little Rock, AK 72205
Phone: (501)257-5560
Fax: (501) 257-5578
E-Mail: rjreis@life.uams edu

Richard Marcotte, PhD.
Bloomfield Center for Research in Aging
Lady Davis Institute
Jewish General Hospital
3755, ch. de la Cote Ste.-Catherine St.
Montreal, Quebec, Canada

Martin Poot
Department of Pathology
Box 357705, HSB K-081
University of Washington
1959 NE Pacific Avenue
Seattle, WA 98195
Phone: (206) 543-5088
Fax: (206) 685-9256
E-Mail: mpoot@u.washington.edu

Peter Rabinovitch
Department of Pathology
Box 357705, HSB K-081
University of Washington
1959 NE Pacific Avenue
Seattle, WA 98195
Phone: (206) 543-5088
Fax: (206) 685-9256
E-mail: petersr@u.washington.edu

Cynthia M. Simbulan-Rosenthal, PhD
Basic Science Building, Room 348347
Dept of Biochemistry and Molecular Biology
Georgetown University School of Medicine
3900 Reservoir Road NW
Washington, DC 20007
Phone: (202)687-1708
Fax: (202) 687-7186
E-Mail: simbulan@bc.georgetown.edu

Mark Smulson, PhD.
Basic Science Building, Room 351
Dept of Biochemistry and Molecular Biology
Georgetown University School of Medicine
3900 Reservoir Road NW

Washington, DC 20007
Phone: (202)687-1718
Fax: (202) 687-7186
E-Mail: smulson@bc.georgetown.edu

Eugenia Wang, PhD.
Bloomfield Center for Research in Aging
Lady Davis Institute
Jewish General Hospital
3755, ch de la Cote Ste. Catherine St.
Montreal, Quebec, Canada
Phone: (514)340-8260
Fax: (514) 340-8295
E-Mail: cznu@musica.mcgill.ca

Junko Oshima,
Department of Pathology
Box 357470, HSB K-543
University of Washington
1959 NE Pacific Avenue
Seattle, WA 98195
Phone: (206) 543-5088
Fax: (206) 685-9256
E-Mail: picard@u.washington.edu

George Martin, M.D.
Department of Pathology
Box 357470, HSB K-543
University of Washington
1959 NE Pacific Avenue
Seattle, WA 98195
Phone: (206) 543-5088
Fax: (206) 685-9256
E-Mail: gmmartin@u.washington.edu

Matthew Gray
Department of Pathology
Box 357470, HSB K-543
University of Washington
1959 NE Pacific Avenue
Seattle, WA 98195
Phone: (206) 543-5088
Fax: (206) 685-9256
E-Mail: mgray@u.washington.edu

Mark S. Eller ,Ph.D.
Boston University School of Medicine
Dept of Dermatology
609 Albany Street
Boston, MA 02118
Phone: (617)638-5535
Fax: (617) 638-5515
E-mail: mseller@bu.edu

Advances in
Cell Aging and Gerontology

Series Editor: Mark P. Mattson

URL: http://www.elsevier.nl/locate/series/acag

Aims and Scope:

Advances in Cell Aging and Gerontology (ACAG) is dedicated to providing timely review articles on prominent and emerging research in the area of molecular, cellular and organismal aspects of aging and age-related disease. The average human life expectancy continues to increase and, accordingly, the impact of the dysfunction and diseases associated with aging are becoming a major problem in our society. The field of aging research is rapidly becoming the niche of thousands of laboratories worldwide that encompass expertise ranging from genetics and evolution to molecular and cellular biology, biochemistry and behavior. ACAG consists of edited volumes that each critically review a major subject area within the realms of fundamental mechanisms of the aging process and age-related diseases such as cancer, cardiovascular disease, diabetes and neurodegenerative disorders. Particular emphasis is placed upon: the identification of new genes linked to the aging process and specific age-related diseases; the elucidation of cellular signal transduction pathways that promote or retard cellular aging; understanding the impact of diet and behavior on aging at the molecular and cellular levels; and the application of basic research to the development of lifespan extension and disease prevention strategies. ACAG will provide a valuable resource for scientists at all levels from graduate students to senior scientists and physicians.

Books Published:

1. P.S. Timiras, E.E. Bittar, *Some Aspects of the Aging Process,* 1996, 1-55938-631-2
2. M.P. Mattson, J.W. Geddes, *The Aging Brain,* 1997, 0-7623-0265-8
3. M.P. Mattson, *Genetic Aberrancies and Neurodegenerative Disorders,* 1999, 0-7623-0405-7
4. B.A. Gilchrest, V.A. Bohr, *The Role of DNA Damage and Repair in Cell Aging,* 2001, 0-444-50494-X

Printed and bound by CPI Group (UK) Ltd, Croydon, CR0 4YY

08/05/2025

01865006-0001